Recent Advances in Biotechnology

(Volume 6)

Algal Biotechnology for Fuel Applications

Edited by

Hüseyin Karaca

Chemical Engineering Department, Engineering Faculty, Inonu University, Elazig Road 15th km, 44280-Campus, Malatya, Turkey

&

Cemil Koyunoğlu

Energy Systems Engineering Department, Engineering Faculty, Yalova University, Yalova, Turkey

Recent Advances in Biotechnology

Volume # 6

Algal Biotechnology for Fuel Applications

Editors: Hüseyin Karaca & Cemil Koyunoğlu

ISSN (Online): 2468-5372

ISSN (Print): 2468-5364

ISBN (Online): 978-981-5051-00-1

ISBN (Print): 978-981-5051-01-8

ISBN (Paperback): 978-981-5051-02-5

First published in 2022.

need for a court order if at any point you breach any terms of this License Agreement. In no event will any delay or failure by Bentham Science Publishers in enforcing your compliance with this License Agreement constitute a waiver of any of its rights.

3. You acknowledge that you have read this License Agreement, and agree to be bound by its terms and conditions. To the extent that any other terms and conditions presented on any website of Bentham Science Publishers conflict with, or are inconsistent with, the terms and conditions set out in this License Agreement, you acknowledge that the terms and conditions set out in this License Agreement shall prevail.

Bentham Science Publishers Pte. Ltd.
80 Robinson Road #02-00
Singapore 068898
Singapore
Email: subscriptions@benthamscience.net

BENTHAM SCIENCE

CONTENTS

PREFACE

Algae are a solution to reduce carbon emissions, especially in today's world where the global climate has become important. The United Nations and the European Union suggested that algae can be used in this regard. With the genetic engineering done in the past, it has been observed that algae can be reproduced with synthetic DNA. It is even said that algae will evolve into a separate life form from other living things. Our book offers a wide range of environment-friendly algae technologies that reduce harmful emissions. In future editions, criticism from valuable readers will allow us to further develop new issues.

Hüseyin Karaca
Chemical Engineering Department,
Engineering Faculty, Inonu University,
Elazig Road 15th km, 44280-Campus,
Malatya, Turkey

Cemil Koyunoğlu
Energy Systems Engineering Department,
Engineering Faculty, Yalova University,
Yalova, Turkey

List of Contributors

Cemil Koyunoğlu   Energy Systems and Engineering Department, Faculty of Engineering, Yalova University, Yalova, Turkey

Ece Polat   Environmental Engineering Department, Faculty of Engineering and Architecture, Sinop University, Sinop, Turkey

Fevzi Yaşar   Chemistry and Chemical Process Technology Department, Vocational School of Technical Sciences, Batman University, 72100, Batman, Turkey

Hüseyin Karaca   Chemical Engineering Department, Engineering Faculty, Inonu University, Malatya, Turkey

Leyla Uslu   Marine Biology Department, Fisheries Faculty, Cukurova University, 01330, Adana, Turke

Mesut Yılmazoğlu   Chemical Engineering Department, Engineering Faculty, Yalova University, 77100, Turkey

Oya Işık   Marine Biology Department, Fisheries Faculty, Cukurova University, 01330, Adana, Turkey

CHAPTER 1

Introduction

Cemil Koyunoğlu[1,*] and **Hüseyin Karaca**[2]

[1] *Energy Systems Engineering Department, Faculty of Engineering, Yalova University, Yalova, Turkey*

[2] *Chemical Engineering Department, Engineering Faculty, Inonu University, Malatya, Turkey*

Abstract: The purpose of writing this book is to justify the production of biofuels from algae to minimize the emissions of fossil fuel technologies to reduce their environmental effects. Moreover, the use of algae, to reduce the amount of CO_2 emissions from the global CO_2 cycle is an alternative to existing biomass conversion technologies. The book covers the most efficient algae-to-oil conversion technologies, fuel characterization, and their reflection on different technologies. It is our hope that the topics here will not only help the scientific community for a more thorough understanding of alternatives to fossil fuels but also the civil society at large as well as policymakers at national and international level.

Keywords: Algae, Algae to oil, Biofuel production, Energy Consumption, Harvesting, Species.

INTRODUCTION

Many algae conversion technologies can convert raw bio materials to liquid biofuels (Fig. **1**). Liquid biofuels are preferred mostly in the transportation industry and many of these technologies produce various intermediates during biomass to liquid fuel conversation (Fig. **2**) [1 - 24].

Industry actors, definitions, and the current status of the biomass process are given in Figs. (**3** - **5**) [1 - 24].

---

* **Correspondence author Cemil Koyunoğlu:** Energy Systems Engineering Department, Faculty of Engineering, Yalova University, Yalova, Turkey; Tel: +90 226 8155378; E-mail: cemil.koyunoglu@yalova.edu.tr

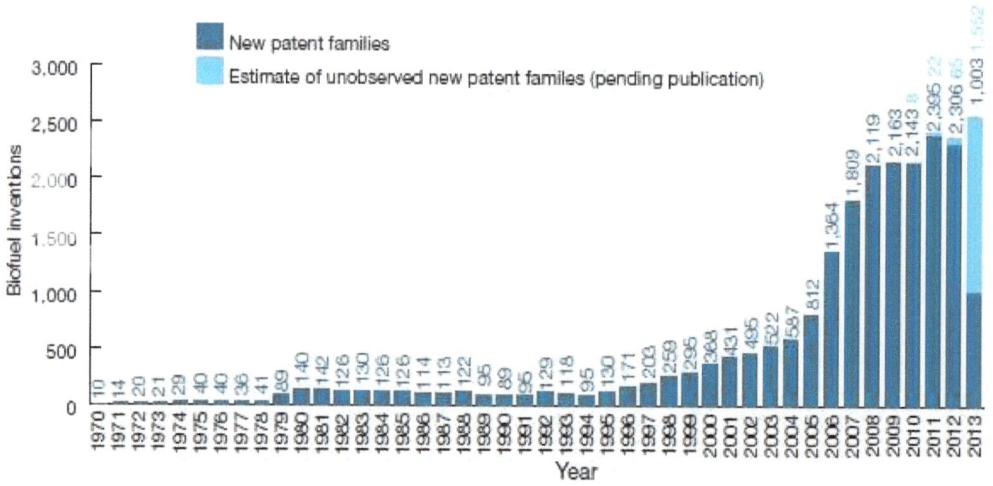

Fig. (1). Patented biofuel technologies [25].

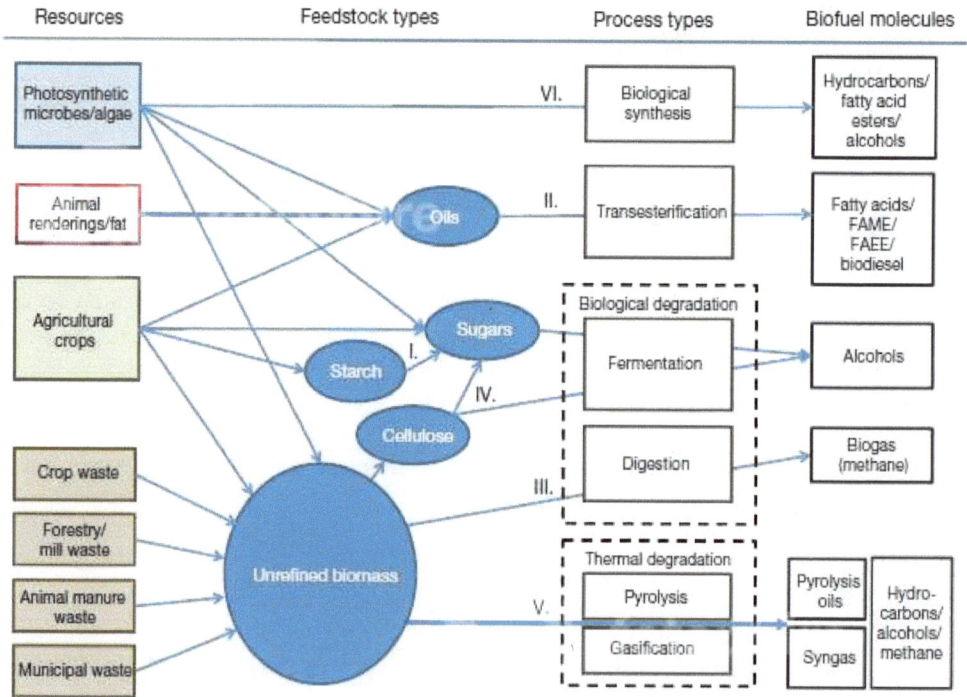

Fig. (2). Biofuel pathways [25].

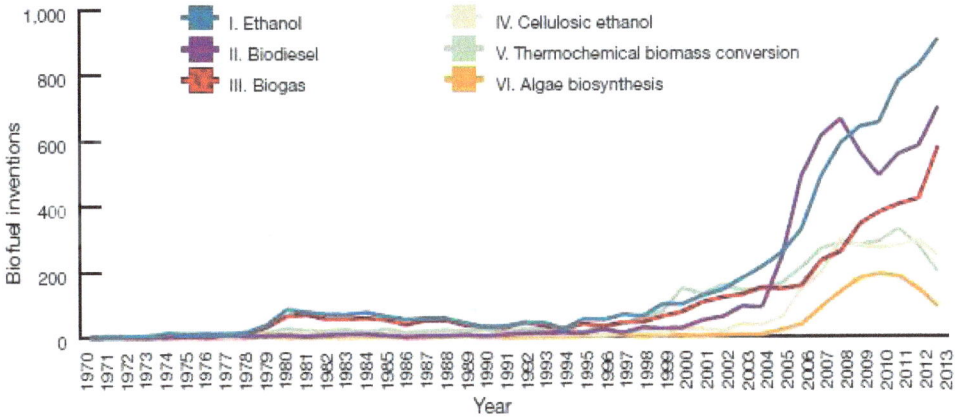

Fig. (3). Biofuel inventions [25].

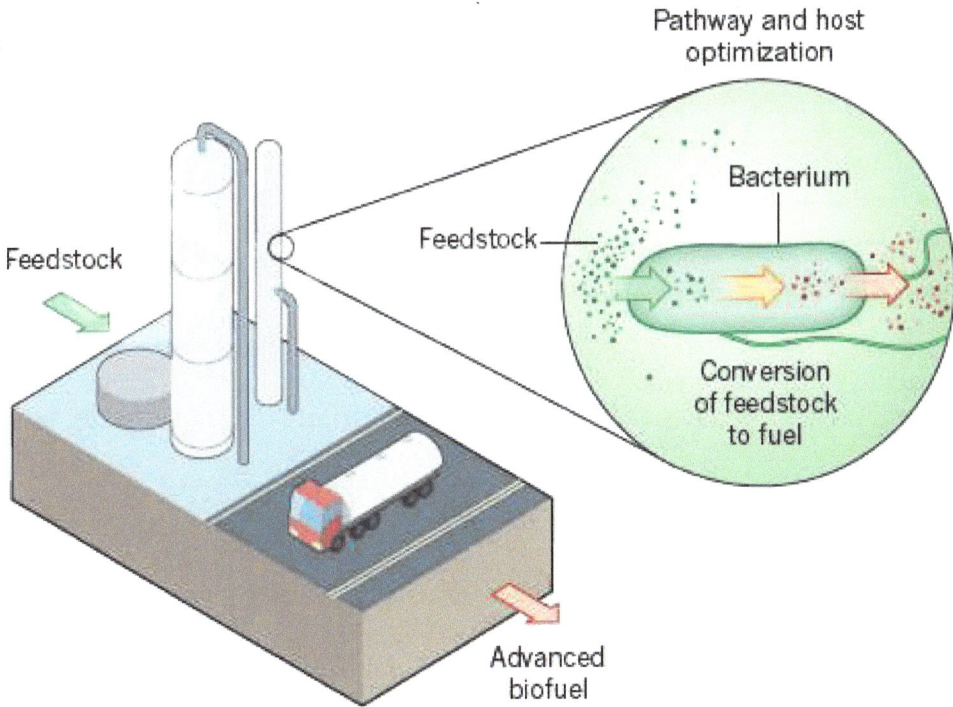

Fig. (4). Advanced biofuel technology basic definition [26].

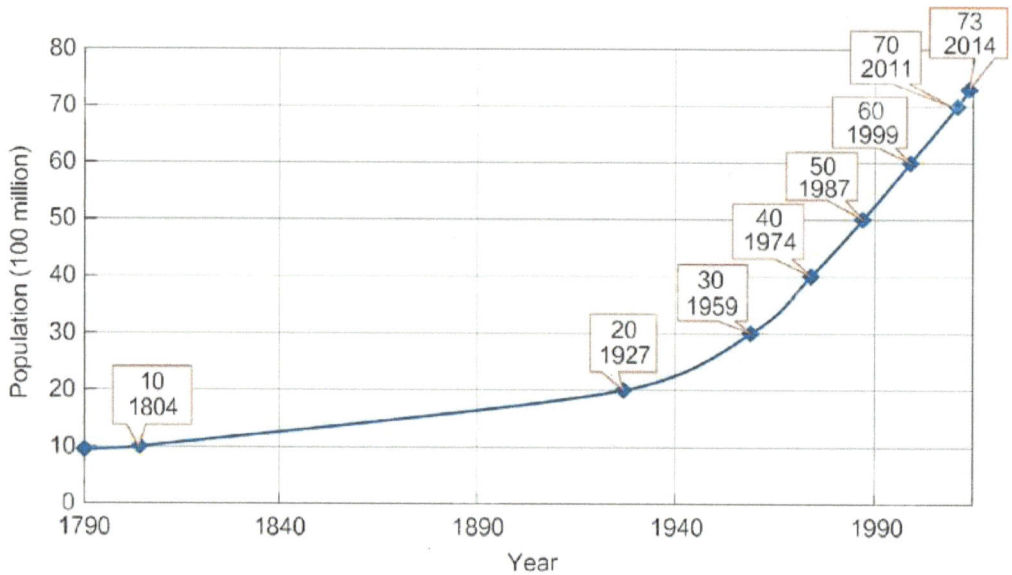

Fig. (5). World population growth rate [27].

Beyond the technology used to transform wastes into biofuel biotechnologically, it is seen that the type of raw material used is more critical [1 - 24].

Agricultural wastes: They include post-harvest biomass such as corn cobs and straw (corn cobs, nutshell, sugarcane meal) [1 - 24].

- Biogenic wastes: Biogenic wastes are in the solid phase, including parks such as garden wastes, municipal solid wastes, industrial wastes, roadside, and all wastes released from household consumption [1 - 24].
- Energy plants other than food consumption: These are grass plants such as willow, poplar, Miscanthus, switchgear, canary grass, cane, and energy herbs such as wood plants [1 - 24].
- Forest residues such as woody biomass shavings, leaves, branches, cutting chips, sawdust, *etc.* come from forestry activities [1 - 24].
- Seaweeds (Macroalgae), commonly used for biogas production, are obtained from private farms or natural habitats and are potential sources of carbohydrates. They are harvested by the anaerobic method in small volumes, given their high water and ash content [1 - 24].
- Microalgae are grown in photobioreactors and open pools and have high lipid content [1 - 24].

Technical potential: Issues such as harvesting, picking, and subsequent reduction in productivity are the remaining challenges to full potential of algae technology [1 - 24].

- Theoretical potential: It is a usable energy potential of raw material from a specific region [1 - 24]. Economic potential: It is the possible technical component that is economical for collecting the product, taking into account the availability of the resource, the costs of production, or harvesting [1 - 24].
- Sustainable potential: Other ecological and social issues such as cultivation methods, protection of soil quality, food safety, pesticide, or synthetic fertilizer application limits and water use are taken into consideration [1 - 24].

It is essential to know the biomass reserves. To better understand the algae as the potential biomass source for bioenergy, it is necessary to know the current status and classification of other biomass. Since the use of algae is sustainable, the economic and environmental conditions should be accurately compared with the conditions of other competing biomass [1 - 24].

However, it is known that the world's energy demand is essential for determining the biofuel potential of algae. According to the IPCC report, by 2050, the energy from biomass is expected to provide one-fifth of the world's primary energy demand at approximately 100 Exajoules. The cost of energy biomass raw materials determined was 1.6-5.2 USD (2020 cost), excluding algae and biogenic wastes. By 2050, the price of forest and agricultural wastes is expected to remain constant or increase partially. The cost of biomass energy used in food production is expected to decrease [1 - 24].

The cost of solid biogenic waste is high due to the need for landfills [1 - 24].

- Agricultural residues: Agricultural residues with a high potential depend on the possibility of energy crops. Transport and collection costs of raw materials may be high because these residues are limited, given the factors such as the use of straw in horse breeding Some of the agricultural remains are left on the site to preserve the natural structure of the soil, while some crop waste is used in household heating [1 - 24].
- Solid biogenic wastes: The energy conversion is still on the agenda due to the need for disposal and less competitive potential. Increasing population and increasing energy needs suggest that this group of biomass will continue to exist as an alternative energy form (Figs. **5** - **8**). Its disposal affects its cost depending on the type of waste. Power and heat are also involved in energy production. In general, such wastes are used in a direct incineration plant. Recyclable

derivatives such as cardboard and paper can be returned to the fields. Minimizing, handling, recycling, and reusing waste is essential in this sense [1 - 24].

- Non-food energy crops: Especially due to land availability, methodological differences, and environmental constraints, these wastes have been brought to the agenda. Increasing the need for land brought by advanced agricultural practices, sustainable soil management, and water availability keep non-food energy crops on the list [1 - 24].

Raw material costs for other residues are higher than algae. Both the development of supply chains, differences in regions with different crops, and changing energy values are the main obstacles in determining cost estimates [1 - 24].

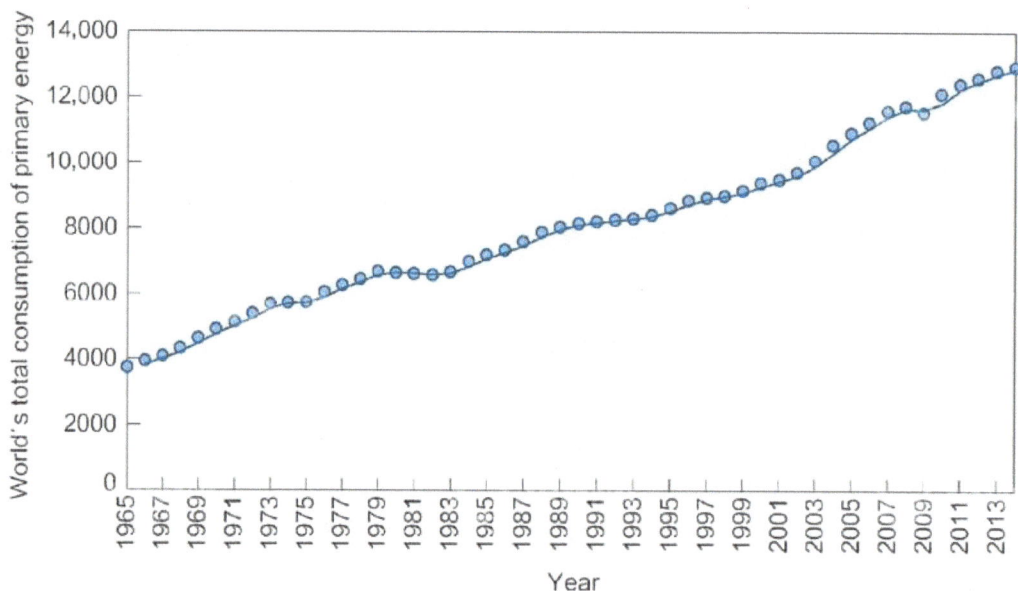

Fig. (6). World energy consumption is due to population growth [27].

- Algae: Determining the amount of raw material is essential in terms of developing breeding systems. Since the parameters such as nutrients, water, and adequate irradiation required for algae cultivation are limited to determine the regional reserves. Although it is known that algae are used in the cosmetic industry today, it seems possible to use them as energy raw materials by collecting, growing, and decreasing harvest costs [1 - 24].
- Forest remains: The determination of their potential is based on the regions where forestry activities are commercial. Low collection costs for heat and electricity generation increase the demand for forest remains. It increases the value of the collection in terms of being accessible to forest remains in non-

commercial areas. The residues occurring in the commercial forest industry required for wood production and animal beds are left on site for the sustainability of soil quality, as mentioned earlier (Fig. **9**) [1 - 24].

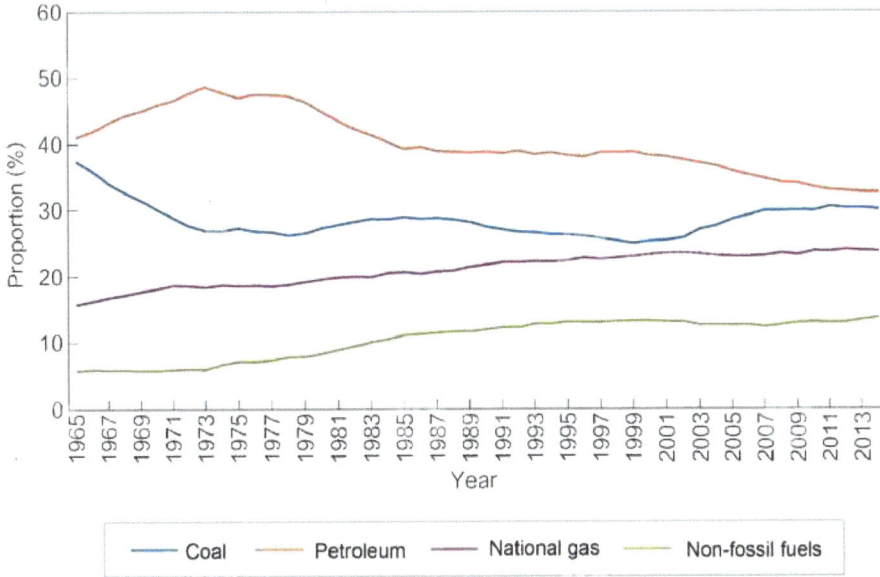

Fig. (7). World primary energy dispersions *via* year.

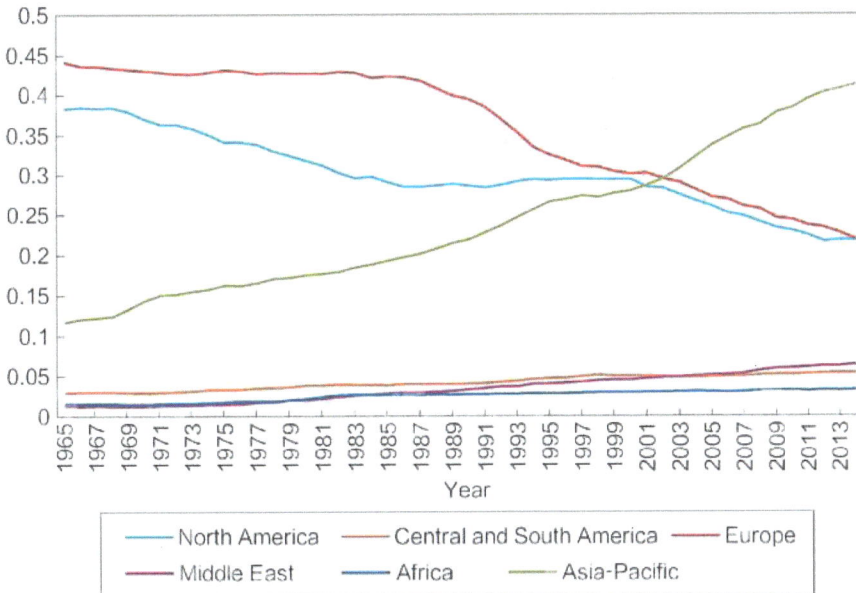

Fig. (8). World geographical regions primary energy consumption *via* year [27].

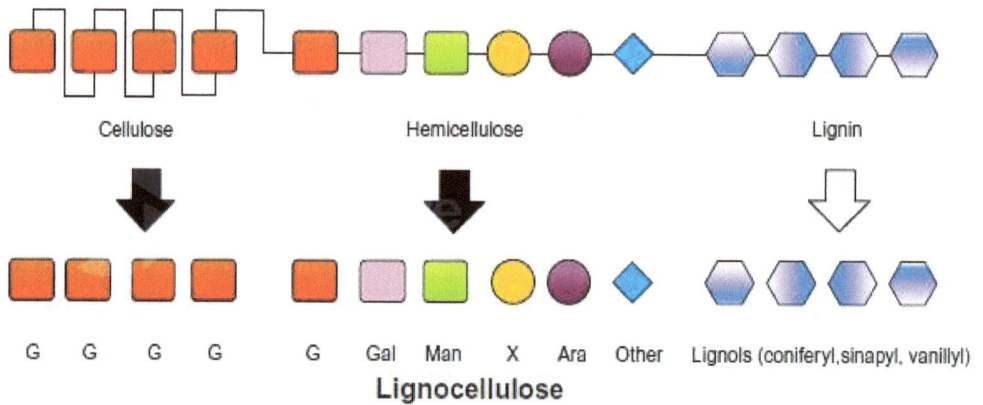

Fig. (9). Lignocellulose structure consists of wood production residue [28].

Various methods are explored, particularly for micro-algae, for their economic production from suitable algae oils (Fig. **10**). These methods include extraction, micro-algae cultivation, and reduction of oil collection costs. Trials of genetically modified microalgae and other species, as well as existing micro-algae species, provide relevant studies in increasing the conversion efficiency of solar energy to organic energy and reducing its cost [1 - 24].

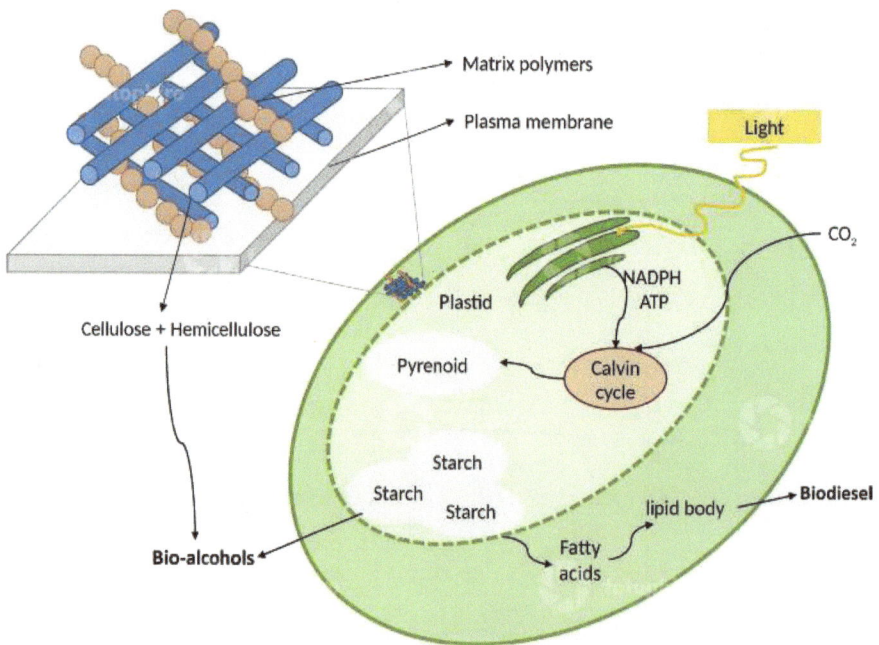

Fig. (10). Algae to biofuel pathway in a cell [29].

- In micro-algae cultivation, it is essential that open pools are practical and economical, and made of durable material that can withstand the growing life in maximum time [1 - 24].
- Determination of the conditions (temperature, pressure, *etc.*) to be carried out in the photobioreactor in a minimum time to grow micro-algae [1 - 24].
- Resolution of situations that negatively affect fertilizer input, nitrogen deficiency, lipid productivity, and quality [1 - 24].
- Mechanical optimization (geometric optimization of the mixer, *etc.*) required to increase the breeding efficiency in the pools, is done to increase the maximum lipid efficiency [1 - 24].
- Study of issues such as separation from solvent separation technologies after oil production (separation with solvent extraction, electromagnetic field separation, and effectiveness of centrifuges) will be mentioned in the following chapters of our book and hydrothermal liquefaction methods draw attention in terms of the efficiency of oil extraction methods [1 - 24]. There are several studies on the effectiveness of the catalyst used in this technology. Another topic of research is the reuse of wastewater, and the oil content obtained is essential [1 - 24].

Microalgae, evaluated in multi and single-celled organism classes, can take CO_2 from the atmosphere and convert it into fuels such as lipid and ethanol. Various algae-type related carbohydrate content list is shown in Table **1** [1 - 24].

Table 1. Carbohydrates content in various microalgae species [29].

Microalgae Species	Carbohydrate Content (Total, % Dry Mass)
Tetraselmis suecica	15–50
aPrymnesium parvum	25–33
Dunaliella salina	32
Dunaliella bioculata	4
Chlorococcum sp.	32.5
Chlorella vulgaris	12–17
Euglena gracilis	14–18
Scenedesmus dimorphus	21–52
Scenedesmus obliquus	15–51.8
Spirogyra sp.	33–64
Nannochloropsis oculata	8
Isochrysis sp.	5.2–16.4
Synechoccus sp.	15
Isochrysis galbana	7.7–13.6

(Table 1) cont.....

Microalgae Species	Carbohydrate Content (Total, % Dry Mass)
Mychonastes afer	28.4
Spirulina maxima	13–16
Porphyridium cruentum	40
Tetraselmis maculate	15
Scenedesmus abundans	41
Tetraselmis sp.	24
Chlorella sp.	19
Spirulina platensis	8–20
Chlorella pyrenoidosa	26
Chlamydomonas reindhardtii	17

Algae feeding stock can be realized in many steps [1 - 24].

Only by determining the types of algae, the variety of species that can produce the most oil and even biofuel is determined. The pool establishes the environment where it will be grown later as closed and open [1 - 24].

Photobioreactors

In algae photobioreactor, which has a tubular design, it is determined as to which type of growing medium (such as CO_2, water, food, and light) will facilitate the growth of algae in minimum time. It also allows micro-algae to grow safely. Photobioreactors provide high efficiency in the absence of water evaporation and oil production. Photobioreactors do not carry the sensitivity of determining the region involved in the installation of pools [1 - 24].

The disadvantages are energy consumption, high capital costs, maintenance of the inner surface of the photobioreactor (algae produce a protein that prevents the permeability of light to the reactor), and effective light transmission [1 - 24].

Heterotrophic Bioreactors

Such algae can be fed in non-light environments, and their carbon sources are sugar. These sugars turn into triglycerides as a result of algae digestion and then fuel. Most of the algae that fit this production method belong to the Chlorella species. Sugar cane can be used as a raw material in these bioreactors. The area required for the installation of pools, which do not need to be installed in specific areas, is less than that of pools [1 - 24].

Outdoor Pools

Open pools may be preferred to increase the contact surface area of microalgae with CO_2 so as to increase oil production. The harvest can thicken in the pond and settle continuously. Its most significant advantage over other algae growing methods is its low installation cost. It is essential to reduce the value of the required components during system design, which is necessary to meet the large-scale fuel demand. These pools have geographical region restrictions. However, the main issues that determine the operating costs are the intensification and drying processes after algae growth. In addition to reducing the energy costs of these stages, it is also on the agenda to use waste sugar sources as raw materials. The methods applied in harvesting microalgae are essential. These are flocculation, sedimentation (H. pluvialis, which separates when it reaches the red color stage), membrane filtration (used for thread algae at intervals> 25 microns), centrifugal water removal, and sieving. For Dunaliella, it is one of the new methods of absorption attached to an iron Filament on a high-grade magnet on polystyrene. There are various methods for separating oil from microalgae such as chemical, enzymatic and physical mechanisms. Micro-algae are mostly grown to produce transportation fuels such as renewable gasoline, FAME biodiesel, jet fuel, and diesel [1 - 24].

Unlike other agricultural biomass species, algae offer environmentally friendly production opportunities. They are grown in bitter, salt, and freshwater in pools set up in uncultivated soils. Micro-algae technology, which is mostly used in cosmetics, with low sales prices of high quantity products, has to be lowered for fuel production. A company called Solazyme has accomplished this and has gone into commercial fuel production and is the largest global manufacturer. This company has 2 million biofuels produced annually, one in Peoria, the second plant in Brazil, which has a biofuel capacity of 100 million liters a year, and a third plant in the US, which has a capacity of 20 million liters. Also, pilot-scale research studies with many funded projects continue in the world. According to the results of the project, production costs should be reduced to a competitive level, and production reliability should be ensured, too. It is crucial to determine values such as high production rates, net energy production, and oil yield, wide range of products (fuel and chemicals) for high-scale production. It can be said that issues such as changing the genetic modifications of algae for low cost and high output of products are on the agenda [1 - 24].

In the following sections, harvesting of algae, production by thermal liquefaction method, biodiesel production, characterization, and finally fuel cell developments are explained in detail [1 - 24].

CONCLUSION

Today, the intense use of fossil fuels in energy production causes significant environmental problems at the global level. Therefore, it is important both to reduce these environmental problems and to research the production ways of new alternative fuels to oil and natural gas. In this context, it is important to evaluate biomass, one of the renewable energy sources, both as an energy source and as a basic chemical raw material source by using various conversion technologies. Algae, which is one of the biomass types, is more advantageous than other biomass types, both to reduce carbon dioxide causing the greenhouse effect and using it as an input to produce new alternative fuels or basic chemical raw materials. Therefore, it is important to produce basic chemical raw materials by using algae directly or indirectly and to develop evaluation technologies as an energy source.

CONSENT FOR PUBLICATION

Not applicable.

CONFLICT OF INTEREST

The author declares no conflict of interest, financial or otherwise.

ACKNOWLEDGEMENT

Declared none.

REFERENCES

[1] Hiloidhari M, Bhuyan N, Gogoi N, *et al.* 16 - Agroindustry wastes: Biofuels and biomaterials feedstocks for sustainable rural development. In: Kumar RP, Gnansounou E, Raman JK, Baskar G, Eds. Refining Biomass Residues for Sustainable Energy and Bioproducts. Academic Press 2020; pp. 357-88.
[http://dx.doi.org/10.1016/B978-0-12-818996-2.00016-8]

[2] Chua SY, Periasamy LAP, Goh CMH, *et al.* Biodiesel synthesis using natural solid catalyst derived from biomass waste — A review. J Ind Eng Chem 2020; 81: 41-60.
[http://dx.doi.org/10.1016/j.jiec.2019.09.022]

[3] Ziolkowska JR. Chapter 1 - Biofuels technologies: An overview of feedstocks, processes, and technologies. In: Ren J, Scipioni A, Manzardo A, Liang H, Eds. Biofuels for a More Sustainable Future. Elsevier 2020; pp. 1-19.
[http://dx.doi.org/10.1016/B978-0-12-815581-3.00001-4]

[4] Patel A, Matsakas L, Sartaj K, Chandra R. Chapter 2 - Extraction of lipids from algae using supercritical carbon dioxide. In: Inamuddin , Asiri AM, Isloor AM, Eds. Green Sustainable Process for Chemical and Environmental Engineering and Science. Elsevier 2020; pp. 17-39.

[5] Lan K, Park S, Yao Y. Chapter 10 - Key issue, challenges, and status quo of models for biofuel supply chain design. In: Ren J, Scipioni A, Manzardo A, Liang H, Eds. Biofuels for a More Sustainable Future. Elsevier 2020; pp. 273-315.

[http://dx.doi.org/10.1016/B978-0-12-815581-3.00010-5]

[6]   Dahiya A. Chapter 13 - Algae biomass cultivation for advanced biofuel production. In: Dahiya A, Ed. Bioenergy. 2nd ed. Academic Press 2020; pp. 245-66.
[http://dx.doi.org/10.1016/B978-0-12-815497-7.00013-0]

[7]   Sadeghbeigi R, Ed. Chapter 17 - Biofuels. Fluid Catalytic Cracking Handbook. 4th ed. Butterworth-Heinemann 2020; pp. 309-27.
[http://dx.doi.org/10.1016/B978-0-12-812663-9.00017-5]

[8]   Makareviciene V, Sendzikiene E. Chapter 22 - Application of microalgae for the production of biodiesel fuel. In: Konur O, Ed. Handbook of Algal Science, Technology and Medicine. Academic Press 2020; pp. 353-65.
[http://dx.doi.org/10.1016/B978-0-12-818305-2.00022-X]

[9]   Katiyar R, Arora A. Health promoting functional lipids from microalgae pool: A review. Algal Res 2020; 46: 101800.
[http://dx.doi.org/10.1016/j.algal.2020.101800]

[10]   Singh D, Sharma D, Soni SL, Sharma S, Kumar Sharma P, Jhalani A. A review on feedstocks, production processes, and yield for different generations of biodiesel. Fuel 2020; 262: 116553.
[http://dx.doi.org/10.1016/j.fuel.2019.116553]

[11]   Prabha SP, Nagappan S, Rathna R, Viveka R, Nakkeeran E. 21 - Blue biotechnology: A vision for future marine biorefineries. In: Kumar RP, Gnansounou E, Raman JK, Baskar G, Eds. Refining Biomass Residues for Sustainable Energy and Bioproducts. Academic Press 2020; pp. 463-80.
[http://dx.doi.org/10.1016/B978-0-12-818996-2.00021-1]

[12]   Behera BK, Prasad R. Chapter 2 - Greenhouse gas capture and conversion. In: Behera BK, Prasad R, Eds. Environmental Technology and Sustainability. Elsevier 2020; pp. 41-71.
[http://dx.doi.org/10.1016/B978-0-12-819103-3.00002-0]

[13]   Dahiya A. Chapter 31 - Cutting-edge biofuel conversion technologies to integrate into petroleum-based infrastructure and integrated biorefineries. In: Dahiya A, Ed. Bioenergy. 2nd ed. Academic Press 2020; pp. 649-70.
[http://dx.doi.org/10.1016/B978-0-12-815497-7.00031-2]

[14]   Bhadani A, Kafle A, Ogura T, *et al.* Current perspective of sustainable surfactants based on renewable building blocks. Curr Opin Colloid Interface Sci 2020; 45: 124-35.
[http://dx.doi.org/10.1016/j.cocis.2020.01.002]

[15]   Rosemann A, Molyneux-Hodgson S. Industrial Biotechnology: To What Extent Is Responsible Innovation on the Agenda? Trends Biotechnol 2020; 38(1): 5-7.
[http://dx.doi.org/10.1016/j.tibtech.2019.07.006] [PMID: 31399264]

[16]   Bhattacharya M, Goswami S. Microalgae – A green multi-product biorefinery for future industrial prospects. Biocatal Agric Biotechnol 2020; 25: 101580.
[http://dx.doi.org/10.1016/j.bcab.2020.101580]

[17]   Rajesh Banu J, Preethi , Kavitha S, Gunasekaran M, Kumar G. Microalgae based biorefinery promoting circular bioeconomy-techno economic and life-cycle analysis. Bioresour Technol 2020; 302: 122822.
[http://dx.doi.org/10.1016/j.biortech.2020.122822] [PMID: 32007307]

[18]   Gholizadeh M, Hu X, Liu Q. A mini review of the specialties of the bio-oils produced from pyrolysis of 20 different biomasses. Renew Sustain Energy Rev 2019; 114: 109313.
[http://dx.doi.org/10.1016/j.rser.2019.109313]

[19]   Rebello S, Anoopkumar AN, Aneesh EM, Sindhu R, Binod P, Pandey A. Sustainability and life cycle assessments of lignocellulosic and algal pretreatments. Bioresour Technol 2020; 301: 122678.
[http://dx.doi.org/10.1016/j.biortech.2019.122678] [PMID: 31982298]

[20]   Sekoai PT, Ouma CNM, du Preez SP, *et al.* Application of nanoparticles in biofuels: An overview.

Fuel 2019; 237: 380-97.
[http://dx.doi.org/10.1016/j.fuel.2018.10.030]

[21]   Tran NH, Bartlett JR, Kannangara GSK, Milev AS, Volk H, Wilson MA. Catalytic upgrading of biorefinery oil from micro-algae. Fuel 2010; 89(2): 265-74.
[http://dx.doi.org/10.1016/j.fuel.2009.08.015]

[22]   Leiva-Candia DE, Pinzi S, Redel-Macías MD, Koutinas A, Webb C, Dorado MP. The potential for agro-industrial waste utilization using oleaginous yeast for the production of biodiesel. Fuel 2014; 123: 33-42.
[http://dx.doi.org/10.1016/j.fuel.2014.01.054]

[23]   Wei H, Liu W, Chen X, Yang Q, Li J, Chen H. Renewable bio-jet fuel production for aviation: A review. Fuel 2019; 254: 115599.
[http://dx.doi.org/10.1016/j.fuel.2019.06.007]

[24]   Sudan Reddy Dandu M, Nanthagopal K. Tribological aspects of biofuels – A review. Fuel 2019; 258: 116066.
[http://dx.doi.org/10.1016/j.fuel.2019.116066]

[25]   Albers SC, Berklund AM, Graff GD. The rise and fall of innovation in biofuels. Nat Biotechnol 2016; 34(8): 814-21.
[http://dx.doi.org/10.1038/nbt.3644] [PMID: 27504772]

[26]   Peralta-Yahya PP, Zhang F, del Cardayre SB, Keasling JD. Microbial engineering for the production of advanced biofuels. Nature 2012; 488(7411): 320-8.
[http://dx.doi.org/10.1038/nature11478] [PMID: 22895337]

[27]   Zhang Y, Zhang F, Wu S, Eds., *et al.* Chapter 1 - Review and outlook of world energy development. Non-Fossil Energy Development in China. Oxford: Academic Press 2019; pp. 1-36.
[http://dx.doi.org/10.1016/B978-0-12-813106-0.00001-5]

[28]   Roy L, Chakraborty S, Bera D, Adak S. Chapter 5 - Application of metabolic engineering for elimination of undesirable fermentation products during biofuel production from lignocellulosics. In: Kuila A, Sharma V, Eds. Genetic and Metabolic Engineering for Improved Biofuel Production from Lignocellulosic Biomass. Elsevier 2020; pp. 63-80.
[http://dx.doi.org/10.1016/B978-0-12-817953-6.00005-1]

[29]   Martín-Juárez J, Markou G, Muylaert K, Lorenzo-Hernando A, Bolado S. 8 - Breakthroughs in bioalcohol production from microalgae: Solving the hurdles.Microalgae-Based Biofuels and Bioproducts. Woodhead Publishing 2017; pp. 183-207.
[http://dx.doi.org/10.1016/B978-0-08-101023-5.00008-X]

CHAPTER 2

Anaerobic Algal Biotechnology

Ece Polat[1,*]

[1] *Environmental Engineering Department, Faculty of Engineering and Architecture, Sinop University, Sinop, Turkey*

Abstract: Biogas is produced with an anaerobic method, which involves live digestion of biomass in an oxygen-free environment. The second part of our book gives information about algae technology of the anaerobic process, which produces biogas by a biological process using animal fertilizers, food waste, and bioenergy products. In general, biogas can be used to produce heat and electricity, and its addition to the natural gas network is even considered as a vehicle fuel. It consists of 30-40% CO_2 as content, 45-65% CH_4. Conversion of CH_4, which is 20 times more harmful than CO_2 as a greenhouse gas, into energy is essential for the protection of environmental impact. In this sense, the burning of biogas emerges as a greenhouse gas reduction strategy.

Keywords: Algae, Algae to oil, Biofuel production, Energy Consumption, Harvesting, Species.

INTRODUCTION

In anaerobic digestion, which is a complex biochemical reaction that is carried out in several steps by the algae in an oxygen-free environment, the products are carbon dioxide and methane, biogas occurs from their mixture (Table **1**) [1 - 29].

In multiple stages, proteins, fats, and carbohydrates turn into water-soluble amino acids, fat subordinates, and sugars (Fig. **1**). The general composition of biogas is presented in Table **2**. The stages of biogas production from algae are hydrolysis, acidogenesis, acetogenesis, and methanogenesis. It is assumed that these stages combine acidogenesis and acetogenesis. In digesting the algae in an oxygen-free environment, all four steps take place inside the reactor. Biogas production takes place almost at the end of stage 3 [1 - 29].

---

[*] **Correspondence author Ece Polat:** Environmental Engineering Department, Faculty of Engineering and Architecture, Sinop University, Sinop, Turkey; Tel: +90 368 2714395; E-mail: epolat@sinop.edu.tr

Table 1. Elemental biogas composition [8].

Basic Components	Biogas Composition (%)
Carbon dioxide (CO_2)	30-40
Methane (CH_4)	45-65
Ammonia (NH_3)	0-1
Hydrogen sulfide (H_2S)	0.3-3
Nitrogen (N_2)	0-5
Moisture (H_2O)	0-10
Oxygen (O_2)	0-2

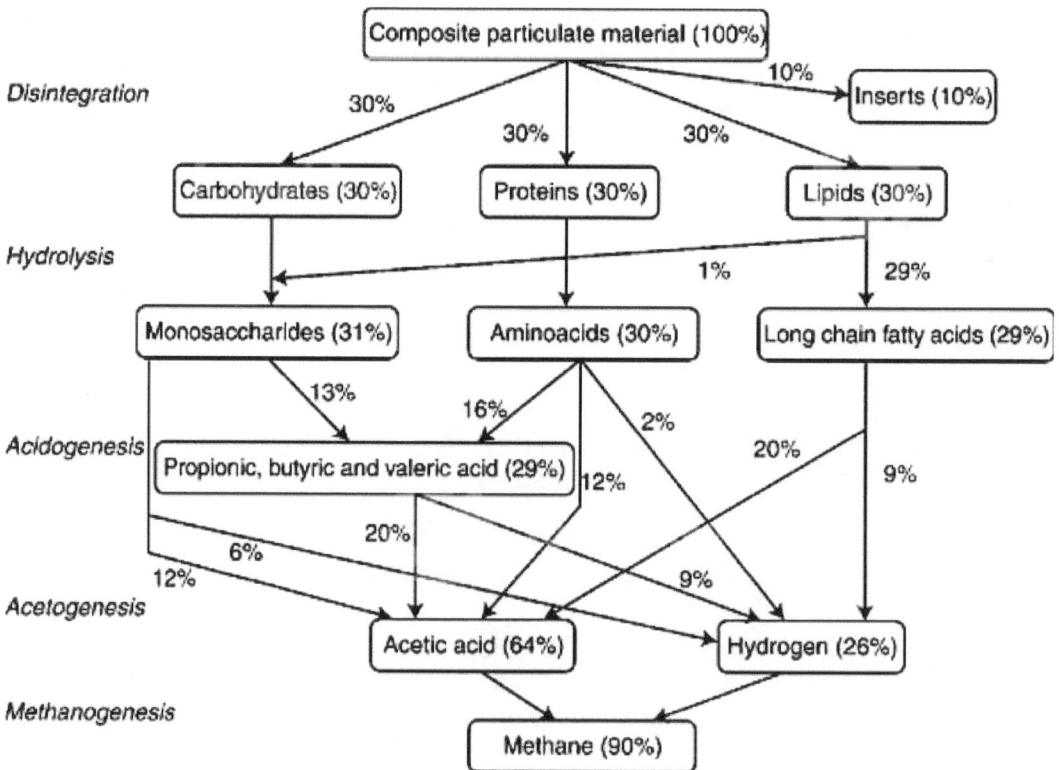

Fig. (1). Primary anaerobic digestion steps [8].

Step 1: Hydrolysis

In hydrolysis, the first step of the anaerobic process, the organic structure with a sizeable polymeric structure breaks down into simpler molecules such as amino acids, carbohydrates, fats, proteins, fatty acids, and simple sugars.

The major hydrolysis products are acetate and hydrogen. In the next process, methanogenic bacteria come into play. Methanogenic bacteria are expected to be large enough to convert the disrupted molecule into methane (Fig. **2** and Table **2**) [1 - 29].

Step 2: Fermentation or Acidogenesis

Other types of biomass break down after hydrolysis break down into simpler compounds during the acidogenesis phase. In an acidic environment, CO_2, H_2, ammonia, H_2S, alcohols, volatile fatty acids, carbonic acids, and by-products are formed. Biomass, which cannot reach sufficient size for methane production, reacts with the acetogenesis bacteria in the environment for the next stage [1 - 29].

Step 3: Acetogenesis

Acetogenesis bacteria use organic molecules of level 2 as energy and carbon source from acetate. CO_2, acetic acid, and H_2 are the essential ingredients for methane production in other products [1 - 29].

Step 4: Methanogenesis

In methanogenesis (the last stage of anaerobic digestion), methanogenic bacteria produce by-products from acidogenesis and hydrolysis, such as methane from products of steps 3 [1 - 29].

The two main reactions known at this stage are the use of carbon dioxide and acetic acid as reactants for the production of methane. From products, methane and CO_2 can turn into water. The primary known CH_4 raw material at this stage is acetic acid [1 - 29].

While CO_2 can be converted to methane and water by the reaction, the primary mechanism for forming methane in methanogenesis is acetic acid. This pathway creates two main products of anaerobic digestion, methane, and CO_2. It is observed that anaerobic digestion technology is applied in three categories [1 - 29].

- Dry continuous system: It is a system based on vertical or horizontal flow reactors working with 15-45% dry matter.
- Wet continuous system: technology in which raw material containing more than 20% dry matter is processed in continuously stirred tanks [1 - 29].
- Dry intermittent operation: These are systems where dry matter solids with 28-50% are processed at stages such as multi-stage irritants and filtration. The final product is used as a fertilizer [1 - 29].

To use biogas in the transportation sector, the engine should be developed keeping in mind the following instructions. One stage of this development is purification. It is based on the removal of pollutants such as water vapor and hydrogen sulfide from biogas. Also, siloxanes must be removed in volatile harmful organic compounds and dangerous trace compounds. It is essential to remove hydrogen sulfide. In the anaerobic digestion process, it can be extracted directly from the digester during the oxidation step (Fig. **3**) [13]. It is possible to remove hydrogen sulfide before CO_2 removal. Some additives will add oxygen to the digester or react with hydrogen sulfide during the process.

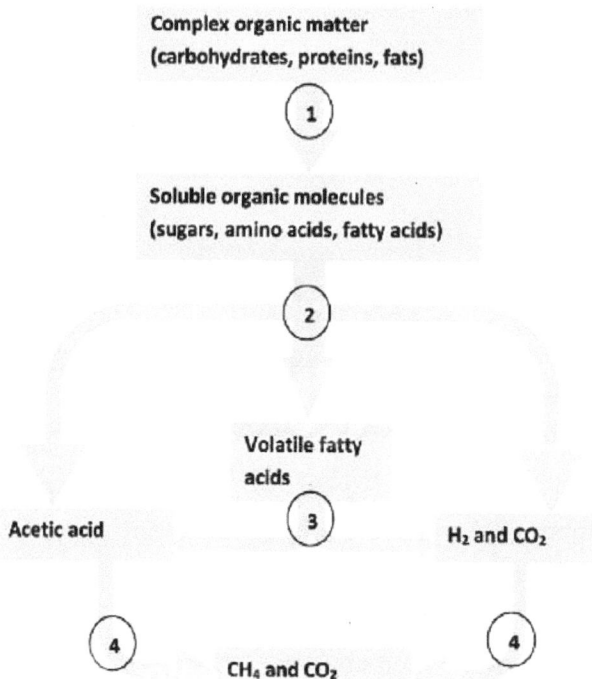

Complex organic matter
(carbohydrates, proteins, fats)

1

Soluble organic molecules
(sugars, amino acids, fatty acids)

2

Volatile fatty
acids

3

Acetic acid H_2 and CO_2

4   **4**

CH$_4$ and CO$_2$

Fig. (2). Main biogas products [13].

With these processes, the amount of hydrogen sulfide is reduced from 3000-5000 ppm to 50-100 ppm. The H_2S content required for the use of biogas as vehicle fuel should be less than five ppm. It is used to remove H_2S in the use of impregnated activated carbon.

Sodium hydroxide is another alternative. It is used as an aluminum or silica gel drying agent to remove water vapor from biogas. Physically, the water vapor is separated from the compressed biogas and cooled in the heat exchanger by adsorption. Often VOC and siloxanes consist of landfills and wastewater treatment. Lubricants and defoamers are a source of siloxanes in biogas [1 - 29].

Table 2. Four steps reactions for anaerobic digestion [13].

Hydrolysis	Acidogenesis	Acetogenesis	Methanogenesis
PO → Glucose	Glucose → VFA	-	Acetate → CH_4, CO_2
PO → Glucose	Glucose → propionate, butyrate, acetate	Propionate, butyrate → acetate	Acetate → CH_4, CO_2
PO → amino acids, FA, sugars	Sugars, amino acids, FA → acetate, propionate	Propionate → acetate	Acetate → CH_4 H_2, CO_2 → acetate
PO → amino acids, FA, sugars	Glucose → propionate, butyrate, acetate	Butyrate, propionate → acetate	Acetate → CH_4 H_2, CO_2 → acetate
EBB → SO	SO → acetate	-	VFA → CH_4
PC → SC	SC → propionate, butyrate, acetate	Propionate, butyrate → acetate	Acetate → CH_4
PC, fats, proteins → sugars, amino acids, FA	SC → propionate, acetate, butyrate	Propionate, butyrate → acetate	Acetate → CH_4
PO → proteins, carbohydrates, fats → sugars, amino acids, FA	Sugars and amino acids → propionate, butyrate, acetate FA → acetate	Propionate, butyrate → acetate	Acetate → CH_4 H_2, CO_2 → acetate

PO: Particulate organics SO: Soluble organics FA: Fatty acids VFA: Volatile fatty acids EBB: Easily biodegradable biomass

The inspection of various properties of the product obtained in anaerobic treatment determines the product quality [1 - 29]. Therefore, a control chart conforming to Table **3** should be followed [1 - 29].

Table 3. Biogas characterization methods [13].

Characterization Method	Properties
Manometric	Biogas flow
Volumetric displacement	Biogas flow
Hydrogen	Exhale hydrogen monitor
Hydrogen	Mercury-mercuric oxide detector cell
Hydrogen	Palladium metal oxide semiconductors
Hydrogen	Thermistor thermal conductivity
Methane	Treatment of biogas with soda lime
Methane	Gas chromatography infrared analyzer
VFA	Gas chromatography and online sampling
VFA	Gas chromatography (off-line)
VFA	Indirectly *via* titration
VFA	Gas phase extraction at pH < 2

(Table 3) cont.....

Characterization Method	Properties
Dissolved hydrogen	Hydrogen/air fuel cell
Dissolved hydrogen	Amperometric probe
Dissolved hydrogen	Teflon or Silicon membrane tubing to transfer dissolved hydrogen to a gas phase
TS, VS	Drying
TOC	Infrared analyzer
COD	Spectrometry and oxidation
BMP	Bioassay

TS: Total solids VS: Volatile solids TOC: Total organic carbon COD: Chemical oxygen demand BMP: Biochemical methane potential

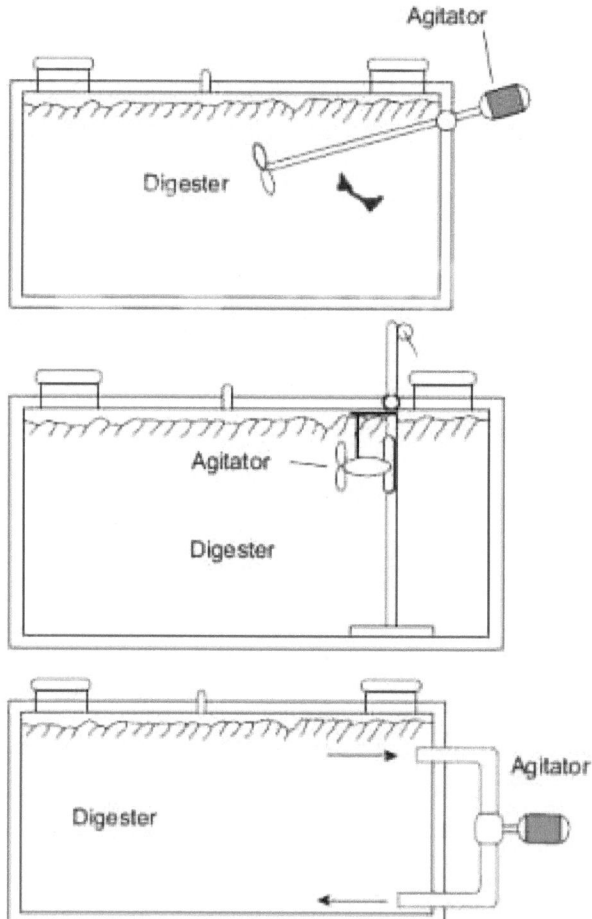

Fig. (3). A simple digester for anaerobic process **a:** impellers in different points, **b:** a pumping force agitation [8].

Activated carbon filters may be required to separate because these contaminants are both water-soluble and insoluble. For better separation, a second filter is installed with the activated carbon filter. Whether the excretion is pure or not, the analysis could be done with periodic impurity of the downstream gas flow. CO_2 removal is also essential for the use of biogas in vehicle engines. It also increases the calorific value of biogas. The energy amount of biogas is almost the same as methane. However, the transfer of CO_2 to certain liquids, the process of separation by using membranes, and the removal from solid material can be realized. The applicability of these technologies depends on capital costs. It also brings up process parameters and techniques for separating energy demand. The plant capacities, which are purified from the impurities of biogas and converted to biomethane, are generally significant. This purification takes place in these plants, which produce approximately 200 m^3/hour of raw biogas. The capacities of the raw biogas purification plants vary between 20 m^3-200 m^3. While the biogas is purified, it is compressed under a suction column at 5-10 bar pressure and injected into the washer while spraying solvent or pure water [1 - 29].

Methane is absorbed while transferring H_2S and CO_2 to solvent or water to increase the contact surface. Because these two molecules dissolve more in water than methane, the biogas is dried as methane in a column, and an extracted biomethane can be recovered. Under the column, solvent or water containing methane and H_2S also includes CO_2. It is injected into a tank by reducing the pressure. Meanwhile, a dissolved gas emission containing methane and CO_2 occurs. This released gas is sent back to the raw biogas inlet at the beginning of the purification. The water in the flash tank, where pressure drop occurs, is renewed at the outlet. H_2S and CO_2 are removed by air desorption in a separation tank. It can be heated to provide the desorption property of the solvent. The solvent or water recovered in the system is fed back into the suction column. It is possible to provide the separation between the liquid medium and CO_2 with a chemical (amine solution). More separation from water is provided in the organic solvent. Besides, the selectivity of the separation between H_2S, CO_2, and CH_4 is ensured. This causes methane to be separated more effectively. The process takes place partially above the ambient pressure. The less electrical energy consumed in the process (for pressure generation), the higher the thermal energy (160 °C) required to recover the solvent. Pressure swing adsorption is carried out for the pretreatment of biogas and removal of H_2S.

The removal of CO_2 with adsorbents takes place with adsorbent materials in a reactor at 4-7 bar temperatures, and the biomethane is recovered at the inlet of the reactor. CO_2 poisoned adsorbent material is removed from the reactor to release CO_2 into the atmosphere. Adsorption of methanol under pressure may occur again. Membrane technology is also used in the development of biogas. Biogas

H_2S purified by this technology is removed from the particulate matter and then dried. It is passed through the membrane after being compressed under 4-16 bar pressure. Methane cannot pass through the membrane at this time, but CO_2 can pass. It is possible in cryogenic separation. Gas behavior varies at -120 °C. Membrane and cryogenic separation technologies are not yet in the industrialization stage [1 - 29].

BIOGAS PRODUCTION *VIA* ANAEROBIC DIGESTION

Biogas Production Global Overview

The technological and later industrial progress of biogas in the world has shown significant development globally. China, Europe, and the USA make up 90% of the biogas production. The biggest biogas producer of today is Europe. The biggest producer of Europe is Germany (2 thirds of the European plant capacity, see Fig. (**4**). The reason for Germany to stand out so much was the right raw material selection (Energy plants) [1 - 10, 20 - 30]. Fig. (**5**) shows the biogas capacities of different production yield, "Food and green" represents leaves and grass call as MSW. Wastewater is excluded *via* high variability of yields, based on wastewater and treatment technologies in different regions. 1 toe = 11.63 MWh = 41.9 gigajoules, 1 toe = 1 tonne of oil equivalent) [30].

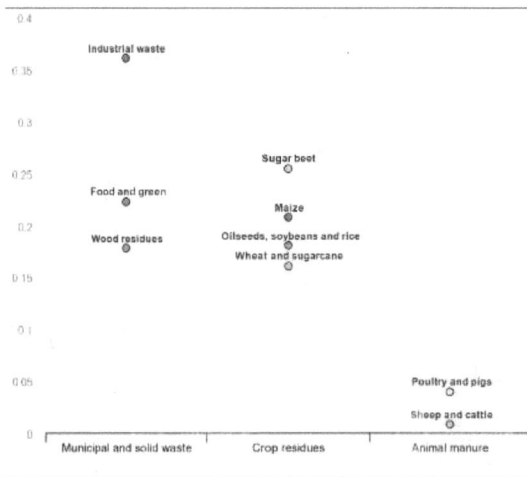

Fig. (4). Global biogas installed power generation capacity between 2010-2018 [30].

The size of the cost varies due to conditions such as animal waste, industrial, agricultural, and forest product residues, and methane storage in landfills. After Germany, biogas production has become active in Italy, France, Denmark, and the Netherlands [1 - 29]. Fig. (**6**) shows the biogas production technologies cost. This figure includes biodigesters constructed in place using sand, cement, and gravel which are traditional construction materials (HH = household; "HH Basic"), "HH Advanced" consists of more expensive composite material pre-manufactured biodigesters. Maintenance and operating costs include labor costs, ordinary and extraordinary maintenance. Capital costs have been levelised for each Technologies production lifetime: 25 years for landfill advanced household biodigesters and gas recovery; 20 years for wastewater digesters and centralized biodigesters (medium, small, and large) and; 15 years for basic household biodigesters. 1 MBtu = 0.29 MWh.) [30].

Fig. (**8**) shows the potential global biogas supply cost curve for 2018. Crops include only subsequent crops (not dedicated energy crops) and crop residues. The curve integrates feedstock costs and technology. Technology costs consist of the biodigester only, *i.e.*, excluding any costs for equipment to transform biogas into heat and power (1 MBtu = 0.29 MWh) [30]. Fig. (**9**) presents the potential global biogas supply for 2040 [30]. Fig. (**10**) shows the potential global biomethane supply cost curve for 2018. It is noted that C & S America means Central and South America. In this Figure, the curve integrates feedstock and technology costs whilst injection costs are not included. The figure presents all the biogas potential that can be upgraded to biomethane. Fig. (**11**) depicts the potential global biomethane supply cost curve for 2040.

Fig. (**7**) shows the global consumption values in 2018.

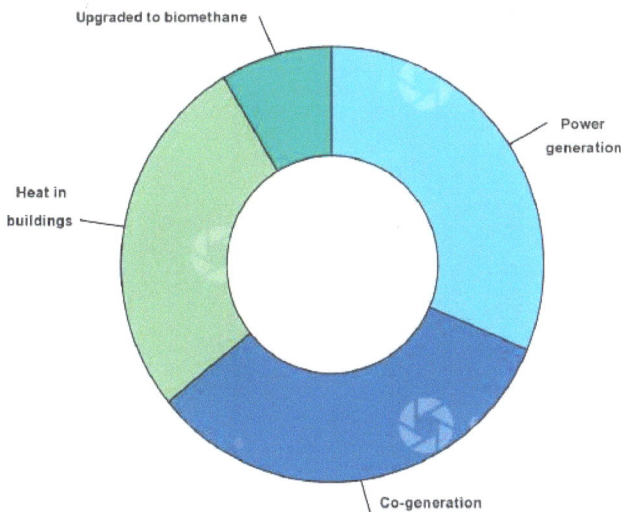

Fig. (5). Average biogas production yield by a tonne of feedstock type.

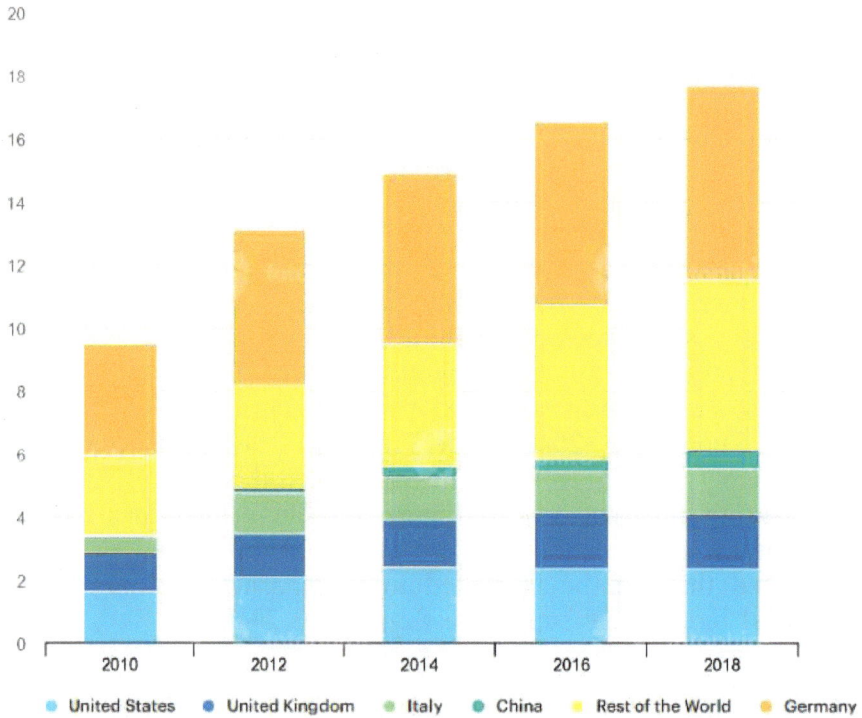

Fig. (6). Average costs of biogas production technologies per unit of energy produced (excluding feedstock).

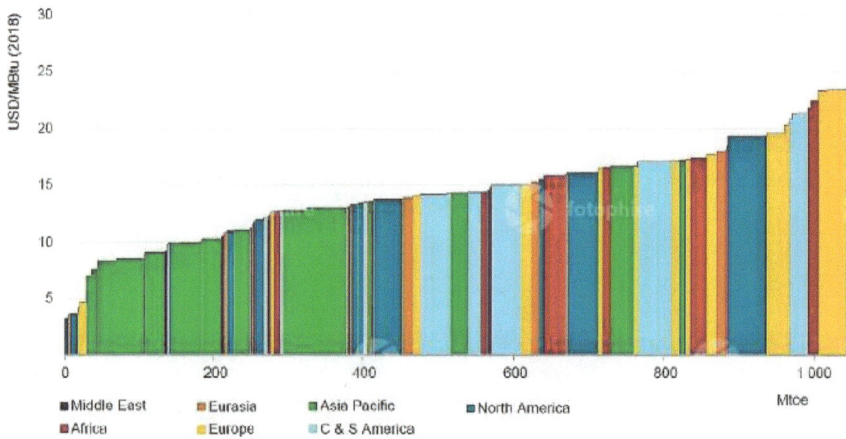

Fig. (7). Global consumption values in 2018.

In China, rural biogas converters are spreading in rural areas due to the government's policies on clean cooking fuels and modern energy. These small facilities account for about 70% of the installed power. Investment in larger-scale facilities for both power and heat-generating plants has increased in China. China has published a guide to popularize the use of biomethane as a vehicle fuel in the transportation industry as well as residential use [1 - 29].

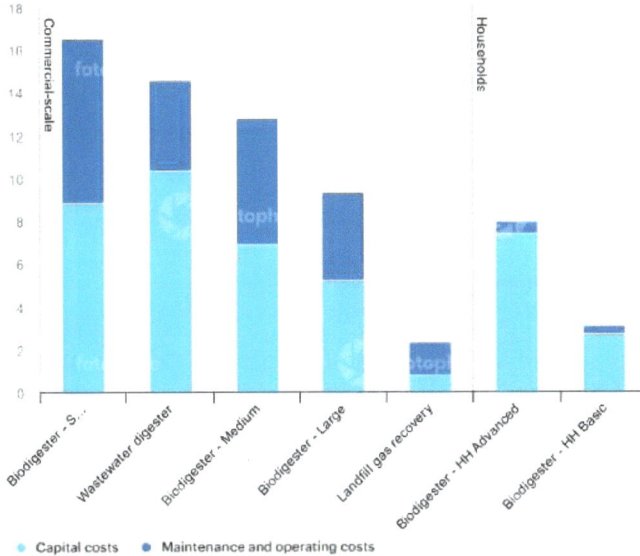

Fig. (8). Global biogas supply by feedstock potential cost curve for 2018.

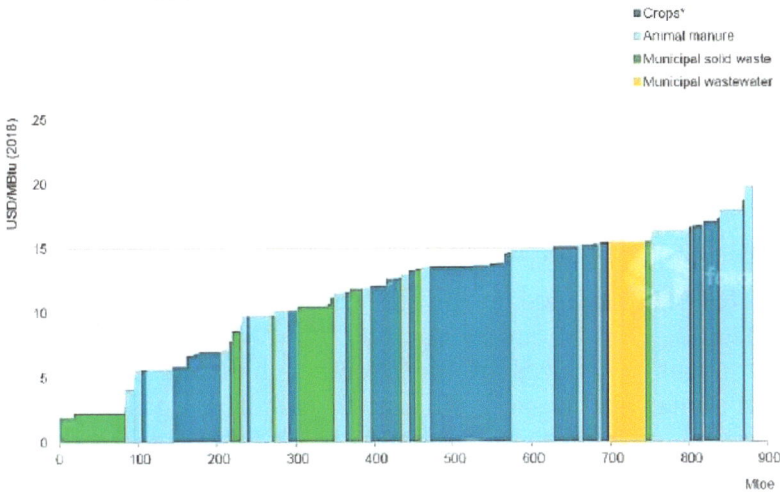

Fig. (9). Global biogas supply by feedstock potential cost curve for 2040.

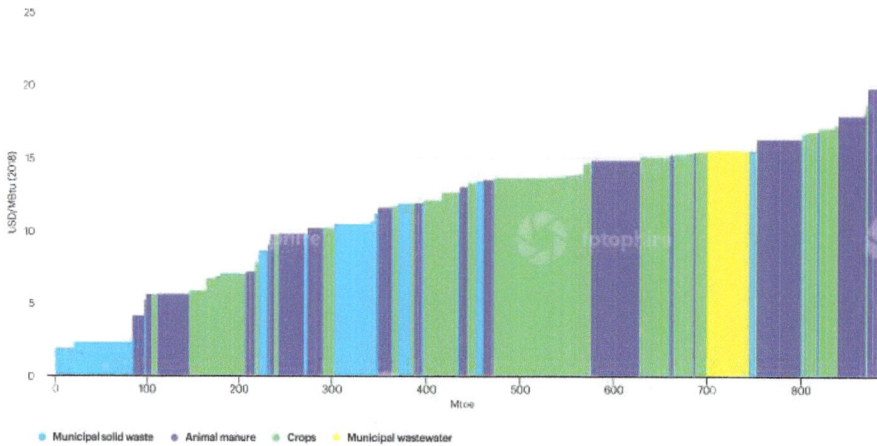

Fig. (10). Global biomethane supply by region potential cost curve for 2018 [30].

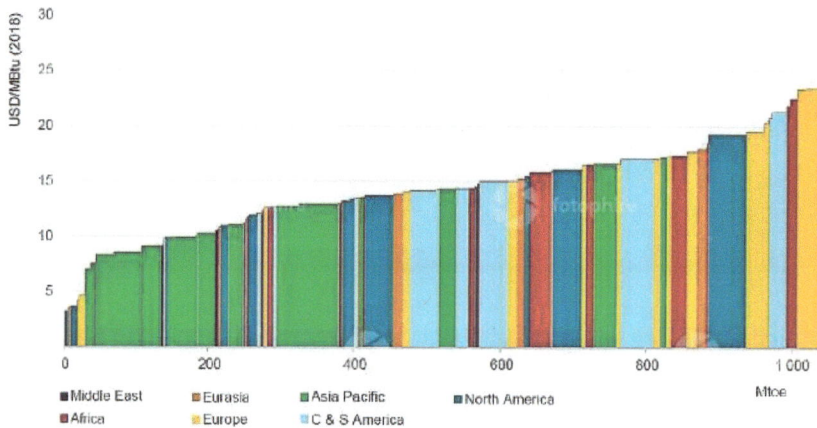

Fig. (11). Global biomethane supply by region potential cost curve for 2040 [30].

The United States has managed to quickly realize its biogas production with natural raw material storage areas. As the livestock industry of Ululsa is responsible for about two-thirds of the country's methane emissions, biogas production from agricultural waste is still on the agenda. Biogas production from agriculture is responsible for approximately 90% of America's total biogas production. According to China, the United States is the global leader in the use of biomethane for vehicle fuel [1 - 29]. Fig. (**12**) presents global biomethane production.

India and Thailand are the other countries where biogas is produced the most. The use of biogas became widespread between 2007 and 2011 thanks to the clean development movement in these two countries. However, since 2011, the amount of these loans has decreased, so biogas projects also have fallen [1 - 20].

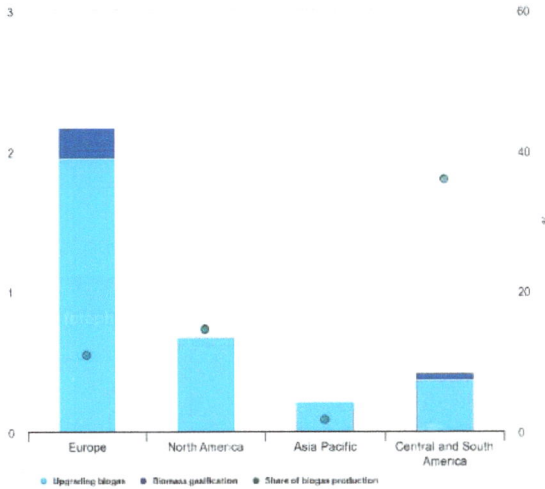

Fig. (12). Biomethane production capacities of the world [30].

The biofuel industries in Thailand produce biogas with the remnants of pig farms and the starch sector [1 - 20].

India aims to develop 5000 new compressed biogas plants. It has issued government supports for biogas production in Brazil and Argentina. Especially ethanol production is typical in Brazil [1 - 29].

In the African continent, many projects are provided with the establishment of the necessary data bank to complete the lack of information for biogas production, with the incentives of some non-governmental organizations, with 50% grant support from some governments such as Burkina Faso, Benin, and Ethiopia. Financial investments, including loan subsidies that cover 50% of the cost of biogas plants, have become widespread in Kenya. The global biogas consumption is shown in Fig. (**7**) [1 - 29].

A few of the factors that affect the cost most in biogas production with the oxygen-free method are plant capacity and raw material price. Both operation and maintenance and the amount of capital decrease partially as the facility scale increases. If the raw materials are ready for the storage and transportation of organic wastes, the establishment of the facility may be more comfortable. In the installation of the biogas plant, the cost of the digester was determined as 55-60% for industrial and agricultural wastes and 40-45% for energy plants. Apart from this, it should be explained in detail in components such as construction costs, machinery and equipment costs, and the amount of investment [1 - 29].

   i. conversion to biomethane
  ii. anaerobic digestion
 iii. distribution

Also, it can be considered to add costs to the use of biomethane as a vehicle fuel in biogas production [1 - 29].

Biogas production from Algae is a direct and easy way method (Fig. **13**). There are lots of algae types that have high lipid contents (Table **4** and Fig. **14**) and different algae to product methods are known (Fig. **15**) [1 - 29]. Table **5** shows the various types of feedstocks biomethane capacities in detail [1 - 29].

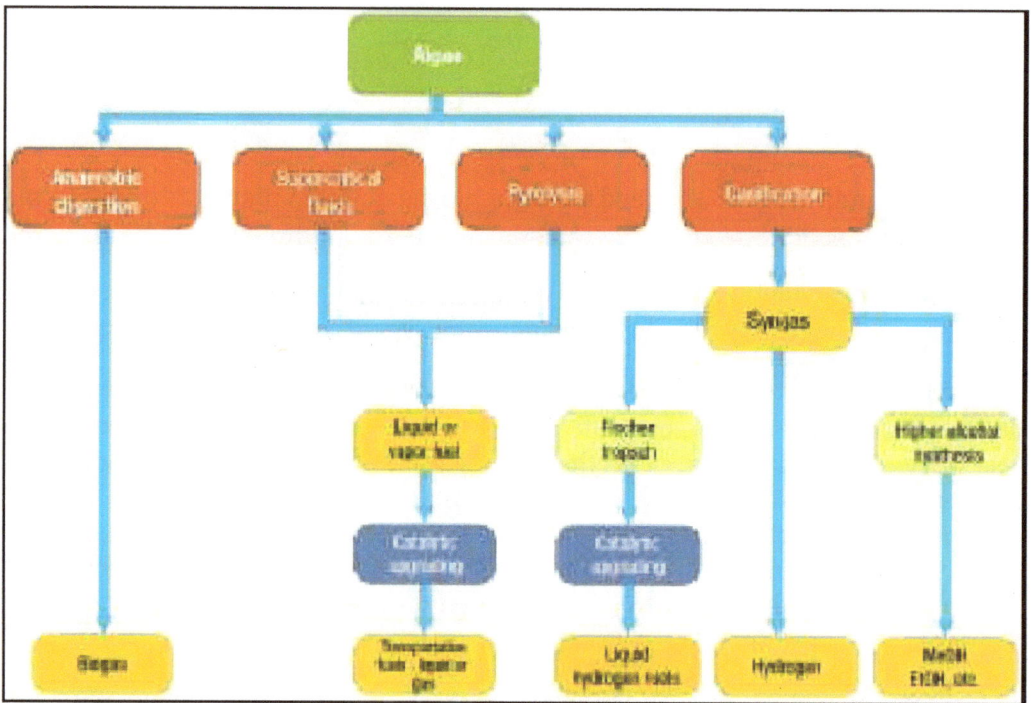

Fig. (13). Biogas production *via* anaerobic digestion has advantages.

Table 4. Algae type *versus* lipid content [31].

Algae Type	Lipid Content (%)
Schizochytrium sp.	50-77
Nitzschia palea	40
Nitzschia dissipate	66
Boekelovia hooglandii	59
Monallantus salina	41-72

(Table 4) cont.....

Algae Type	Lipid Content (%)
Navicula saprophila	58
Navicula accept	47
Navicula pelliculosa	45
Navicula pseudotenelloides	42
Chlorella minutissima	57
Chlorella pyrenoidosa	36
Chlorella vulgaris	41
Dunaliella sp.	45-55
Neochloris oleoabundans	35-54
Nannochloris sp.	48
Amphora sp.	51
Ourococcus sp.	50
Scenedesmus sp.	45
Scenedesmus obliquus	41
Nannochloropsis salina	46
Ankitodesmus sp.	40
Chaetoceros calcitrans	40
Chaetoceros muelleri	39
Amphiprora hyalina	37
Cyclotella cryptica	37
Cylindrotheca sp.	16-37
Amphiprora hyalina	37
Pavlova lutheri	36

Table 5. Different feedstocks have biomethane capacities [8].

Feedstock	TS: VS (%)	VS (%)	TS (%)	Max CH$_4$ (%)	Biogas Potential (L/biogas/kg Initial VS)
Manure, swine	52	1.4	2.7	70	248 ± 32
Manure, dairy, freestall, wood shaving bedding	82	6.8	8.3	56	510 ± 90
Coffee grounds	98	39.3	40.2	46	417 ± 97
Manure, dairy, dry lot, daily scrape	72	15.1	21.1	56	429 ± 76
Fruit waste, pressed grape residuals	74	29.9	40.2	53	489 ± 106

(Table 5) cont.....

Feedstock	TS: VS (%)	VS (%)	TS (%)	Max CH$_4$ (%)	Biogas Potential (L/biogas/kg Initial VS)
Cellulose, microcrystalline	88	86.6	98.6	38	514 ± 70
FOG	88	3.5	4.0	67	870 ± 250
Brewer grain	96	25.6	26.5	96	739 ± 49
Glycerin	95	68.4	71.7	65	1010 ± 10
Switchgrass, fresh	96	57.0	59.6	50	540 ± 30
Cheese whey	83	3.9	4.7	56	710 ± 6
Fruit waste, pineapple	98	11.5	11.8	63	990 ± 410
Primary solids/waste-activated sludge	80	3.5	4.3	554 ± 76	48

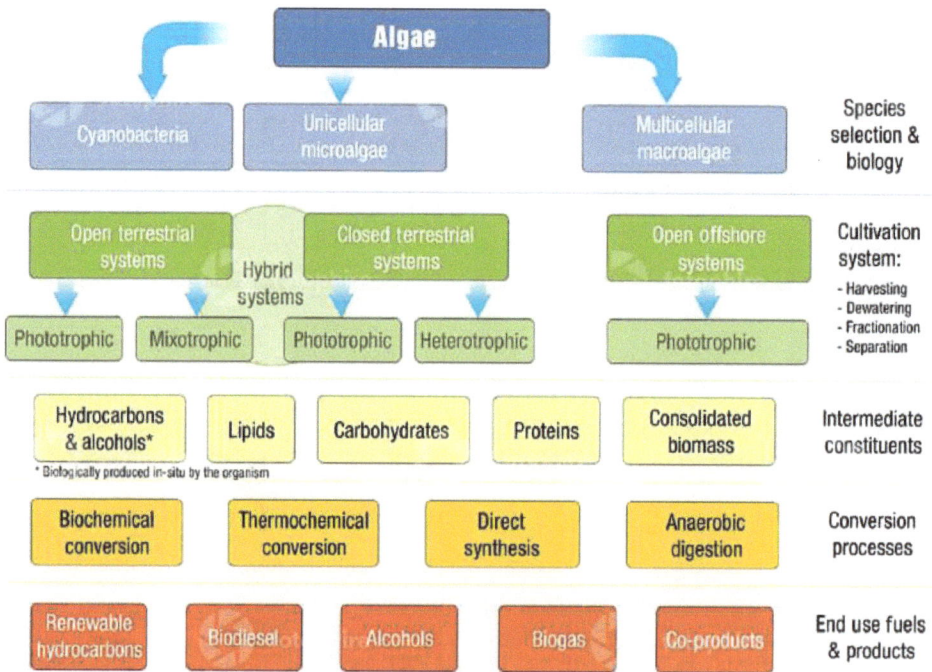

Fig. (14). Different types of microscopic algae cells [31].

While the interest in biofuels produced from lags has increased recently, companies such as Solazyme have decided to invest since 2014. The Solazyme company, which has partnered with the Bunge company, produces 100,000 mt of fuel per year. However, Sapphire Energy company has a 2200-acre algae farm to produce 10000 barrels of fuel in New Mexico. The development of these technologies is that fuels produced from algae can compete with fossil fuels in the market. The main effective parameters that should be considered are ensuring the operational and cost efficiency of algae production facilities (appropriate logistics network, reducing algae cultivation cost, etc.), preferring the oil-rich ones of algae raw materials, making oil production efficient, and finally supplying the produced oil easily and quickly in this competition (Fig. **4**) [1 - 29, 32 - 36].

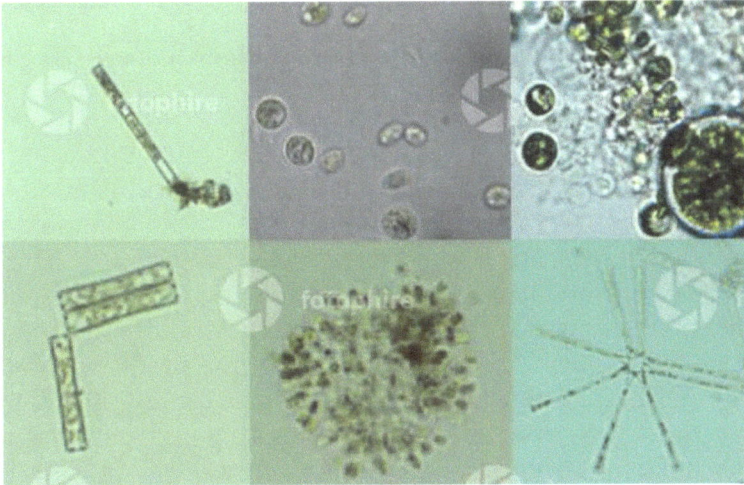

Fig. (15). Various algae to biofuel methods [31].

Costs

Raw material storage costs (energy costs, logistics costs of agricultural and industrial residues)

- Production (Biogas production)
- Pretreatment
- Feeding equipment
- Electric control system
- Buildings
- Labor costs [1 - 29],

Raw material costs also depend on factors such as soil fertility and global agricultural commodity rates, and regional climatic conditions. If waste products are provided free of charge, it can partially reduce the cost of raw materials. Even on-demand, biogas plants can earn money from waste treatment [1 - 29].

The following costs (or revenues reflected at negative costs) are common values for biogas production from these types of raw materials in Central Europe [11, 10, 20 - 29]:

- Energy crops (*e.g.*, corn silage, sugar beet, wheat grain): 9 USD / GJ - 14 USD / GJ,

Agricultural Waste Products (Mostly Slurry and Fertilizer): 0 USD / GJ - 6 USD / GJ

- Industrial waste products (*e.g.*, distilled distillation, grain vinasse): 3 USD / GJ - 6 USD / GJ
- Municipal waste (municipal solid waste or sewage sludge): -20 USD / GJ - -9 USD / GJ

Heat is required to reach the optimum temperature for digestion. With relatively dry raw materials (*e.g.*, energy crops), the heat requirements are relatively low, and consequently, heat is costly because less water should be heated compared to the use of manure or industrial wastewater. Also, heat demand depends on the climate conditions of the facility and the location of the biogas plant. Typically, 5-10% of the total energy from biogas produced to heat digesters is needed. It is estimated that the electricity consumed by the pumps that perform the mass transfers required during biogas production is between 20-30 kWh for 1 MWh energy production. These values vary according to the type of raw material. For example, this value for fertilizer biogas production is approximately $ 0.23 per m^3 methane production and $ 0.55 per m^3 for energy crops. industrial waste methane production is $ 0.12 per m^3. As the capacity of the facility increases, the number of raw materials increases, so the transfer will be difficult, and this can affect the cost negatively. In the case of raw material mixtures, it may lead to an increase in production costs due to situations such as raw material inventory. Therefore, the operating cost of small capacity facilities can be called more than large-scale facilities. To use the produced biogas in the transportation sector, formalization is required. It can be said that the technology, facility capacity, and environmental regulations used in addition to the factors evaluated with the increase in fuel quality affect the total cost. The operating and capital costs of the facilities, where the use of raw materials increases, is lower. However, amine scrubbers, water, membranes, and pressure swing adsorption systems considered in intermediate stage processes are another issue affecting the cost. In the meantime, operating costs can be affected by system optimization over time. Upgrades can be reduced with some ease. Cost savings can be achieved by sending biomethane to gas grills at high pressure before they are obtained. It is important to position the biomethane at fuel stations where it will be stored. Apart from storage, the costs in its distribution do not bring any costs in the presence of a fuel station. It can be said that the distribution of biomethane, biogas, or compressed biomethane gas liquefied during the distribution of the biogas from the production facility to the stage, where they are used by customers, will be effective in reducing the cost of the product [1 - 29].

If there is a gas network of less than 5 km from the biogas production facility, the cost of the biogas plant is not significantly affected. Costs related to the gas network can be classified under two headings as follows [1 - 29].

1. Effective gas transportation due to the state's ability to transfer to the natural gas network
2. Additional costs related to operation and capital due to the transfer to the network.

Considering the transfer costs of the transfer to the network, which are explained in the two items above, both the installation of the required network for the transfer and the additional costs brought by the organization, it has become necessary to make a comparison between the output fuel prices from the existing fuel stations and the product output price of the biogas fuel station. There is a situation where supply costs are at the forefront, in addition to the fuel price produced. However, there are individual operational costs, that are different between countries, such costs include sales tax, and customs [1 - 29]. It means that there is a cost related to the pressure of the gas network and the injection of the produced biogas into the network, together with the additional costs. The cost of a biogas production plant located 10 km from the main connection point can determine the installation cost of the facility. Increasing the injection capacity is important in terms of decreasing the cost. The maximum cost item is 5-8 bar facility outlet pressure and 55 bar operating pressure, while the minimum cost is 300-800 millibars working pressure. The cost can be reduced when used by mixing the produced biomethane with liquefied petroleum gas depending on the situation. Mixing biomethane with natural gas or petroleum gas is determined according to the desired energy value [1 - 29].

Depending on the injection capacities, the cost of biomethane injection into the gas network is important. The average injection cost varies from country to country, but it is known as $ 0.06-0.047/1 m^3. Taxes on transfer to the public gas network are added to the additional cost. The additional cost items above vary from one country to another, and even from one country to the next. These items differentiate issues regarding maximum output capacity and total annual production. On average, the gas network usage fee of a gas filling station varies between $ 0.05-0.26/m^3. These figures do not change much for biomethane and natural gas. The distribution cost of produced biomethane is 0.09-0.22 $ per m^3. If the cylinder made of high-pressure steel is used in trucks, there is a distribution cost of $ 0.17-0.22 per m^3. A certain cost is required for the manufacture and operation of the biogas filling unit. The gas transfer rate of the filling station is another factor that determines the cost. The total cost for operating and financing a fast-filling station with an average capacity of 3500 m³ is $ 0.5 per m³. The produced biogas is liquefied at 200 bars by increasing its quality as a result of a chemical conversion at a low cost, and it is transported at a lower cost by reaching the liquid natural gas quality. This process takes place at product-network station distances of more than 100 km. Transport of 3-300 m³/h can take place at an average cost of $ 0.07-0.35 [1 - 29].

CONCLUSION

In our study, biomethane production stages from algae, fermentation, acetogenesis, and methanogenesis stages are summarized. Then, biofuel production stages; biomethane conversion, anaerobic digestion, and distribution. Economic analyses of the above-mentioned stages were made. It is mentioned that the produced biomethane can be transferred to the existing natural gas pipeline, but the transfer cost will gain importance at this stage.

CONSENT FOR PUBLICATION

Not applicable.

CONFLICT OF INTEREST

The author declares no conflict of interest, financial or otherwise.

ACKNOWLEDGEMENT

Declared none.

REFERENCES

[1]   Ometto F, Karlsson A, Ejlertsson J, Björn AV, Shakeri SY. 5 - Anaerobic digestion: An engineered biological process.Substitute Natural Gas from Waste. Academic Press 2019; pp. 63-74.
        [http://dx.doi.org/10.1016/B978-0-12-815554-7.00005-2]

[2]   Muthudineshkumar R, Anand R. 6 - Anaerobic digestion of various feedstocks for second-generation biofuel production.Advances in Eco-Fuels for a Sustainable Environment. Woodhead Publishing 2019; pp. 157-85.
        [http://dx.doi.org/10.1016/B978-0-08-102728-8.00006-1]

[3]   Stamatelatou K, Antonopoulou G, Michailides P. 15 - Biomethane and biohydrogen production via anaerobic digestion/fermentation.Advances in Biorefineries. Woodhead Publishing 2014; pp. 476-524.
        [http://dx.doi.org/10.1533/9780857097385.2.476]

[4]   Oppong G, O'Brien M, McEwan M, Martin EB, Montague GA. Advanced control for anaerobic digestion processes: Volatile solids soft sensor development.Computer Aided Chemical Engineering. Elsevier 2012; Vol. 30: pp. 967-71.

[5]   de la Rubia MA, Villamil JA, Mohedano AF. Chapter 4 - Anaerobic digestion for methane and hydrogen production. In: Olivares JA, Puyol D, Melero JA, Dufour J, Eds. Wastewater Treatment Residues as Resources for Biorefinery Products and Biofuels. Elsevier 2020; pp. 67-83.

[http://dx.doi.org/10.1016/B978-0-12-816204-0.00004-7]

[6] Labatut RA, Pronto JL. Chapter 4 - sustainable waste-to-energy technologies: Anaerobic digestion. In: Trabold TA, Babbitt CW, Eds. Sustainable Food Waste-To-energy Systems. Academic Press 2018; pp. 47-67.
[http://dx.doi.org/10.1016/B978-0-12-811157-4.00004-8]

[7] Jankowska E, Zieliński M, Dębowski M, Oleśkowicz-Popiel P. Chapter 15 - Anaerobic digestion of microalgae for biomethane production. In: Basile A, Dalena F, Eds. Second and Third Generation of Feedstocks. Elsevier 2019; pp. 405-36.
[http://dx.doi.org/10.1016/B978-0-12-815162-4.00015-X]

[8] Kirk DM, Gould MC. Chapter 17 - Bioenergy and anaerobic digestion. In: Dahiya A, Ed. Bioenergy. 2nd ed. Academic Press 2020; pp. 335-60.
[http://dx.doi.org/10.1016/B978-0-12-815497-7.00017-8]

[9] Gould MC. Chapter 18 - Bioenergy and Anaerobic Digestion. In: Dahiya A, Ed. Bioenergy. Boston: Academic Press 2015; pp. 297-317.
[http://dx.doi.org/10.1016/B978-0-12-407909-0.00018-3]

[10] Nguyen D, Nitayavardhana S, Sawatdeenarunat C, Surendra KC, Khanal SK. Chapter 31 - Biogas production by anaerobic digestion: Status and perspectives. Biofuels: Alternative Feedstocks and Conversion Processes for the Production of Liquid and Gaseous Biofuels. Second Edition. Pandey A, Larroche C, Dussap C-G, Gnansounou E, Khanal SK, Ricke S. Academic Press 2019; pp. 763-78.

[11] Murphy JD, Thanasit T. 5 - Fundamental science and engineering of the anaerobic digestion process for biogas production. In: Wellinger A, Murphy J, Baxter D, Eds. The Biogas Handbook. Woodhead Publishing 2013; pp. 104-30.
[http://dx.doi.org/10.1533/9780857097415.1.104]

[12] Banks CJ, Heaven S. 6 - Optimisation of biogas yields from anaerobic digestion by feedstock type. In: Wellinger A, Murphy J, Baxter D, Eds. The Biogas Handbook. Woodhead Publishing 2013; pp. 131-65.
[http://dx.doi.org/10.1533/9780857097415.1.131]

[13] Stamatelatou K, Antonopoulou G. 12 - Production of biogas *via* anaerobic digestion. In: Luque R, Campelo J, Clark J, Eds. Handbook of Biofuels Production. Woodhead Publishing 2011; pp. 266-304.

[14] Draa KC, Zemouche A, Alma M, Voos H, Darouach M. Chapter 4 - Control of anaerobic digestion process. In: Boubaker O, Zhu Q, Mahmoud MS, Ragot J, Karimi HR, Dávila J, Eds. New Trends in Observer-based Control. Academic Press 2019; pp. 99-135.

[15] Córdova O, Chamy R. Chapter 15 - Microalgae to biogas: Microbiological communities involved. In: Yousuf A, Ed. Microalgae Cultivation for Biofuels Production. Academic Press 2020; pp. 227-49.
[http://dx.doi.org/10.1016/B978-0-12-817536-1.00015-1]

[16] Guangyin Z, Youcai Z. Chapter Five - Harvest of Bioenergy From Sewage Sludge by Anaerobic Digestion. In: Guangyin Z, Youcai Z, Eds. Pollution Control and Resource Recovery for Sewage Sludge. Butterworth-Heinemann 2017; pp. 181-273.
[http://dx.doi.org/10.1016/B978-0-12-811639-5.00005-X]

[17] Sahoo K, Mani S. Economic and environmental impacts of an integrated-state anaerobic digestion system to produce compressed natural gas from organic wastes and energy crops. Renew Sustain Energy Rev 2019; 115: 109354.
[http://dx.doi.org/10.1016/j.rser.2019.109354]

[18] Zhang L, Loh KC, Zhang J. Enhanced biogas production from anaerobic digestion of solid organic wastes: Current status and prospects. Bioresour Technol Rep 2019; 5: 280-96.
[http://dx.doi.org/10.1016/j.biteb.2018.07.005]

[19] Pérez-Camacho MN, Curry R, Cromie T. Life cycle environmental impacts of substituting food wastes for traditional anaerobic digestion feedstocks. Waste Manag 2018; 73: 140-55.

[http://dx.doi.org/10.1016/j.wasman.2017.12.023] [PMID: 29291908]

[20]   Harun N, Othman NA, Zaki NA, Mat Rasul NA, Samah RA, Hashim H. Simulation of Anaerobic Digestion for Biogas Production from Food Waste Using SuperPro Designer. Mater Today Proc 2019; 19: 1315-20.
[http://dx.doi.org/10.1016/j.matpr.2019.11.143]

[21]   Tranter RB, Swinbank A, Jones PJ, Banks CJ, Salter AM. Assessing the potential for the uptake of on-farm anaerobic digestion for energy production in England. Energy Policy 2011; 39(5): 2424-30.
[http://dx.doi.org/10.1016/j.enpol.2011.01.065]

[22]   Raheem A, Hassan MY, Shakoor R. Bioenergy from anaerobic digestion in Pakistan: Potential, development and prospects. Renew Sustain Energy Rev 2016; 59: 264-75.
[http://dx.doi.org/10.1016/j.rser.2016.01.010]

[23]   Shrestha S, Fonoll X, Khanal SK, Raskin L. Biological strategies for enhanced hydrolysis of lignocellulosic biomass during anaerobic digestion: Current status and future perspectives. Bioresour Technol 2017; 245(Pt A): 1245-57.
[http://dx.doi.org/10.1016/j.biortech.2017.08.089] [PMID: 28941664]

[24]   Thomas P, Soren N, Rumjit NP, George James J, Saravanakumar MP. Biomass resources and potential of anaerobic digestion in Indian scenario. Renew Sustain Energy Rev 2017; 77: 718-30.
[http://dx.doi.org/10.1016/j.rser.2017.04.053]

[25]   Wilkinson KG. A comparison of the drivers influencing adoption of on-farm anaerobic digestion in Germany and Australia. Biomass Bioenergy 2011; 35(5): 1613-22.
[http://dx.doi.org/10.1016/j.biombioe.2011.01.013]

[26]   Farangi M, Asl Soleimani E, Zahedifar M, Amiri O, Poursafar J. The environmental and economic analysis of grid-connected photovoltaic power systems with silicon solar panels, in accord with the new energy policy in Iran. Energy 2020; 202: 117771.
[http://dx.doi.org/10.1016/j.energy.2020.117771] [PMID: 32367905]

[27]   Holm-Nielsen JB, Al Seadi T, Oleskowicz-Popiel P. The future of anaerobic digestion and biogas utilization. Bioresour Technol 2009; 100(22): 5478-84.
[http://dx.doi.org/10.1016/j.biortech.2008.12.046] [PMID: 19217772]

[28]   Edwards J, Othman M, Burn S. A review of policy drivers and barriers for the use of anaerobic digestion in Europe, the United States and Australia. Renew Sustain Energy Rev 2015; 52: 815-28.
[http://dx.doi.org/10.1016/j.rser.2015.07.112]

[29]   Kumaran P, Hephzibah D, Sivasankari R, Saifuddin N, Shamsuddin AH. A review on industrial scale anaerobic digestion systems deployment in Malaysia: Opportunities and challenges. Renew Sustain Energy Rev 2016; 56: 929-40.
[http://dx.doi.org/10.1016/j.rser.2015.11.069]

[30]   IEA. Energy Supply and Investment Outlook (ESIO) Division of the Directorate of Sustainability Outlook for biogas and biomethane 2020; 1-93.

[31]   Dahiya A. Chapter 14 - Algae biomass cultivation for advanced biofuel production. In: Dahiya A, Ed. Bioenergy. Boston: Academic Press 2015; pp. 219-38.
[http://dx.doi.org/10.1016/B978-0-12-407909-0.00014-6]

[32]   BASF. Focus on Surfactants 2015; 2015(10): 3.

[33]   Behera BK, Prasad R. Chapter 2 - Greenhouse gas capture and conversion. In: Behera BK, Prasad R, Eds. Environmental Technology and Sustainability. Elsevier 2020; pp. 41-71.
[http://dx.doi.org/10.1016/B978-0-12-819103-3.00002-0]

[34]   Bhadani A, Kafle A, Ogura T, *et al.* Current perspective of sustainable surfactants based on renewable building blocks. Curr Opin Colloid Interface Sci 2020; 45: 124-35.
[http://dx.doi.org/10.1016/j.cocis.2020.01.002]

[35]   Bhattacharya M, Goswami S. Microalgae – A green multi-product biorefinery for future industrial prospects. Biocatal Agric Biotechnol 2020; 25: 101580.
[http://dx.doi.org/10.1016/j.bcab.2020.101580]

[36]   Solazyme reveals new identity TerraVia with high focus on algae innovation platform. Focus on Surfactants 2016; 2016(5): 7.
[http://dx.doi.org/10.1016/j.fos.2016.05.032]

CHAPTER 3

Thermal Liquefaction Based Algal Biotechnology

Cemil Koyunoğlu[1,*] and **Hüseyin Karaca**[2]

[1] *Energy Systems Engineering Department, Engineering Faculty, Yalova University, Yalova, Turkey*

[2] *Chemical Engineering Department, Engineering Faculty, Inonu University, Malatya, Turkey*

Abstract: Large amounts of residues are obtained after lipid extraction when producing biodiesel from microalgae. From these residues, animal feed or bioethanol production may be obtained. Another alternative biofuel that can be obtained from microalgae biomass residues is bio-oil from pyrolysis or hydrothermal processes. Out of these, microalgae for biofuel production stand out due to the high thermal value of Algal biomass of around 24 MJ / kg. The organic components of the biological mass can be thermally decomposed in the production of fuel by thermochemical applications, such as direct combustion, gasification, pyrolysis, and liquefaction. With the hydrothermal liquefaction process, microalgae are converted into liquid crude oil with or without a catalyst. The reaction takes place at 280-370 °C and 10-25 MPa pressure on wet biomass in water. Biological oil production by hydrothermal liquefaction method from microalgae has gained considerable attention in recent years. Compared to the biodiesel obtained mainly due to the lipid content, hydrothermal liquefaction converts not only the lipid content but also carbohydrates and proteins.

Keywords: Algae, Algae to oil, Biofuel production, Catalytic process, Liquefaction.

INTRODUCTION

The purpose of converting biomass into liquid fuels is to use biomass that is difficult to use, has low energy contents, and takes up much space. The production of oils allows them to be stored, used directly, and pumped in combustion furnaces or used to obtain certain chemicals and fuels. As the essential product is liquid, the word liquefaction has been accepted to describe this process. A hydrothermal conversion process is a process involving water at high pressures and temperatures. Although high-temperature water (HTW) indicates that the liquid water is below the critical pressure and temperature

---

[*] **Correspondence author Cemil Koyunoğlu:** Energy Systems Engineering Department, Engineering Faculty, Yalova University, Yalova, Turkey; Tel: +90 226 8155378; E-mail: cemil.koyunoglu@yalova.edu.tr

Hüseyin Karaca and Cemil Koyunoğlu (Eds.)

(374.1 °C, 221 bar), over this point, it becomes a highly compressible liquid called supercritical water (SCW). An advantage of hydrothermal treatment for biomass is that hot water can act as a catalyst, a solvent, or a catalyst precursor. While many biomass compounds (*e.g.*, cellulose, lignin) are insoluble in ambient conditions, most are readily soluble in supercritical water (SCW) or HTW. The hydrothermal treatment has unique advantages for wet biomass related to both engineering (*e.g.*, higher energy efficiency) and chemistry (*e.g.*, rapid hydrolysis). The most crucial advantage of hydrothermal conversion is that hydrothermal treatment eliminates the need for feed water drying and removal [1 - 50].

Hydrothermal liquefaction is one of the thermochemical processes used to convert biomass into liquid products with high-energy content. It is generally performed under subcritical water conditions, an organic liquid called bio-oil, and high pressures are obtained (Fig. **1**). The hydrothermal liquefaction process resembles the way fossil fuels are formed. Nevertheless, fossil fuels are formed by the exposure of biomass underground with high pressures and temperatures for many years, while hydrothermal liquefaction results in liquid fuel in much shorter times in hours, even minutes. The fact that water is environmentally friendly and a unique solvent, the process can be applied to wet biomass, and therefore the biomass is not necessarily dried. It can be performed with high-energy efficiency, making hydrothermal liquefaction attractive and at much lower temperatures than pyrolysis. The goal of hydrothermal liquefaction is to obtain a commercial fossil fuel with high enthalpy density after the conversion of biomass. The essential point of the process is the risk and cost of using high pressure [1 - 50].

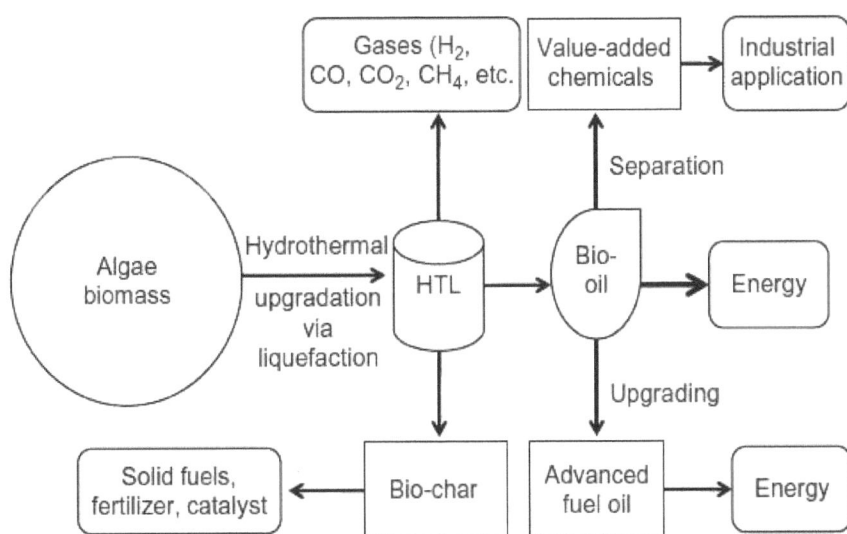

Fig. (1). A typical hydrothermal liquefaction process of algae.

There are also different types of production methods of algae to oil in current technology (Figs. **2** and **3**).

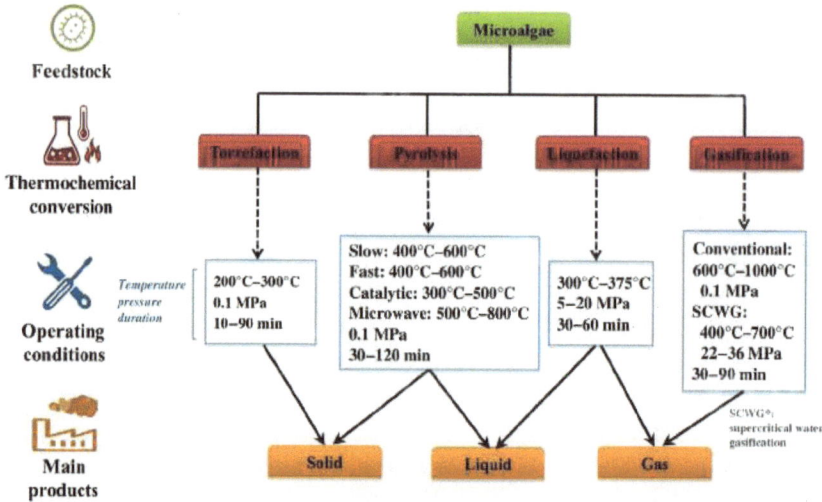

Fig. (2). Various techniques are used for algae to oil technologies [51].

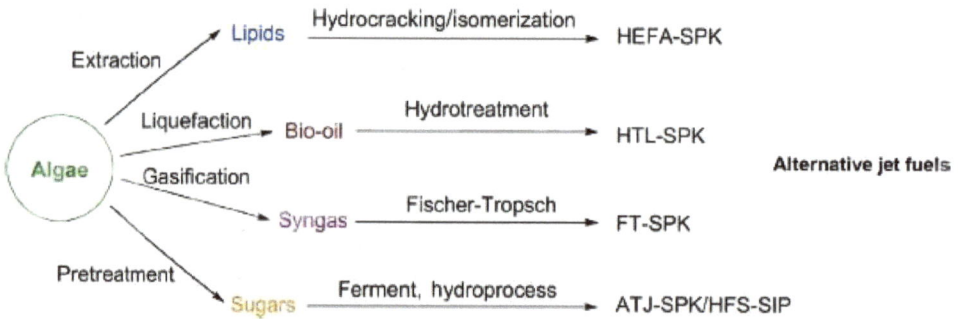

Fig. (3). Algae to Jet fuel pathway (HEFA-SPK: Hydro-processed Esters and Fatty Acids-Synthetic Paraffinic Kerosene, HTL-SPK: Hydrothermal processed-Synthetic Paraffinic Kerosene, FT-SPK: Fischer Tropsch processed-Synthetic Paraffinic Kerosene, ATJ-SPK/HFS-SIP: Alcohol to Jet Synthetic Paraffinic Kerosene/ Hydroprocessed Fermented Sugars to Synthetic Isoparaffins) [52].

The water used in the hydrothermal liquefaction process is an environmentally friendly and unique solvent; this process can be applied to wet biomass; it is performed with high energy efficiency at lower temperatures than pyrolysis. Water, which is the most critical solvent in nature, has very different properties in the above-critical conditions. Water cannot only dissolve organic compounds under normal conditions; it can also dissolve organic compounds under critical conditions. Under normal conditions, while the water has a high dielectric constant ($\varepsilon \sim 80$) according to the effect of hydrogen bonds, the dielectric constant ($\varepsilon \sim 5$) reduces when it is heated towards the critical point. Water with a dielectric constant in this range can also dissolve hydrophobic substances. This reduction in the dielectric constant changes the structure of the water from polar to apolar, making it a suitable solvent for gases, organic compounds, carbohydrates, and lignin [1 - 50].

The hydrothermal process is based on the unique properties of water at high pressure and temperature. Most importantly, at high temperatures, the hydrogen bond reduces and decreases the dielectric constant of water. Therefore, many organic compounds become completely miscible in HTW. In terms of solubility, HTW behaves like a polar organic solvent. Accordingly, HTW has higher concentrations of OH^{-1} and H^{+1} than liquid water in the environment, increasing the effectiveness of dehydration and hydrolysis known to be base or acidic base catalysts [1 - 50].

Hydrothermal liquefaction produces high-energy-density organic liquid by converting the biomass to liquid water at high temperatures. The effect of catalysts (*e.g.*, metals, alkali salts), co-solvents (*e.g.*, glycerol, acetone, recycled bio-oil), and reducing gases (*e.g.*, CO and H_2), on composition and product yield were investigated.

Generally, feedstocks containing about 80% water are subjected to subcritical temperatures (250-350 °C) to form a hydrophobic bio-oil with reduced oxygen content (10-18%) compared to the parent material (about 40%). Bio oils have been found to contain more than 400 compounds depending on processing conditions and raw materials. These bio-oils are a good option for upgrading to high-quality distilled fuels (*e.g.*, diesel and gasoline), and can be used as a direct heavy oil replacement for cooking with coal (Fig. **4**). Oxygenated hydrocarbons have higher melting points and lower energy content, viscosity, and boiling points compared to hydrocarbons with a similar molecular weight; hence oxygen removal by decarboxylation is critical to ensure fuel quality. The breakdown of biomass mainly consists of dehydration where water molecules are removed, decarboxylation where CO_2 molecules are removed, and deamination steps where amino acid molecules are removed from the environment. Decarboxylation and Dehydration reactions facilitate the removal of O_2 from the biomass as CO_2 and H_2O. Biomass, which consists of a large part of macromolecules (Biopolymer), is hydrolysis and turns into polar monomers and oligomers.

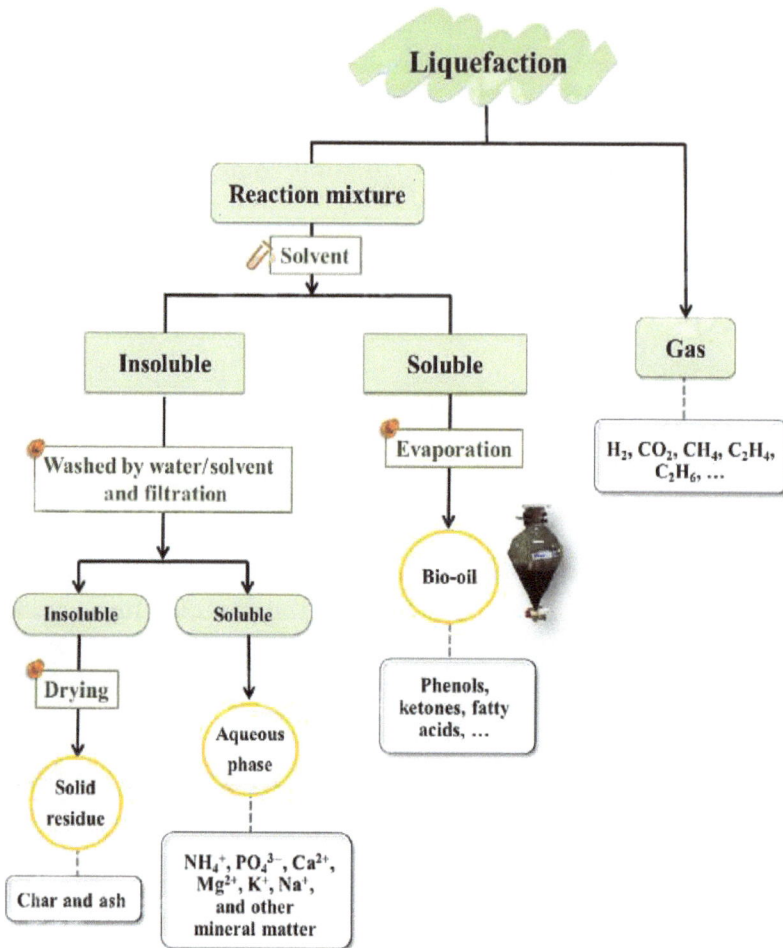

Fig. (4). Microalgae liquefaction pathways [51].

Its high pressure and temperature water solvent feature lead to the formation of glucose monomers and it breaks hydrogen bonds in the cellulose structure. Fructose is more reactive than glucose; they turn into products of the small structure by isomerization, dehydration, hydrolysis, alternating-aldol defragmentation [1 - 50].

In the hydrothermal liquefaction process, the water-soluble part of the biomass begins to dissolve into the water at 100 °C, followed by the hydrolysis event above 150 °C. Breakdown occurs in the monomeric chains of the cellulosic and hemicellulosic biomass fractions. At a pressure of 1 MPa, the mechanism proceeds towards liquefaction or gasification, and a slurry is formed at a temperature of 200 °C [1 - 50].

In the hydrothermal liquefaction process, the water-soluble part of the biomass begins to dissolve into the water at 100 °C, followed by a hydrolysis event above 150 °C. The disintegration occurs in the monomeric chains of hemicellulose and cellulosic biomass fractions. At a pressure of 1 MPa, a slurry is formed at a temperature of 200 °C, and the mechanism proceeds towards liquefaction or gasification. Cellulose contributes to liquid product formation with its moderate polymerization feature, and hemicellulose can be easily disintegrated *via* its amorphous structure, while lignin is more effective in solid formation with its complex structure and limited depolymerization ability [1 - 50].

The raw material particle size used as a biomass source is essential for higher degrees of depolymerization. Nevertheless, considering the reactions taking place in the hydrothermal liquefaction environment, it was seen that particle size is not a very useful parameter. Considering the cost of the grinding process, trials for this inactive parameter are usually done at optimum particle size. Biomass consists of many different compounds. Due to the chemical structures and different molecules of these compounds, the main components, such as hemicellulose, lignin, cellulose in the biomass can behave differently in the hydrothermal environment. As a consequence, various products may emerge depending on the raw material structure [1 - 50].

Many biomass sources, such as forest waste, domestic waste, fruit kernels, herbaceous plants, and sewage waste, have been converted into solid and liquid products without and with catalyst hydrothermal liquefaction and pyrolysis. Analysis methods such as 1H NMR, FT-IR, GC-MS, elemental, SEM are used in the examination of the products obtained [1 - 50].

THERMAL LIQUEFACTION TECHNOLOGY

The critical pressure of the water is 22.1 MPa, and the critical temperature is 374 °C. When it reaches the critical point, the properties of water such as density, reactivity, dielectric constant, and solubility change significantly. As a result, the water, which has become a high reactivity medium, has a high temperature and pressure. The solubility of the water below the critical point increases, and its abrasiveness decreases. Therefore, it is easier to convert microalgae (with 80-90% water content) into liquid fuels in the liquefaction process. Microalgae, which are suitable for heat liquefaction because of their small size, do not need to be dried before the process. During the process, an effective heat transfer takes place inside the reactor due to its small size. Therefore, microalgae are very suitable energy raw materials for heat liquefaction (Figs. **4** and **5**) [1 - 50].

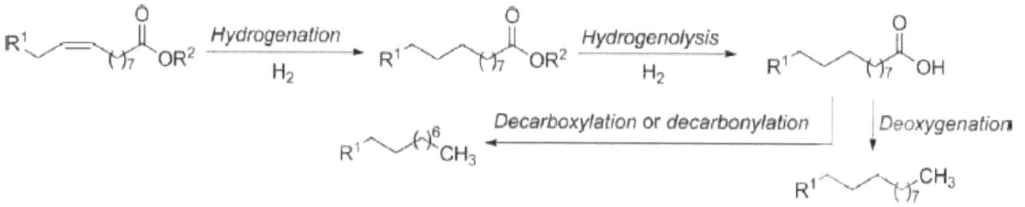

Fig. (5). Fatty acids to triglyceride pathways [52].

During the heat liquefaction process, the pressure and temperature range vary from 5-20 MPa to 300-350 °C. The reason for the high pressure is to keep the water in the liquid phase during the process [1 - 50].

The use of catalysts may be preferred. Recreation time is kept between 5-60 minutes. In general, the microalgae mass percentage of the product obtained in the liquefaction process in an autoclave is between 5-50%. After the treatment, the reactor is cooled to room temperature. The products obtained as a result of gas product analysis are determined. It may contain solvents such as dichloromethane and chloroform in liquid and solid products. The liquid product obtained is firstly evaporated to recover and purify bio-oil soluble products, after that the remaining products which dissolve in the aqueous phase are filtered after washing with water. Intermediate stages such as hydrolysis, depolymerization, and polymerization may occur in the liquefaction reaction. In the first stage, carbohydrates, lipids, and proteins in microalgae decompose. In the later stages, bio-oil, gas, and solid compounds, where organic oils are the leading products, polymerize [1 - 50]. Due to the advantages mentioned above, more bio-oil can be produced by using the Thermo liquefaction method compared to other methods.

The oils obtained from the thermal liquefaction method, which produces higher quality oils than pyrolysis, are energy-intensive liquids. It is dark in color and viscous. Bio-oil has a higher calorific value and carbon content (Fig. **6**, Table **1**) [1 - 50].

Table 1. Bio-crudes from different sources elemental analysis [53].

Properties	Biocrude	-	-	-	-
-	Vegetable Oil	Wood	-	Microalgae	-
-	-	Pyrolysis	Liquefaction	Pyrolysis	Liquefaction
C (%)	77,9	56,4	64,5	61,5	74,0
H (%)	11,7	6,2	7,7	8,5	8,7
H/C	1,8	1,3	1,4	1,7	1,4

(Table 1) cont.....

Properties	Biocrude	-	-	-	-
-	Vegetable Oil	Wood	-	Microalgae	-
O (%)	10,4	37,3	22,5	20,2	10,4
N (%)	0,04	<0,1	<0,1	9,8	6,9
Density (kg/l)	0,89	1,21	1,12	1,2	1,1

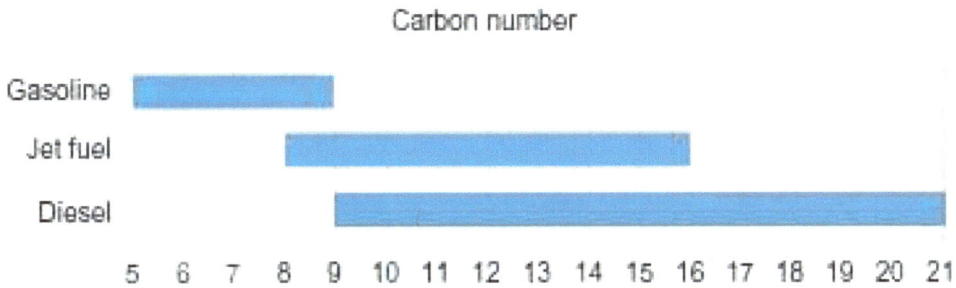

Fig. (6). Gasoline, Diesel, and Jet fuels carbon content [52].

While producing bio-oil, degradation of microalgae, lipids, protein, and carbohydrates takes place, respectively. For example, lipids are produced from fatty acids (Fig. 7), proteins, pyrroles, nitrogen heterocycles, and indoles, cyclic ketones, and phenols from carbohydrates (Figs. 8 & 9). While the bio-oil yield varies between 30-65%, the calorific value varies between 30-50 MJ (Table 2). The calorific value (HHV) of petroleum oil is between 40-44 MJ/kg. Elemental oxygen value is low in products at the end of liquefaction, but it has higher oxygen content compared to petroleum oil. The nitrogen content in products obtained in liquefaction biofuel is approximately 6%. This is because microalgae have high protein content. The sulfur content of the biomass obtained is 1% by weight in terms of releasing low SO_x emissions, especially when burned. This value is generally lower than the sulfur content (0.05-5%) obtained from fossil fuels (Table 3) [1 - 50].

Table 2. Different bio-oil yields values from various algae to oil techniques [52].

Technique	Conditions	Species	Bio-oil yields
Slow pyrolysis	300-500 °C, 10 °C/min	Nannochloropsis	31,1
Slow pyrolysis	Max temp. = 500 °C, 10 °C/min.	Tetraselmis. chui	43
Slow pyrolysis	300-700 °C, 15 min.	Blue-Green algae	54,9
Fast pyrolysis	400-600 °C, 600 °C/s	M. aeruginosa	24
MW pyrolysis	200-300 W	Macroalgae	21
MW pyrolysis	500-1250 W, 460-630 °C, 20 min.	Chlorella	28,6

(Table 2) cont.....

Technique	Conditions	Species	Bio-oil yields
HTL	300 °C, NaCO$_3$ (cat.)	B. braunii	64
HTL	340 °C, 10 MPa, 60 min.	D. tertiolecta	37
HTL	200-300 °C, 8,9-10,3 MPA, 30-120 min.	C. pyrenoidosa	24-39,4
HTL	350 °C, 60 min.	Nannochloropsis	43
HTL	Spirulina	300 °C, 10-12 MPa, 30 min.	32,6
HTL	Scenedesmus	300 °C, 10-12 MPa, 30 min.	24-45

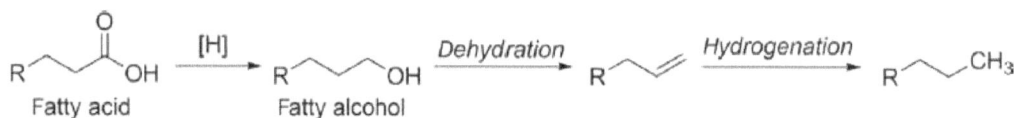

Fig. (7). Fatty acids to n-alkanes pathways [52].

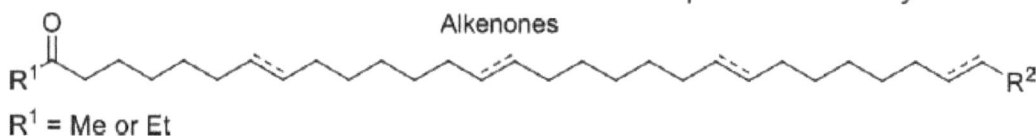

Fig. (8). Botryococcene structure [52].

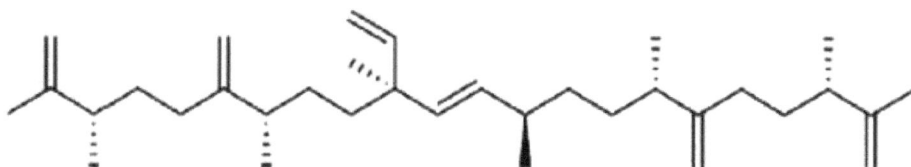

Fig. (9). Structures of Alkenone [52].

Table 3. Conventional petrodiesel and jet fuels ASTM specifications [52].

Diesel fuel (ASTM D9 75)	Jet fuel (ASTM D 1655)	Specification
1,9-4,1 at 40 °C (No.2 DF)	8.0 max at -20 °C	Kinematic viscosity (mm^2/s)
-*	-40 max (Jet A); -47 max (Jet A-1)	Freezing Point (°C)
-	-	Cloud point (°C) or LTFT/CFPP (°C)
-	775-840	Density (40 °C; kg/m^3)
38 (No.1 DF); 52 min (No.2 DF)	38 min.	Flash Point (°C)
-	42,8 min.	Heat of Combustion (MJ/kg)

(Table 3) cont.....

Diesel fuel (ASTM D9 75)	Jet fuel (ASTM D 1655)	Specification
35 max (D 1319) (as aromaticity)**	25 max (ASTM D 1319), 26,6 max (D 6379)	Aromatics (% by vol.)
-	25 min. 18 min (smoke point and naphtalenes, vol%)	Smoke point (mm)
15 (for low-sulfur diesel S15)	-	Sulfur (ppm; ug/g)
-	0,30	Sulfur (mass-%)
-	0,003	Sulfur (mercaptan) (mass %)

* Not specified ** Given as one of two properties to be met with cetane index being the other property.

The content of the gas product obtained in the liquefaction process generally includes gas products such as H_2, N_2, C_2H_4, CH_4, C_2H_6, CO_2. Among them, the percentage of gas is the most gaseous. On the other hand, CH_4 and H_2 concentrations are also high. The concentration of other gases and CO is lower compared to the gasification process. This is due to the presence of more methanation reactions during the process than gasification. The obtained aqueous product can be reused for microalgae cultivation because it contains metallic cations such as Na^{+1}, K^{+1}, Mg^{+2}, Ca^{+2}, phosphorus such as PO_4^{-3}, nitrogen like NH_4^{+1}, and other necessary minerals. However, there is a high proportion of organic carbon in the aqueous phase. This can lead to the use of heterotrophic strains as a source of carbon. It is an indication that a high-carbon aqueous product is producing biocarbon oil with less carbon content. The solid products obtained are ash and bell. The energy density of the tsar is low. The efficiency of liquid fuel production from microalgae is directly affected by reaction time, temperature, raw material ratio, and catalyst additive (Figs. **10 - 12**) [1 - 50].

Temperature is one of the most essential parameters in bio-oil production. It is known that more proteins and lipids are produced in a reaction at 250 °C. When the reaction temperature is increased up to 300-375 °C, the formation of protein and carbohydrates is more intense. The reason for this is that the temperature increases both the carbon content and calorific value in bio-oils, and the hydrogen and oxygen content decreases. However, it increases as more protein is produced due to the high reaction efficiency in nitrogen content. As the liquid conversion efficiency increases, the mole percentages of molecules such as CH_4 increase. CO_2 and N_2 mole ratios increase. It generally increases liquid and gas efficiency during the temperature increase (Figs. **13 - 16**). In Fig. (**13**), it is the reaction zone between two red dashed lines. Relationships between bio-crude oil, gases, aqueous organics, solid products, hydrothermal carbonization, hydrothermal gasification, and hydrothermal liquefaction have been linked to the dominant products of processes. All of these processes consist of aqueous organics. In Fig. (**13**), arrows indicate the shifts of O/C, H/C, and N/C from algae feedstock to algae bio-oil due to the HTL [1 - 50].

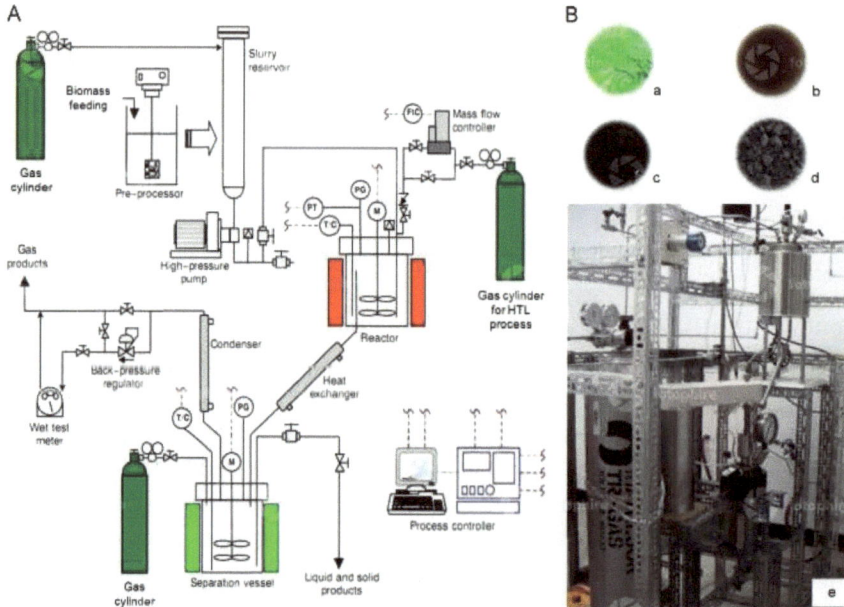

Fig. (10). A: **a)** HTL flowcart, **b)** algae pictures, B: **a)** Slurry of algae, **b)** aqueous products, **c)** bio-crude oil, **d)** solid residue, **e)** autoclave [54].

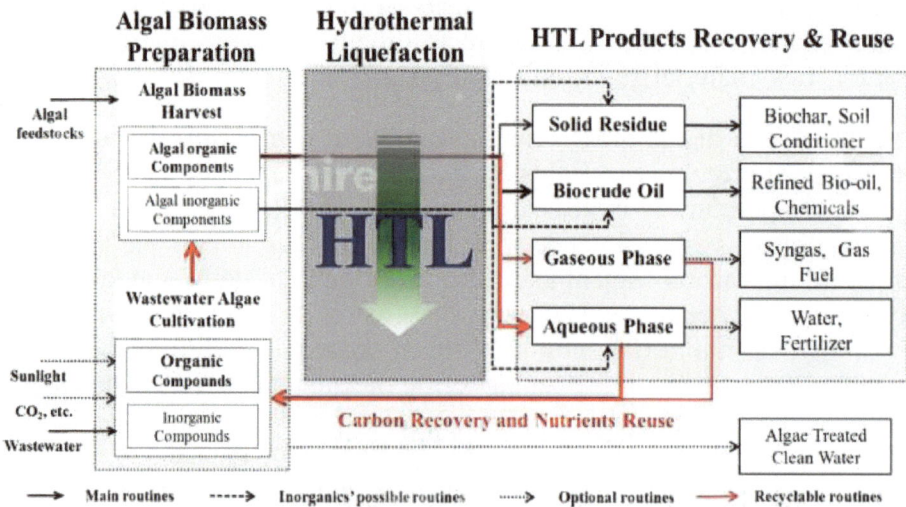

Fig. (11). Includes algae cultivation and harvesting, HTL, product recovery, and reuse [54].

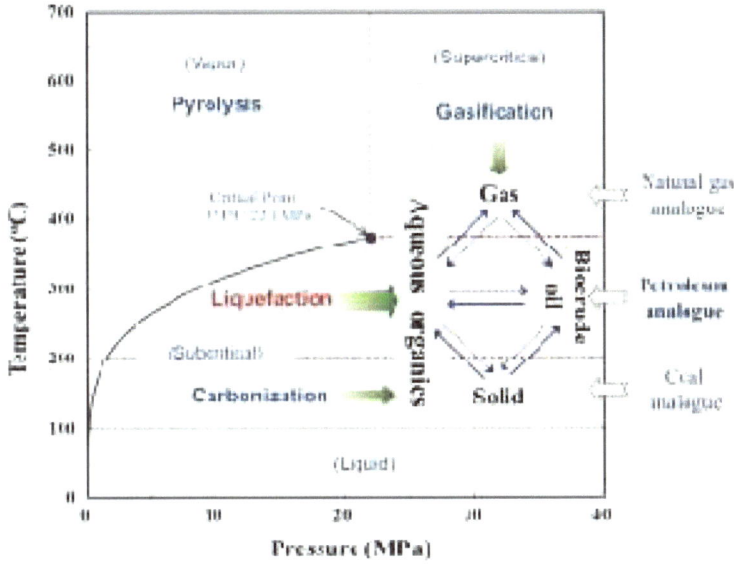

Fig. (12). A simple HTL process [54].

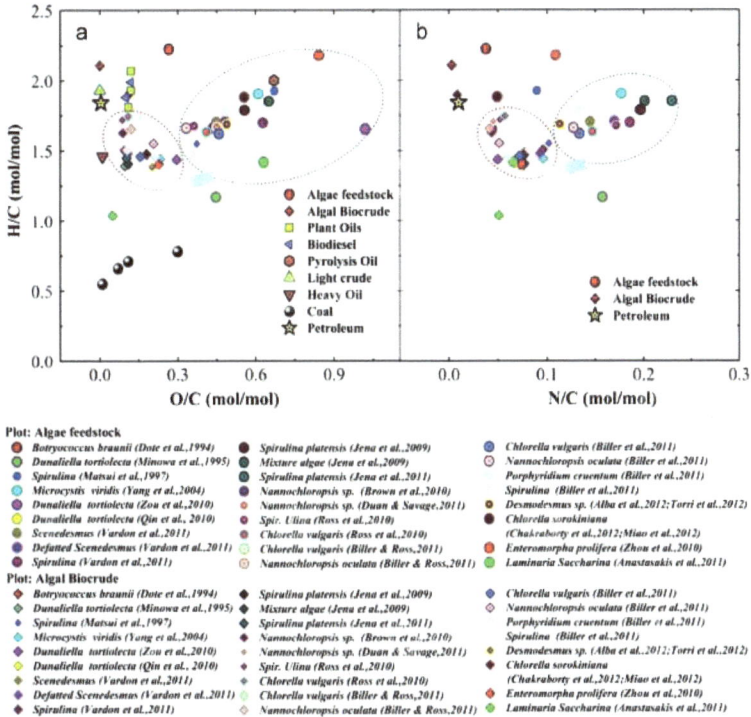

Fig. (13). Bio-crude oils and Algaes Van Krevelen diagram [54].

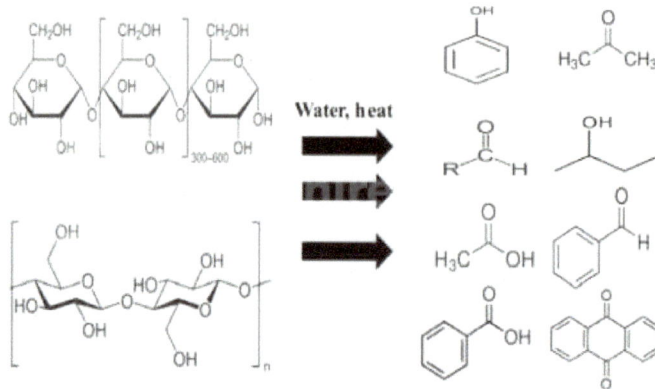

Fig. (14). Carbohydrates to oxygenate compounds obtainable during algae liquefaction [55].

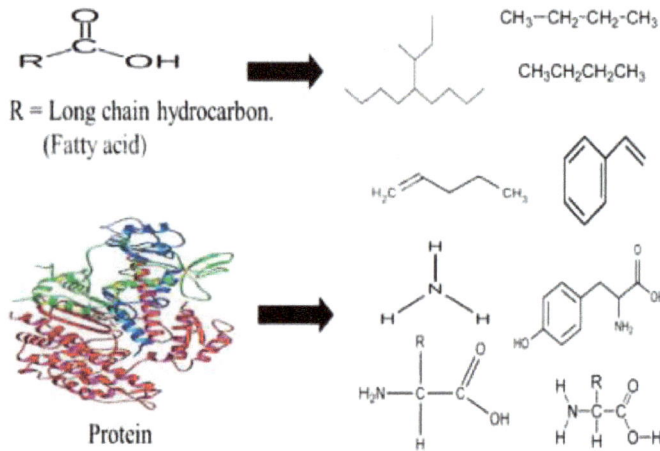

Fig. (15). Fatty acids/proteins degradation during liquefaction process [55].

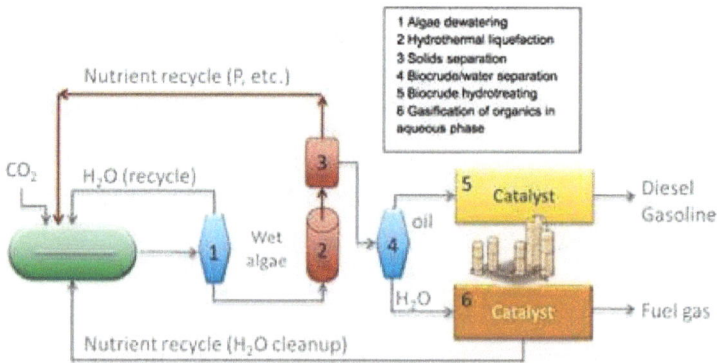

Fig. (16). Process flow diagram for HTL with catalyst used [56].

Determination of reaction time is carried out in order to effectively transform microalgae to bio-oil. This value is usually 60 minutes. It is known to extend up to 120 minutes. However, high temperature determines low reaction temperature, and low temperature determines a high reaction time [1 - 50].

It is an important parameter affecting liquefaction efficiency, such as temperature and time in raw material ratio. The increase in gas and solid yields is usually made possible by increasing the low feed rates by twice [1 - 50].

In the case of catalyst used, alkali salts and metals (heterogeneous and homogeneous catalyst) can be used. Catalysts such as Na_2CO_3 play an active role in the conversion of carbohydrates into bio-oil, while it is more concentrated in the absence of catalyst in the conversion of proteins and lipids (Table **4**, Fig. **11**). When using a heterogeneous catalyst, the calorific value (HHV) and conversion of bio-oil may increase, but the oxygen content decreases (Fig. **17**). Therefore, the addition of catalyst increases both product conversion and affects the elemental content. Microalgae pre-treatment before liquefaction also affects the bio-oil yield (Table **5**) [1 - 50].

Table 4. Influence of homogenous catalysts on algae liquefaction [55].

Conditions	Algae types	Catalyst	Bio-oil yield	% Hydrocarbon
350 °C, 3 g algae + 27 ml of catalyst solution	Spirulina	KOH	15,2	Mainly oxygenates were produced
350 °C, 3 g algae + 27 ml of catalyst solution	Spirulina	Na_2CO_3	20,0	Mainly oxygenates were produced
350 °C, 3 g algae + 27 ml of catalyst solution	Spirulina	CH_3COOH	16,6	Mainly oxygenates were produced
350 °C, 3 g algae + 27 ml of catalyst solution	Spirulina	HCOOH	14,2	Mainly oxygenates were produced
300 °C, 30 min, 20 g algae + 150 ml water + 5 wt% catalyst.	E. prolifera	Na_2CO_3	23	20,0
350 °C, algae/water = 1:10, 15 min.	Laminariasaccharina	KOH	67	Not reported

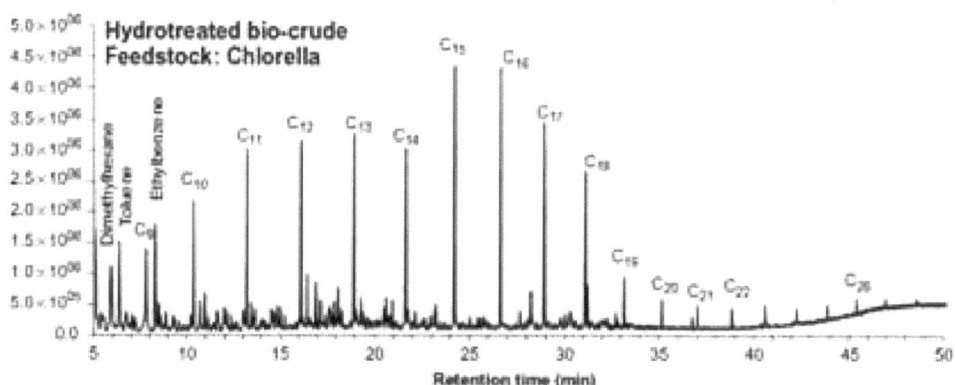

Fig. (17). Bio-crude oil's carbon-related products are produced from chlorella liquefaction [57].

Table 5. Influence of some heterogeneous catalyst for algae liquefaction [55].

Algae Type	Catalyst	Conditions	Bio-oil Yield, %	Hydrocarbon Composition, %
Nannochloropsis sp.	Non-catalytic	350 °C, 1 h, 0.384 g of catalyst, 95% water volüme	35	Mainly oxygenates
Nannochloropsis sp.	Pd/C	350 °C, 1 h, 0.384 g of catalyst, 95% water volüme	57	Mainly C15 to C18, 20%.
Nannochloropsis sp.	Pt/C	350 °C, 1 h, 0.384 g of catalyst, 95% water volume	49	Mainly C15 to C18.
Nannochloropsis sp.	Ru/C	350 °C, 1 h, 0.384 g of catalyst, 95% water volume	50	Mainly C15 to C18.
Nannochloropsis sp.	Ni/SiO$_2$-Al$_2$O$_3$	350 °C, 1 h, 0.384 g of catalyst, 95% water volume	50	Mainly C15 to C18.
Nannochloropsis sp.	CoMo/Al$_2$O$_3$	350 °C, 1 h, 0.384 g of catalyst, 95% water volume	55	Mainly C15 to C18.
Nannochloropsis sp.	Zeolite	350 °C, 1 h, 0.384 g of catalyst, 95% water volume	48	Mainly C15 to C18.
Nannochloropsis sp.	Non-catalytic	350 °C, 1 h, 0.384 g of catalyst, 95% water volume	35	Oxygenates as dominant products
Nannochloropsis sp.	Pd/C	350 °C, 1 h, 0.384 g of catalyst, 95% water volume	40	C5 to C20 alkanes, Aromatics
Nannochloropsis sp.	Pd/C	350 °C, 1 h, 0.384 g of catalyst, 95% water volume	38	C5 to C20 alkanes, Aromatics
C. pyrenoidosa	Non-catalytic	7 g algae, 70 ml water, 0.35 g catalyst, 300 °C, 20 min	32	Oxygenates

(Table 4) cont.....

Algae Type	Catalyst	Conditions	Bio-oil Yield, %	Hydrocarbon Composition, %
C. pyrenoidosa	H-ZSM-5	7 g algae, 70 ml water, 0.35 g catalyst, 300 °C, 20 min	38	Mainly C6, 5–10%
C. pyrenoidosa	Ce/H-ZSM-5	7 g algae, 70 ml water, 0.35 g catalyst, 300 °C, 20 min	52	C6 to C16, > 20%

Recent research shows that two-stage hydrothermal liquefaction has more advantages in more bio-oil production at lower operating conditions and less production cost [58, 59].

CONCLUSION

The main issues affecting the characteristics of thermal methods in biofuel production are summarized below.

1. Presence of free fatty acid and water causes saponification, catalyst depletion, and reduced catalyst effectiveness.
2. Alkali catalysts are much faster than acid catalysts. NaOH, which has a moderate catalytic effect, is cheaper than others and is widely used in industry due to its lower molar mass and better properties.
3. Methanol is generally used due to its low price, physical and chemical advantages (short-chain and polarity).
4. In high molar ratio reactions, a higher rate of ester conversion takes place in a much shorter time.

CONSENT FOR PUBLICATION

Not applicable.

CONFLICT OF INTEREST

The author declares no conflict of interest, financial or otherwise.

ACKNOWLEDGEMENT

Declared none.

REFERENCES

[1]   Biswas B, Kumar J, Bhaskar T. Chapter 10 - Advanced hydrothermal liquefaction of biomass for bio-oil production. In: Pandey A, Larroche C, Dussap C-G, Gnansounou E, Khanal SK, Ricke S, Eds. Biofuels: Alternative Feedstocks and Conversion Processes for the Production of Liquid and Gaseous Biofuels. Second Edition. Academic Press 2019; pp. 245-66.

[2]   Gu Y, Zhang X, Deal B, Han L. Biological systems for treatment and valorization of wastewater generated from hydrothermal liquefaction of biomass and systems thinking: A review. Bioresour Technol 2019; 278: 329-45.
[http://dx.doi.org/10.1016/j.biortech.2019.01.127] [PMID: 30723025]

[3]   Hu Y, Gong M, Feng S, Xu CC, Bassi A. A review of recent developments of pre-treatment technologies and hydrothermal liquefaction of microalgae for bio-crude oil production. Renew Sustain Energy Rev 2019; 101: 476-92.
[http://dx.doi.org/10.1016/j.rser.2018.11.037]

[4]   Kumar M, Oyedun AO, Kumar A. Chapter 29 - Biohydrogen production from bio-oil *via* hydrothermal liquefaction. In: Pandey A, Larroche C, Dussap C-G, Gnansounou E, Khanal SK, Ricke S, Eds. Biofuels: Alternative Feedstocks and Conversion Processes for the Production of Liquid and Gaseous Biofuels. Second Edition. Academic Press 2019; pp. 715-32.

[5]   Matayeva A, Basile F, Cavani F, Bianchi D, Chiaberge S. Chapter 12 - Development of upgraded bio-oil *via* liquefaction and pyrolysis. In: Albonetti S, Perathoner S, Quadrelli EA, Eds. Studies in Surface Science and Catalysis. Elsevier 2019; 178: pp. 231-56.

[6]   Mathimani T, Mallick N. A review on the hydrothermal processing of microalgal biomass to bio-oil - Knowledge gaps and recent advances. J Clean Prod 2019; 217: 69-84.
[http://dx.doi.org/10.1016/j.jclepro.2019.01.129]

[7]   Ong HC, Chen WH, Farooq A, Gan YY, Lee KT, Ashokkumar V. Catalytic thermochemical conversion of biomass for biofuel production: A comprehensive review. Renew Sustain Energy Rev 2019; 113: 109266.
[http://dx.doi.org/10.1016/j.rser.2019.109266]

[8]   Perkins G, Batalha N, Kumar A, Bhaskar T, Konarova M. Recent advances in liquefaction technologies for production of liquid hydrocarbon fuels from biomass and carbonaceous wastes. Renew Sustain Energy Rev 2019; 115: 109400.
[http://dx.doi.org/10.1016/j.rser.2019.109400]

[9]   Taghipour A, Ramirez JA, Brown RJ, Rainey TJ. A review of fractional distillation to improve hydrothermal liquefaction biocrude characteristics; future outlook and prospects. Renew Sustain Energy Rev 2019; 115: 109355.
[http://dx.doi.org/10.1016/j.rser.2019.109355]

[10]   Tao Y, You F. Consequential life cycle analysis for food-water- energy-waste nexus. In: Kiss AA, Zondervan E, Lakerveld R, Özkan L, Eds. Computer Aided Chemical Engineering. Elsevier 2019; 46: pp. 1705-10.

[11]   Ali M, Saleem M, Khan Z, Watson IA. 16 - The use of crop residues for biofuel production. In: Verma D, Fortunati E, Jain S, Zhang X, Eds. Biomass, Biopolymer-Based Materials, and Bioenergy. Woodhead Publishing 2019; pp. 369-95.
[http://dx.doi.org/10.1016/B978-0-08-102426-3.00016-3]

[12]   Biswas B, Bhaskar T. Chapter 17 - Hydrothermal upgradation of algae into value-added hydrocarbons. In: Pandey A, Chang J-S, Soccol CR, Lee D-J, Chisti Y, Eds. Biofuels from Algae. Second Edition. Elsevier 2019; pp. 435-59.

[13]   Chisti Y. Chapter 1 - Introduction to algal fuels. In: Pandey A, Chang J-S, Soccol CR, Lee D-J, Chisti Y, Eds. Biofuels from Algae. Second Edition. Elsevier 2019; pp. 1-31.

[14]   de Farias Silva CE, Barbera E, Bertucco A. Chapter 17 - Biorefinery as a promising approach to promote ethanol industry from microalgae and cyanobacteria. In: Ray RC, Ramachandran S, Eds. Bioethanol Production from Food Crops. Academic Press 2019; pp. 343-59.
[http://dx.doi.org/10.1016/B978-0-12-813766-6.00017-5]

[15]   Lam MK, Khoo CG, Lee KT. Chapter 19 - Scale-up and commercialization of algal cultivation and biofuels production. In: Pandey A, Chang J-S, Soccol CR, Lee D-J, Chisti Y, Eds. Biofuels from

Algae. Second Edition. Elsevier 2019; pp. 475-506.

[16]   Mathimani T, Baldinelli A, Rajendran K, *et al.* Review on cultivation and thermochemical conversion of microalgae to fuels and chemicals: Process evaluation and knowledge gaps. J Clean Prod 2019; 208: 1053-64.
[http://dx.doi.org/10.1016/j.jclepro.2018.10.096]

[17]   Obeid F, Chu Van T, Brown R, Rainey T. Nitrogen and sulphur in algal biocrude: A review of the HTL process, upgrading, engine performance and emissions. Energy Convers Manage 2019; 181: 105-19.
[http://dx.doi.org/10.1016/j.enconman.2018.11.054]

[18]   Okoro OV, Sun Z, Birch J. 10 - Thermal depolymerization of biogas digestate as a viable digestate processing and resource recovery strategy. In: Azad K, Ed. Advances in Eco-Fuels for a Sustainable Environment. Woodhead Publishing 2019; pp. 277-308.
[http://dx.doi.org/10.1016/B978-0-08-102728-8.00010-3]

[19]   Srifa A, Chaiwat W, Pitakjakpipop P, Anutrasakda W, Faungnawakij K. Chapter 6 - Advances in bio-oil production and upgrading technologies. In: Rai M, Ingle AP, Eds. Sustainable Bioenergy. Elsevier 2019; pp. 167-98.

[20]   Yang J, He Q, Yang L. A review on hydrothermal co-liquefaction of biomass. Appl Energy 2019; 250: 926-45.
[http://dx.doi.org/10.1016/j.apenergy.2019.05.033]

[21]   Boboescu IZ, Chemarin F, Beigbeder JB, *et al.* Making next-generation biofuels and biocommodities a feasible reality. Curr Opin Green Sustain Chem 2019; 20: 25-32.
[http://dx.doi.org/10.1016/j.cogsc.2019.07.005]

[22]   Cao Y, Chen SS, Zhang S, *et al.* Advances in lignin valorization towards bio-based chemicals and fuels: Lignin biorefinery. Bioresour Technol 2019; 291: 121878.
[http://dx.doi.org/10.1016/j.biortech.2019.121878] [PMID: 31377047]

[23]   Kim JY, Lee HW, Lee SM, Jae J, Park YK. Overview of the recent advances in lignocellulose liquefaction for producing biofuels, bio-based materials and chemicals. Bioresour Technol 2019; 279: 373-84.
[http://dx.doi.org/10.1016/j.biortech.2019.01.055] [PMID: 30685133]

[24]   Marulanda VA, Gutierrez CDB, Alzate CAC. Chapter 4 - Thermochemical, biological, biochemical, and hybrid conversion methods of bio-derived molecules into renewable fuels. In: Hosseini M, Ed. Advanced Bioprocessing for Alternative Fuels, Biobased Chemicals, and Bioproducts. Woodhead Publishing 2019; pp. 59-81.
[http://dx.doi.org/10.1016/B978-0-12-817941-3.00004-8]

[25]   Pang S. Advances in thermochemical conversion of woody biomass to energy, fuels and chemicals. Biotechnol Adv 2019; 37(4): 589-97.
[http://dx.doi.org/10.1016/j.biotechadv.2018.11.004] [PMID: 30447327]

[26]   Ubando AT, Rivera DRT, Chen WH, Culaba AB. A comprehensive review of life cycle assessment (LCA) of microalgal and lignocellulosic bioenergy products from thermochemical processes. Bioresour Technol 2019; 291: 121837.
[http://dx.doi.org/10.1016/j.biortech.2019.121837] [PMID: 31353166]

[27]   Walmsley TG, Ong BHY, Klemeš JJ, Tan RR, Varbanov PS. Circular Integration of processes, industries, and economies. Renew Sustain Energy Rev 2019; 107: 507-15.
[http://dx.doi.org/10.1016/j.rser.2019.03.039]

[28]   Why ESK, Ong HC, Lee HV, Gan YY, Chen WH, Chong CT. Renewable aviation fuel by advanced hydroprocessing of biomass: Challenges and perspective. Energy Convers Manage 2019; 199: 112015.
[http://dx.doi.org/10.1016/j.enconman.2019.112015]

[29]   Wu W, Chang JS. Integrated algal biorefineries from process systems engineering aspects: A review.

Bioresour Technol 2019; 291: 121939.
[http://dx.doi.org/10.1016/j.biortech.2019.121939] [PMID: 31400827]

[30]   Xie L-F, Xu Y-P, Shi X-L, Wang F, Duan P-G, Li S-C. Hydrotreating the distillate fraction of algal biocrude with used engine oil over Pt/C for production of liquid fuel. Catal Today 2020; 355: 65-74.

[31]   Ahmad FB, Zhang Z, Doherty WOS, O'Hara IM. The outlook of the production of advanced fuels and chemicals from integrated oil palm biomass biorefinery. Renew Sustain Energy Rev 2019; 109: 386-411.
[http://dx.doi.org/10.1016/j.rser.2019.04.009]

[32]   Chowdhury H, Loganathan B. Third-generation biofuels from microalgae: a review. Curr Opin Green Sustain Chem 2019; 20: 39-44.
[http://dx.doi.org/10.1016/j.cogsc.2019.09.003]

[33]   de Morais MG, de Freitas BCB, Moraes L, Pereira AM, Costa JAV. Chapter 18 - Liquid biofuels from microalgae: recent trends. In: Hosseini M, Ed. Advanced Bioprocessing for Alternative Fuels, Biobased Chemicals, and Bioproducts. Woodhead Publishing 2019; pp. 351-72.
[http://dx.doi.org/10.1016/B978-0-12-817941-3.00018-8]

[34]   Kumar D, Singh B. Algal biorefinery: An integrated approach for sustainable biodiesel production. Biomass Bioenergy 2019; 131: 105398.
[http://dx.doi.org/10.1016/j.biombioe.2019.105398]

[35]   McNutt J, He QS. Spent coffee grounds: A review on current utilization. J Ind Eng Chem 2019; 71: 78-88.
[http://dx.doi.org/10.1016/j.jiec.2018.11.054]

[36]   Morales M, Collet P, Lardon L, Hélias A, Steyer J-P, Bernard O. Chapter 20 - Life-cycle assessment of microalgal-based biofuel. In: Pandey A, Chang J-S, Soccol CR, Lee D-J, Chisti Y, Eds. Biofuels from Algae (Second Edition). Elsevier 2019; pp. 507-50.

[37]   Nayak A, Bhushan B. An overview of the recent trends on the waste valorization techniques for food wastes. J Environ Manage 2019; 233: 352-70.
[http://dx.doi.org/10.1016/j.jenvman.2018.12.041] [PMID: 30590265]

[38]   Torres-Mayanga PC, Lachos-Perez D, Mudhoo A, *et al.* Production of biofuel precursors and value-added chemicals from hydrolysates resulting from hydrothermal processing of biomass: A review. Biomass Bioenergy 2019; 130: 105397.
[http://dx.doi.org/10.1016/j.biombioe.2019.105397]

[39]   Wang L, Chang Y, Li A. Hydrothermal carbonization for energy-efficient processing of sewage sludge: A review. Renew Sustain Energy Rev 2019; 108: 423-40.
[http://dx.doi.org/10.1016/j.rser.2019.04.011]

[40]   Zhang J, Zhang X. 15 - The thermochemical conversion of biomass into biofuels. In: Verma D, Fortunati E, Jain S, Zhang X, Eds. Biomass, Biopolymer-Based Materials, and Bioenergy. Woodhead Publishing 2019; pp. 327-68.
[http://dx.doi.org/10.1016/B978-0-08-102426-3.00015-1]

[41]   Choi HI, Hwang SW, Sim SJ. Comprehensive approach to improving life-cycle CO_2 reduction efficiency of microalgal biorefineries: A review. Bioresour Technol 2019; 291: 121879.
[http://dx.doi.org/10.1016/j.biortech.2019.121879] [PMID: 31377048]

[42]   Dunn JB. Biofuel and bioproduct environmental sustainability analysis. Curr Opin Biotechnol 2019; 57: 88-93.
[http://dx.doi.org/10.1016/j.copbio.2019.02.008] [PMID: 30928828]

[43]   Khoo CG, Dasan YK, Lam MK, Lee KT. Algae biorefinery: Review on a broad spectrum of downstream processes and products. Bioresour Technol 2019; 292: 121964.
[http://dx.doi.org/10.1016/j.biortech.2019.121964] [PMID: 31451339]

[44]   Markou G, Monlau F. Chapter 6 - Nutrient recycling for sustainable production of algal biofuels. In:

Pandey A, Chang J-S, Soccol CR, Lee D-J, Chisti Y, Eds. Biofuels from Algae (Second Edition). Elsevier 2019; pp. 109-33.

[45]   Mishra S, Roy M, Mohanty K. Microalgal bioenergy production under zero-waste biorefinery approach: Recent advances and future perspectives. Bioresour Technol 2019; 292: 122008.
[http://dx.doi.org/10.1016/j.biortech.2019.122008] [PMID: 31466819]

[46]   Mohan SV, Rohit MV, Subhash GV, *et al*. Chapter 12 - Algal oils as biodiesel. In: Pandey A, Chang J-S, Soccol CR, Lee D-J, Chisti Y, Eds. Biofuels from Algae (Second Edition). Elsevier 2019; pp. 287-323.

[47]   Pereira da Silva P, Ribeiro LA. Chapter 19 - Assessing microalgae sustainability as a feedstock for biofuels. In: Hosseini M, Ed. Advanced Bioprocessing for Alternative Fuels, Biobased Chemicals, and Bioproducts. Woodhead Publishing 2019; pp. 373-92.
[http://dx.doi.org/10.1016/B978-0-12-817941-3.00019-X]

[48]   Shandilya KK, Pattarkine VM. Chapter 7 - Using microalgae for treating wastewater. In: Hosseini M, Ed. Advances in Feedstock Conversion Technologies for Alternative Fuels and Bioproducts. Woodhead Publishing 2019; pp. 119-36.
[http://dx.doi.org/10.1016/B978-0-12-817937-6.00007-2]

[49]   Sunil Kumar M, Buddolla V. Chapter 12 - Future prospects of biodiesel production by microalgae: a short review. In: Buddolla V, Ed. Recent Developments in Applied Microbiology and Biochemistry. Academic Press 2019; pp. 161-6.
[http://dx.doi.org/10.1016/B978-0-12-816328-3.00012-X]

[50]   Wei H, Liu W, Chen X, Yang Q, Li J, Chen H. Renewable bio-jet fuel production for aviation: A review. Fuel 2019; 254: 115599.
[http://dx.doi.org/10.1016/j.fuel.2019.06.007]

[51]   Chen W-H, Lin B-J. Chapter 13 - Thermochemical conversion of microalgal biomass. In: Basile A, Dalena F, Eds. Second and Third Generation of Feedstocks. Elsevier 2019; pp. 345-82.
[http://dx.doi.org/10.1016/B978-0-12-815162-4.00013-6]

[52]   O'Neil GW, Knothe G, Reddy CM. Chapter 15 - Jet biofuels from algae. In: Pandey A, Chang J-S, Soccol CR, Lee D-J, Chisti Y, Eds. Biofuels from Algae (Second Edition). Elsevier 2019; pp. 359-95.

[53]   Furimsky E. Hydroprocessing challenges in biofuels production. Catal Today 2013; 217: 13-56.
[http://dx.doi.org/10.1016/j.cattod.2012.11.008]

[54]   Tian C, Li B, Liu Z, Zhang Y, Lu H. Hydrothermal liquefaction for algal biorefinery: A critical review. Renew Sustain Energy Rev 2014; 38: 933-50.
[http://dx.doi.org/10.1016/j.rser.2014.07.030]

[55]   Galadima A, Muraza O. Hydrothermal liquefaction of algae and bio-oil upgrading into liquid fuels: Role of heterogeneous catalysts. Renew Sustain Energy Rev 2018; 81: 1037-48.
[http://dx.doi.org/10.1016/j.rser.2017.07.034]

[56]   Elliott DC. Review of recent reports on process technology for thermochemical conversion of whole algae to liquid fuels. Algal Res 2016; 13: 255-63.
[http://dx.doi.org/10.1016/j.algal.2015.12.002]

[57]   Biller P. 4 - Hydrothermal liquefaction of aquatic Feedstocks. In: Rosendahl L, Ed. Direct Thermochemical Liquefaction for Energy Applications. Woodhead Publishing 2018; pp. 101-25.
[http://dx.doi.org/10.1016/B978-0-08-101029-7.00003-5]

[58]   Gu X, Yu L, Pang N, Martinez-Fernandez JS, Fu X, Chen S. Comparative techno-economic analysis of algal biofuel production *via* hydrothermal liquefaction: One stage *versus* two stages. Appl Energy 2020; 259: 114115.
[http://dx.doi.org/10.1016/j.apenergy.2019.114115]

[59] Martinez-Fernandez JS, Chen S. Sequential hydrothermal liquefaction characterization and nutrient recovery assessment. Algal Res 2017; 25: 274-84.
[http://dx.doi.org/10.1016/j.algal.2017.05.022]

CHAPTER 4

Biodiesel Production from Algae Oil

Fevzi Yaşar[1,*]

[1] *Chemistry and Chemical Process Technology Department, Vocational School of Technical Sciences, Batman University, 72100 Batman, Turkey*

Abstract: In this study, biodiesel production was investigated by the transesterification reaction from algae oil. For biodiesel production, the oil obtained from Chlorella protothecoides type algae grown in freshwater with 5% thermal water was added to fully automated closed-loop system high-tech pyramid photobioreactors and adapted for oil production, which had a low acid value (0.23 mg KOH/g). Because of this, base catalyst transesterification was applied. For the transesterification reaction, 99.7% purity of methyl alcohol as alcohol and 99.9% purity of potassium hydroxide (KOH) was used as a catalyst. In order to determine the most suitable conditions for the production of biodiesel from algae oil, a series of laboratory-scale preliminary experiments have been carried out. As a result of the optimization studies, the 6:1 methyl alcohol/oil molar ratio, the use of KOH up to 0.75% of the oil by mass, the reaction temperature of 60 °C, and the reaction time of 60 minutes were determined as the most suitable conditions for biodiesel production. Under these conditions, 96.4% methyl ester yield was obtained, and kinematic viscosity and density values of the final biodiesel product were measured as 4.493 mm^2/s and 882 kg/m^3. As a result of the physical and chemical analysis of the produced biodiesel, it has been determined that it has an ester content of over 96% and that the free and total glycerol content with methanol, mono-, di- and tri-glyceride is well below the maximum values specified in the EN 14214 and ASTM 6751 biodiesel standards. However, properties such as viscosity, density, flash point, cetane number, acid value, sulfur and water content were found to be compatible with the specified standards. In addition, besides having the standard fuel properties of the produced biodiesel, its high cetane number (57) and good cold filter clogging point (-11 °C), makes it an important alternative diesel engine fuel.

Keywords: Algae oil, Biodiesel fuel properties, Transesterification.

---

[*] **Correspondence author Fevzi Yaşar:** Chemistry and Chemical Process Technology Department, Vocational School of Technical Sciences, Batman University, 72100, Batman, Turkey; Tel: +904882173639; E-mail: fevzi.yasar@batman.edu.tr

Hüseyin Karaca and Cemil Koyunoğlu (Eds.)
All rights reserved-© 2022 Bentham Science Publishers

INTRODUCTION

Algae Oil Characterization And Fatty Acid Distribution

Algae have oil contents with different compositions depending on the species types. Some species were identified to have suitable fatty acid values. In the same

way, some algae have more components of fatty acids due to their dry masses. Microalgae can grow in different conditions, even in the availability of fewer nutrients. They are best to be chosen for cultivation. The collection of samples needs care so that the entire biofuel contents could be obtained through careful handling of the instruments. The growth is also affected by different environmental factors that are not explicitly known for every region, so the process needs increasing attention accordingly [1].

The algae oil used for biodiesel production in the study was obtained from a commercial company (Soley Biotechnology, Istanbul, Turkey). The seaweed oil used was obtained from Chlorella protothecoides algae, grown in fresh water with 5% thermal water, and in fully automated closed-loop system high-tech pyramid photobioreactors adapted to produce heterotrophs and oils (Fig. **1** and Table **1**).

Fig. (1). Algae oil is produced from biodiesel.

Table 1. Some properties of algae oil.

Properties	Unit	Algae Oil	Pretreated Algae Oil
Density (15 °C)	kg/m^3	917	914
Cinematic Viscosity (40 °C)	mm^2/s	36,8	33,6
Acidity Value	mgKOH/g	0,29	0,23
Moisture Content	ppm	742,6	228
Calorific Value	kJ/kg	39,287	39,370

Some properties were analyzed before algae oil was converted to biodiesel. In the analyses made, it has been determined that seaweed oil has high water content, viscosity, and density values. To reduce the water content of algae oil, the sample oil was taken into a beaker after being filtered and preheated by mixing it with a magnetic stirrer for 2 hours at a temperature (above 1100 °C) above the boiling point of water. In the preheating process, after the oil was rested in a suitable environment, the same parameters were analyzed again. In Table **1**, some properties of the moss oil used in biodiesel production before and after pretreatment are given. When Table **1** is examined, it is seen that there is a significant decrease in the water content, especially after the seaweed oil has been subjected to a preheating and filtering process. The viscosity of algae oil appears to be too high to use as a fuel in diesel engines. In order for vegetable and animal oils to be used as an alternative fuel in diesel engines, it is necessary to reduce the viscosities that are too high compared to oil-based diesel fuel (diesel oil). For this purpose, dilution, microemulsion forming, pyrolysis, and transesterification methods are used. In the dilution method, the oil is diluted by mixing with a certain amount of diesel fuel. The mixture improves the viscosity, evaporation, and spraying properties of the fuel, depending on the diesel fuel ratio. In microemulsion generation, the oil is intended to form microemulsions with short-chain alcohols such as methanol, ethanol, or 1-butanol. The microemulsion is the equilibrium distribution of optically isotropic liquid microstructures with sizes between 1-150 nm, which is formed by the combination of two ordinarily non-mixing liquids and one or more active substances. In the pyrolysis method, fat molecules are broken down into smaller molecules in an oxygen-free environment at high temperatures. Pyrolysis process: It is divided into three parts; such as hydrocracking, catalytic cracking, and thermal cracking. The amount of product produced depends on the method used and the reaction parameters. With this method, although the fuel properties of the oils are similar to the diesel fuel properties, the high energy consumption is the most crucial disadvantage. Transesterification is the re-esterification reaction of fatty acids (vegetable oils, domestic waste oils, animal fats) with alcohol (methanol, ethanol, *etc.*)

accompanied by a primary catalyst. In this method, fat molecules react chemically with alcohol at a specific temperature with the help of a catalyst. With this chemical reaction, fatty acids separate from their triglycerides and form new esters with alcohols [2 - 4].

The production of biodiesel was performed by algae oil and a transesterification reaction. The acid value of the oil was measured using the AOCS Official Method CD 3A – 63 test methods. During the measurement, titration was carried out using a solution of 0.01 normal potassium hydroxide (KOH). After measurements, the acid value of algae oil was found to be 0.29 mg KOH/g (0.23 mg KOH/g after pretreatment). Thus, it has been decided that transesterification with an alkali catalyst can be achieved.

Another essential feature for biodiesel production is the fatty acid distribution of algae oil. The fuel properties of biodiesel vary greatly depending on the fatty acid distribution [5 - 8]. The most essential features are ignition ability, cold flow properties, and oxidative stability. Although saturation and fatty acid distribution in lipids do not have a significant effect on the production of biodiesel by the transesterification method, it directly concerns the properties of the fuel. For example, biodiesel fuels produced from saturated oils have a high cetane number and extreme oxidative stability while exhibiting lower combustion properties [9 - 11]. Biodiesel fuels produced from saturated oils tend to convert into gels at ambient temperatures. Biodiesel produced from long-chain unsaturated fatty acid-rich raw material has suitable cold flow properties. Nevertheless, the tendency of these fatty acids to oxidize is very high compared to others. Therefore, problems may arise during the long-term storage of biodiesel.

The fatty acid distribution of algae oil used in the study was determined using the Shimadzu brand GC-2030 model GC at Batman University Food Analysis Application and Research Center and is shown in Table **2**.

Table 2. The fatty acid distribution of algae oil in this study.

Fatty Acid	C:D	Closed Formula	MA	% Mass Weighted
Saturated	-	-	-	-
Miristik	C14:0	$C_{24}H_{28}O_2$	228,4	0,07
Palmitic	C16:0	$C_{16}H_{32}O_2$	256,4	6,68
Stearic	C18:0	$C_{18}H_{36}O_2$	284,5	2,68
Arakhidik	C20:0	$C_{20}H_{40}O_2$	312,5	0,97
Behenic	C22:0	$C_{22}H_{44}O_2$	340,6	0,60
Lignokerik	C24:0	$C_{24}H_{48}O_2$	368,63	0,27

(Table 2) cont.....

Fatty Acid	C:D	Closed Formula	MA	% Mass Weighted
Monounsaturated	-	-	-	-
Palmitoleic	C16:1	$C_{16}H_{30}O_2$	254,4	0,50
Cis-10-Heptadekanoik	C17:1	$C_{18}H_{34}O_2$	282,5	0,12
Trans Oleik	C18:1n9t	$C_{19}H_{36}O_2$	295,5	49,32
Cis-11-Eikosenoik	C20:1	$C_{20}H_{38}O_2$	310,5	3,03
Polyunsaturated	-	-	-	-
Linoleic	C18:2	$C_{18}H_{32}O_2$	294,5	21,72
Linolenic	C18:3	$C_{18}H_{30}O_2$	278,4	12,01
Cis-4,7,10,13,16,19	C22:6n3	$C_{18}H_{30}O_2$	1,80	1,80

When Table **2** is examined, it can be seen that algae oil consists of 14 fatty acids. Fatty acids dominating algae oil are trans oleic (49.32% mass) and linoleic acid (21.72% mass). Similar studies have been conducted in the literature, and fatty acid distributions of Chlorella protothecoides algae have been investigated. Xu *et al.* (2006) stated in their study that the oil obtained from Chrotella protothecoides algae from heterotrophic organic origin mostly consists of 9-Octadecenoic (trans oleic) and 9,12-Octadecadienoic acids (Linoleic) [12]. Al-lwayzy *et al.* (2010) stated that Chlorella protothecoides algae oil has the highest acid distribution in its acid distribution analysis ethics studies with 9-octadecenoic acid methyl ester ($C_{19}H_{36}O_2$) (60.8%) and 9.12-octadecadienoic acid methyl ester ($C_{19}H_{34}O_2$) (17.4%) [13]. Campenni *et al.* (2013) stated that Chlorella protothecoides algae grown as heterotrophic are a suitable source for biodiesel production since they contain high amounts of lutein and fatty acids. The 43.4% lipid content of the Chlorella protothecoides algae species used in their studies found that in the analysis of the fatty acid distribution made, it contains a high content of unsaturated fatty acids (Oleic acid 33.64% and Linoleic acid 11.53%) [14]. Krzemińska *et al.* (2015) pointed out that the lipid content of Chlorella protothecoides algae cultivated in high light intensity environment increased and was suitable for biodiesel production. They expressed that light intensity also affects fatty acid composition in their studies. They found that the increase in light intensity increased the monounsaturated fatty acid while reducing the polyunsaturated fatty acid content. They also stated that the increase in light intensity increases the content of C16-C18 fatty acids, but the linolenic acid content decreases with increasing light intensity. The saturation rate of the moss oil we used in the study was measured as 11.27%, monounsaturated by 52.97%, and polyunsaturation rate of 35.75%.

Biodiesel Production from Algae Oil

Before producing the amount of biodiesel to be used in engine tests from algae oil, production was made on a sample basis, and suitable conditions for biodiesel production were specified to determine the production stages and some reaction inputs. These studies were carried out in Batman University Technical Sciences Vocational School Refinery and Petro-Chemistry Technology Program Laboratory. For the pre-experiments, the magnetically stirred heater seen in Fig. (**2**), the bottom flat three-necked glass balloon, digital temperature meter, back cooler, precision scales, glass containers, separatory funnels, paper filters, and necessary auxiliary apparatus were used.

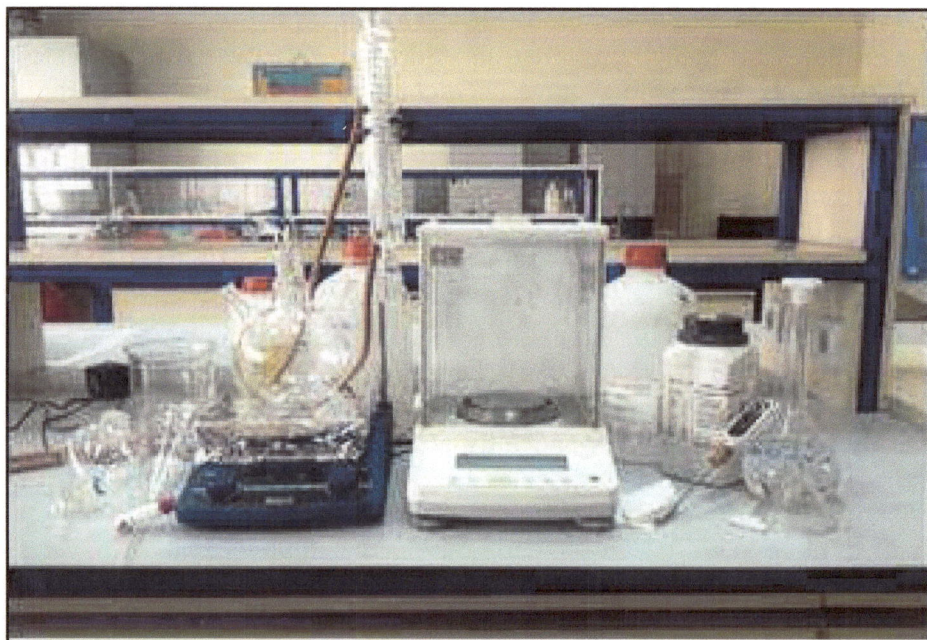

Fig. (2). Materials and tools used in the experiments.

EXPERIMENTAL

Methyl alcohol was chosen as the type of alcohol according to EN 14214 standard. In order to produce fuel suitable for the standard, Sigma-Aldrich's 99.7% purity methyl alcohol was purchased from a commercial company. Potassium hydroxide (KOH) was chosen as the catalyst for alkaline catalyzed transesterification as a result of algae oil-free fatty acid being below 0.5%. Merck brand 99.9% purity KOH was purchased from a commercial company. At this stage, catalyst amount, reaction temperature, and reaction time were optimized first in laboratory-scale experiments. In the study, primarily 500 g algae oil was heated up to 60 °C. The calculation of the reactants to be used in the reaction according to the amount of this raw material was calculated as follows.

The molecular weight of single fatty acid 'i' can be calculated as:

$$MW_i = 14.027C - 2.016d + 31.9988 \tag{1}$$

Where 'C' is the number of carbons and' is the number of double bonds. The average molecular weight of the fatty acid mixture can be calculated by dividing the sum of all reported fatty acid weights by total moles in the mixture. Numerically,

Where fi is the weight fraction of a reported fatty acid, the molecular weight of the triglyceride (oil molecule) containing three fatty acids can be calculated as:

Average molecular weight of fatty acids = $\sum f_i / \sum (f_i / MW_i)$
The average molecular weight of algae oil according to the formula above 883.5 g/mol
Mole amount of oil = 500 g/883.5 g/mol = 0.566 mol
Methyl alcohol amount = 6 x 0.566 mol x 32.04g/mol (methanol molar mass) = 108.807 g
KOH amount = 500 g x 0.75**/100 = 3.75 g
(**: Algae oil studies usually use 0.75% of the amount of oil.)

According to the calculations above, 3.75 g of KOH were dissolved in 108.807 g of methyl alcohol in room conditions in another container. The reaction was started by adding the mixture obtained to the previously heated oil under suitable conditions. In Fig. (**3**), in the transesterification reaction, the transfer of the methyl alcohol-KOH mixture to the reaction vessel and completion of the reaction can be seen.

At the end of the reaction period, the reaction was terminated by turning off the heater and mixer, and the resulting ester-glycerin mixture was transferred to the separatory funnel and allowed to rest for 12 hours (Fig. **4a**). At the end of the period, the glycerin accumulated at the bottom of the separatory funnel was weighed into a separate container due to the difference in density. Then, the ester taken from the glycerin was washed four times with distilled water at 22, 40, 50, and 60 °C to remove unwanted alcohol and solid particles, if any (Fig. **4b**).

Fig. (3). Transfer the methanol-KOH solution to the reaction vessel and complete the reaction.

(a)   (b)

Fig. (4). Taking glycerin from the separatory funnel (**a**), washing the ester (**b**).

As some water and alcohol may remain in the remaining ester at the end of the fourth washing, the methyl ester was heated at 110 °C for about 60 minutes to allow the evaporation of the water and alcohol. Then, the remaining amount of ester is weighed on the sensitive balance and noted. Fig. (**5**) shows the methyl ester obtained as a result of the removal of water and alcohol remaining in the ester and transesterification.

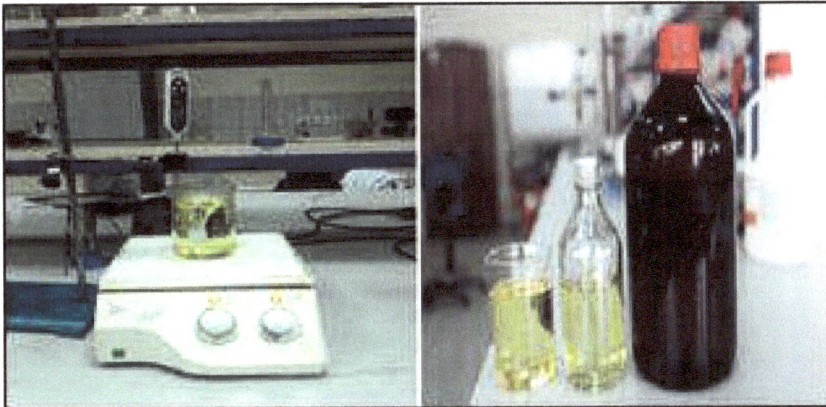

Fig. (5). Removal of water and alcohol remaining in the ester and the resulting methyl ester.

Fig. (**6**) shows the flow chart of algae oil converting to biodiesel.

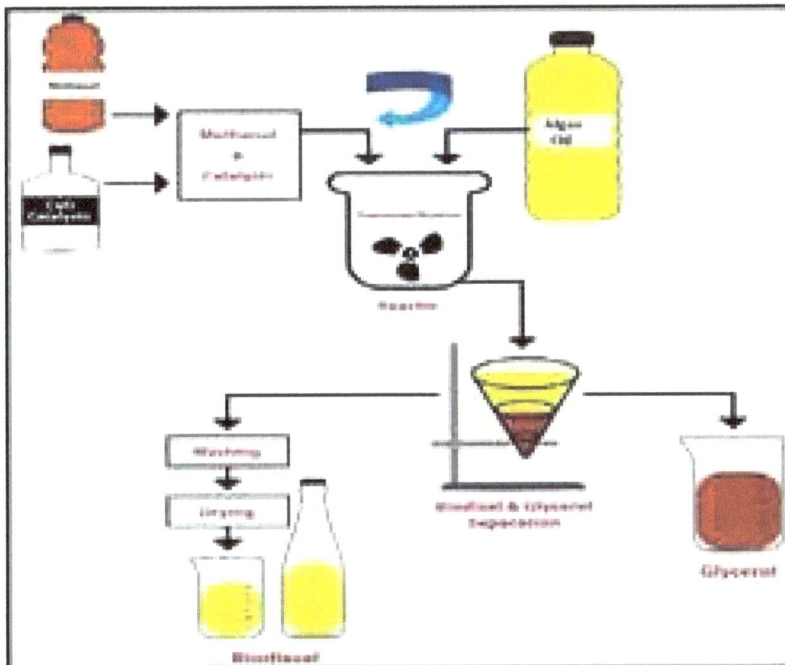

Fig. (6). Flow chart of algae oil conversion to biodiesel.

RESULTS AND DISCUSSIONS

Analysis of Fuel Properties of Algae Oil Biodiesel

In this study, the methods and devices used in the characterization of algae oil and in determining the properties of the methyl ester fuel produced working principles, and sensitivity of the devices are shown in Table **3**.

Table 3. Methods and devices used in oil characterization and determination of fuel properties.

Properties	Unit	Device	Test Method	Precision
Esther Content	% (m/m)	Agilent 6890 C	EN 14103	0,09
Density (15 °C)	kg/m³	DMA38	EN ISO 12185	0,1
Viscosity (40 °C)	mm²/s	HVM 472	EN ISO 3104	0,001
Flash Point	°C	HFP 339	EN ISO 3679	0,5
Sulfur Content	mg/kg	Flash 2000	EN ISO 20846	0,1
Polycyclic Aromatic Hydrocarbons	% weight	HPLC-RID SHIMADZU	EN 12916	-
Cetane Number	-	ZX-440 Analyser	EN ISO 5165	-
Cold Filter Plugging Point	°C	FPP 5Gs	EN 116	0,1
Lower Heating Value	kJ/kg	IKA C5003	TS 1740	1,0
Lubrication (at 60 °C)	μm	HFRR	EN ISO 12156	-
Water Content	mg/kg	KEM Karl-Fischer MKC-501	EN ISO 12937	0,01
Acidity Value	mg KOH/g	Titration	EN 14104	-
Fatty Acid Distribution	% (m/m)	GSMS-TQ8030	EN ISO 5508	-
Methanol Content	% (m/m)	Agilent 6890 GC	EN 14110	-
Monogliserid Content	% (m/m)	Agilent 6890 C	EN 14105	-
Digliserid Content	% (m/m)	Agilent 6890 C	EN 14105	0,02
Trigliserid Content	% (m/m)	Agilent 6890 C	EN 14105	0,005
Free Glycerol	% (m/m)	Agilent 6890 C	EN 14105	0,0006
Total Glycerol	% (m/m)	Agilent 6890 C	EN 14105	0,014

Evaluation of Fuel Properties

The results of the analysis of biodiesel produced in Table **4** are given in comparison with EN14214 and ASTM D6751 biodiesel standard values and petroleum-based diesel fuel.

Table 4. Physical-chemical properties of algae oil used in pilot-scale biodiesel production.

Properties	Unit	EN 14124	ASTM D6751	Biodiesel	Diesel
Esther Content	% (m/m)	96,5 min.	-	96,4	-
Density (15 °C)	kg/m³	860-900	-	882	841
Viscosity (40 °C)	mm²/s	3,5-5,0	1,9-6,0	4,493	3,188
Flash Point	°C	120 min.	130 min.	143	61
Sulfur Content	mg/kg	10 max.	50 max.	6,81	9,81
PAH	% weight	-	-	5,6	4,3
Cetane number	-	51 min.	47 min.	57	55
SFTN	°C	-	-	-11	-14
Lower Heating Value	kJ/kg	-	-	39,820	42,880
Lubrication	μm	-	-	228	458
Water Content	mg/kg	500 max.	-	198	157
Acidity value	mg KOH/g	0,5 max.	0,5 max.	0,22	-
Methanol Content	% (m/m)	0,2 max.	-	<0,01	-
Monogliserid content	% (m/m)	0,8 max.	-	0,33	-
Digliserid content	% (m/m)	0,2 max.	-	0,12	-
Trigliserid content	% (m/m)	0,2 max.	-	0,01	-
Free Glycerol	% (m/m)	0,02 max.	0,02 max.	0,002	-
Total Glycerol	% (m/m)	0,25 max.	0,24 max.	0,016	-
Distillation	-	-	-	-	-
First Boiling Point	°C	-	-	328,8	170
10% achieved temperature	°C	-	-	340,2	202
50% achieved temperature	°C	-	-	339,7	264
90% achieved temperature	°C	-	-	350,3	326

The ester content of algae oil biodiesel was measured as 96.4%. This value appears to comply with EN 14214 standards. Total and free glycerol content, methanol content, and mono-, di- and triglyceride contents appear to comply with EN 14214 standards. The high density of algae oil has significantly decreased after it has been converted to the methyl ester form. The density value of the biodiesel is between the limit values specified in EN 14214 standards. The density of biodiesel is about 4.88% higher when compared to diesel fuel density. The viscosity of algae oil, which is very high compared to diesel fuel, also decreased significantly after it turned into a methyl ester form. The viscosity of algae oil methyl ester remained within the viscosity limit values specified in EN 14214 and

ASTM D6751 biodiesel standards. The flashpoint of the biodiesel has been measured as 141 °C, and this value appears to be following EN 14214 and ASTM D6751 standards. It is seen that the produced biodiesel sulfur content is lower than petroleum-based diesel fuel and complies with the standards. The thermal value of biodiesel was found to be approximately 7.68% lower than the thermal value of diesel fuel. It is seen that the lubricating feature of biodiesel is much better than that of petroleum diesel. The water content and acid values of algae oil biodiesel were also determined by the analyses conducted following the standards. Besides having the standard fuel properties of the produced biodiesel, its properties such as high cetane number (57) and good cold filter clogging point (-11 °C) make it an essential alternative diesel engine fuel.

CONCLUSION

The obtained highlights can be summarized as follows.

1. For vegetable and animal oils to be used as alternative fuels in diesel engines, their viscosity, which is very high compared to petroleum-based diesel fuel (diesel), must be lowered. For this purpose, dilution, microemulsion formation, pyrolysis, and transesterification methods are used.

2. In the dilution method, the oil is diluted by mixing it with diesel fuel at a certain rate. Depending on the mixture diesel fuel ratio, the viscosity, evaporation, and spraying properties of the fuel improve.

3. Microemulsion formation, this method, is aimed to form microemulsions of oil with short-chain alcohols such as methanol, ethanol, or 1-butanol. The microemulsion is the equilibrium distribution of optically isotropic liquid microstructures with dimensions between 1-150 nm and is formed by the combination of two normally immiscible liquids and one or more active substances.

4. In the pyrolysis method, oil molecules are broken down into smaller molecules in an oxygen-free environment at high temperatures. The pyrolysis process is divided into three parts: hydrocracking, catalytic cracking, and thermal cracking. The amount of product produced depends on the method used and the reaction parameters. With this method, although the fuel properties of the oils approach the diesel fuel properties, the high energy consumption is the most important negative.

5. Transesterification is the re-esterification reaction of fatty acids (vegetable oils, household waste oils, animal fats) with alcohol (methanol, ethanol, *etc.*) in the presence of a basic catalyst. In this method, oil molecules enter into a chemical

reaction with alcohol at a certain temperature with the help of a catalyst. With this chemical reaction, fatty acids are separated from their triglycerides and form new esters with alcohols.

CONSENT FOR PUBLICATION

Not applicable.

CONFLICT OF INTEREST

The author declares no conflict of interest, financial or otherwise.

ACKNOWLEDGEMENT

Declared none.

REFERENCES

[1]   Khan S, Siddique R, Sajjad W, *et al*. Biodiesel production from algae to overcome the energy crisis. Hayati J Biosci 2017; 24(4): 163-7.
        [http://dx.doi.org/10.1016/j.hjb.2017.10.003]

[2]   Yaşar F. Biodiesel production from algae oil and its use in a diesel engine as alternative fuel. Elazığ: Fırat Universiry 2016.

[3]   ASTM International: West Conshohocken, PA 2018.

[4]   Vignesh G, Barik D. Chapter 6 - Toxic waste from biodiesel production industries and its utilization. In: Barik D, Ed. Energy from Toxic Organic Waste for Heat and Power Generation. Woodhead Publishing 2019; pp. 69-82.
        [http://dx.doi.org/10.1016/B978-0-08-102528-4.00006-7]

[5]   Liu J, Huang J, Sun Z, Zhong Y, Jiang Y, Chen F. Differential lipid and fatty acid profiles of photoautotrophic and heterotrophic Chlorella zofingiensis: Assessment of algal oils for biodiesel production. Bioresour Technol 2011; 102(1): 106-10.
        [http://dx.doi.org/10.1016/j.biortech.2010.06.017] [PMID: 20591657]

[6]   Lamaisri C, Punsuvon V, Chanprame S, Arunyanark A, Srinives P, Liangsakul P. Relationship between fatty acid composition and biodiesel quality for nine commercial palm oils. Songklanakarin J Sci Technol 2015; 37: 389-95.

[7]   Singh G, Jeyaseelan C, Bandyopadhyay K K, Paul D. Comparative analysis of biodiesel produced by acidic transesterification of lipid extracted from oleaginous yeast Rhodosporidium toruloides. 3 Biotech 2018; 8(10): 434.

[8]   Knothe G. Dependence of biodiesel fuel properties on the structure of fatty acid alkyl esters. Fuel Process Technol 2005; 86(10): 1059-70.
        [http://dx.doi.org/10.1016/j.fuproc.2004.11.002]

[9]   Knothe G, Matheaus AC, Ryan TW III. Cetane numbers of branched and straight-chain fatty esters determined in an ignition quality tester□. Fuel 2003; 82(8): 971-5.
        [http://dx.doi.org/10.1016/S0016-2361(02)00382-4]

[10]   Giakoumis E, Sarakatsanis C. A comparative assessment of biodiesel cetane number predictive correlations based on fatty acid composition. Energies 2019; 12(3): 422.
        [http://dx.doi.org/10.3390/en12030422]

[11]   Knothe G. A comprehensive evaluation of the cetane numbers of fatty acid methyl esters. Fuel 2014; 119: 6-13.
[http://dx.doi.org/10.1016/j.fuel.2013.11.020]

[12]   Xu H, Miao X, Wu Q. High quality biodiesel production from a microalga Chlorella protothecoides by heterotrophic growth in fermenters. J Biotechnol 2006; 126(4): 499-507.
[http://dx.doi.org/10.1016/j.jbiotec.2006.05.002] [PMID: 16772097]

[13]   Al-lwayzy SH, Yusaf T, McCabe B, Pittaway P, Aravinthan V. Microalgae as alternative fuel for compressionignition (CI) engines Southern Region Engineering Conference. 11-12 November 2010; Toowoomba, Australia SREC2010-F2-4.

[14]   Campenni' L, Nobre BP, Santos CA, *et al.* Carotenoid and lipid production by the autotrophic microalga Chlorella protothecoides under nutritional, salinity, and luminosity stress conditions. Appl Microbiol Biotechnol 2013; 97(3): 1383-93.
[http://dx.doi.org/10.1007/s00253-012-4570-6] [PMID: 23160982]

Recent Advances in Biotechnology, 2022, Vol. 6, 73-97 73

CHAPTER 5

Algal Biodiesel Chemical Characterization

Cemil Koyunoğlu[1,*] and **Fevzi Yaşar[2]**

[1] *Energy Systems Engineering Department, Engineering Faculty, Yalova University, Yalova, Turkey*

[2] *Department of Chemistry and Chemical Process Technology, Vocational School of Technical Sciences, Batman University, 72100 Batman, Turkey*

Abstract: Algae have been produced or evaluated as a nutritional supplement in animal husbandry, rather than as an alternative energy source for many years. As a result of biomass energy research, which has accelerated in recent years with the impact of rising oil prices, algae have started to be seen as a promising energy source. Despite being successful in laboratory research, pilot, and small-scale experiments, also called third-generation biofuel technologies and aiming to use many algae species in nature as an energy source, the desired yield cannot be obtained if ideal processes cannot be created in large-scale local productions. In general, algae may contain about 15-77% fat although the volume varies by species. Compared to other oil plants, their high oil content and growth efficiency make algae attractive for biodiesel and biogas production. The production of these fuels from algae has the potential to respond to the increasing global energy need and, in part, to contribute to the prevention of global warming by converting more than enough carbon dioxide in the atmosphere into efficient products through photosynthesis.

Keywords: Algal biodiesel, Chemical Characterization, Environmental issues, Oil quality.

INTRODUCTION

One of the advantages of using algae as raw materials for biofuels is that different types of fuels can be produced through it. In addition to biodiesel and biogas, algae have properties that can meet our needs for ethanol, bio-jet fuel, bio-gasoline, or other fuel. Biodiesel is a diesel fuel derived from plant or animal lipids (oils or fats). Studies show that some algae species contain more than 80% of their total dry weight fat. Algae cells have large-scale biomass production capacities, as they are grown in an aqueous suspension medium in pools and photobioreactors with water, CO_2, and dissolved nutrients. Fats produced from

---

* **Correspondence author Cemil Koyunoğlu:** Energy Systems Engineering Department, Engineering Faculty, Yalova University, Yalova, Turkey; Tel: +902268155378; E-mail: cemil.koyunoglu@yalova.edu.tr

Hüseyin Karaca and Cemil Koyunoğlu (Eds.)

algae can be converted into biodiesel for later use in automobiles. In Fig. (**1**), biodiesel and various biofuels are obtained as a result of the extraction of oil from algae that was established next to the power plant and grown by using wastewater such as agricultural or sewage.

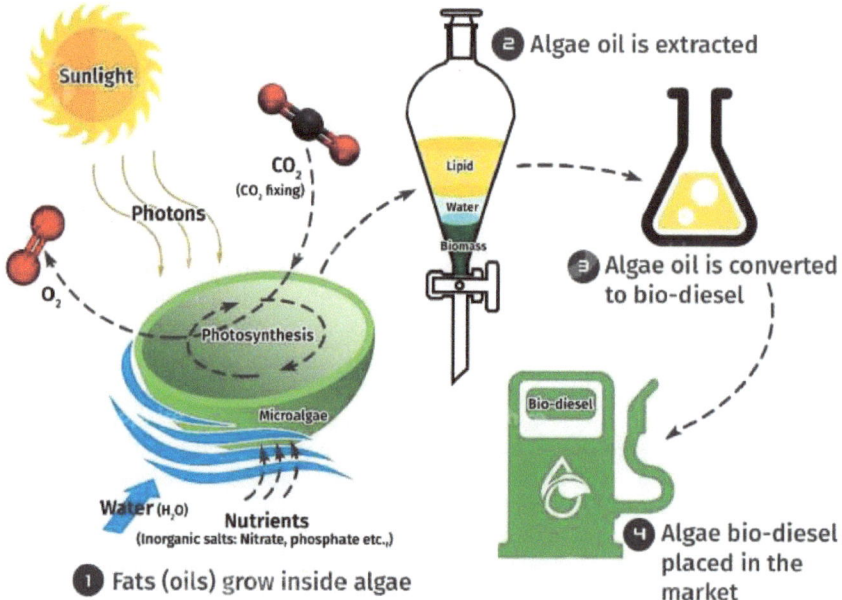

Fig. (1). Flowchart for biotechnology [1].

ALGAE OIL AND ITS PROPERTIES

Due to the quality and quantity of oil produced from algae; the nutrients in the environment vary depending on CO_2, water, light intensity, pH, and temperature. Minerals and vitamins, which are very useful for health, are high in algae oil. Algae oil, which is mostly preferred in the cosmetics industry, is also useful for general health. Algae oil is very effective in the study of thyroids, regeneration of tissues, and acceleration of metabolism. The usage area of moss oil is extensive. However, it is generally preferred for body's beauty. The reason is that the minerals and vitamins contained in algae oil meet the body's needs. Also, as a result of the rapid consumption of fossil fuels in recent years, the pollution of the atmosphere at the same speed and the increase of the effects of greenhouse gases have increased the demand for alternative energy sources. As a result of the studies, the idea that algae oil will be an alternative source of fossil fuels has been strengthened. In addition, fuels such as biodiesel, bio-jet fuel, bio-gasoline, biobutanol, bioethanol, and methane were obtained from algae oils, and it was determined that their fuel properties are suitable (Table **1**) [2].

Table 1. Mandates, targets, and policies of selected countries [2].

Biodiesel and Ethanol	Policies	Country
Obligation by 2030 (% energy), 30% biofuel supply	-	Finland
5% ethanol	To produce ethanol approved feedstocks, biofuels' policy expands	India
25% biodiesel (%vol.) by 2036 and 32% ethanol	-	Thailand
10% biodiesel, 12% ethanol (%vol.)	-	Argentina
10% renewable energy by 2020 (T) in transport with 7% for conventional biofuels (% energy)	In 2030, a Provisional agreement for 14% renewable energy in transport	European Union
Low carbon Fuel Standard in Oregon and California	by 2022 136 billion L.	USA
10% biodiesel (%vol.) and 27% ethanol	RenovaBio signed into law, by 2028 10% GHG reduction	Brazil
10% ethanol (%vol.)	To extend nationwide distribution in 2020, 10% ethanol was mandated	China

Some properties are analyzed before algae oil is converted to biodiesel. In the analyses made, it has been determined that algae oil has high acid value, viscosity, and density values and partially contains water. To reduce the water content of algae oil, the sample oil is first passed through a filter. Then, it is taken to a beaker and subjected to preheating by mixing it with a magnetic stirrer for 2 hours at a temperature over 100 °C (at 110 °C), which is the boiling point of the water. After the preheating process, the oil is rested in a suitable environment and then the acid value, viscosity, density, and water content parameters are analyzed again. In Table **2**, some properties of the moss oil used in biodiesel production before and after pretreatment are given [2].

Table 2. Some properties of algae oil.

Properties	Unit	Algae Oil	Pretreated Algae Oil
Density (at 15 °C)	kg/m^3	915	913
Kinematic viscosity (40 °C)	mm^2/s	35,6	32,4
Acidity value	mg KOH/g	0,28	0,22
Water content	ppm	734,7	220

When Table **2** is examined, it is seen that there is a significant decrease in the water content, especially after the seaweed oil has been subjected to a preheating and filtering process. The viscosity of algae oil appears to be too high to be used as a fuel in diesel engines. Therefore, its viscosity should be reduced in order to be evaluated as diesel engine fuel.

Table **3** shows Scenedesmus almeriensis, Nannochloropis gaditana, and Spirulina platensis oils chemical properties. Table **4** depicts their fame contents.

Table 3. Some selected algae oil production properties [3].

Properties	Scenedesmus Almeriensis	Nannochloropis Gaditana	Spirulina Platensis
Biodiesel, $g_{biodiesel}/g_{biomass}$	0,084	0,119	0,067
Purity, %	64,6	68,1	63,7
FA content, %dry.wt.	6,03	10,64	5,03
Conversion, %	90,3	92,7	84,3

Table 4. Fatty acid properties of biodiesel [3].

FAME/Algae Types	Scenedesmus Almeriensis	Nannochloropis Gaditana	Spirulina Platensis
(C16:0), Methyl palmitate	47,12	32,71	42,25
(C18:0), Methyl stearate	0,64	-	2,52
(C20:4), Methyl arachidonate	-	3,27	-
(C14:0), Methyl tetradecanoate	0,48	5,61	4,38
(C16:3), Methyl hexadecatrienoate	6,35	-	-
(C16:4), Methyl hexadecatetraenoate	2,65	-	-
(C18:3), Methyl linolenate	28,55	-	13,82
(C18:2), Methyl linoleate	7,9	3,74	21,93
(C16:1), Methyl palmioleate	2,04	29,91	8,44
(C20:4), Methyl arachidonate	-	3,27	-
(C20:5), Methyl eicosapentaenoate	-	15,42	-

Table **5** presents the chlorella-based biodiesel properties.

Table 5. Chlorella biodiesel properties [4].

Algae Types	Properties (Experimental)	Standards (ASTM D6751)	Standards (EN14214)
Chlorella vulgaris	-	-	-
Density (15 °C, kgm^{-3})	885	875-900	860-900
Viscosity (40 °C, mm^2 s^{-1})	4,7	1,9-6	3,5-5,0
Cetane number (min)	65,9	min.47	min.51

(Table 5) cont.....

Algae Types	Properties (Experimental)	Standards (ASTM D6751)	Standards (EN14214)
Iodine value (g I2 100 g^{-1})	60,26	-	max.120
CFPP	-5.0863	-	-
Pour point (°C)	-11.6373	-15 to 10	-
Cloud point (°C)	-4.43681	-3 to 12	-
DU (%)	67,04	-	-
LCSF (wt %)	3.62565	-	-
Acid value (mg KOH g^{-1})	0,46	max. 0,8	max. 0,5
Saponification value (mg KOH g^{-1})	164,184	-	-
Chlorella sp.	-	-	-
Density (15 °C, kgm^{-3})	-	875-900	860-900
Viscosity (40 °C, mm^2 s^{-1})	-	1,9-6	3,5-5,0
Cetane number (min)	51,94	min.47	min.51
Iodine value (g I2 100 g^{-1})	98,65	-	max.120
CFPP	-15,30	-	-
Pour point (°C)	-	-15 to 10	-
Cloud point (°C)	-	-3 to 12	-
DU (%)	-	-	-
LCSF (wt %)	0,38	-	-
Acid value (mg KOH g^{-1})	-	max. 0,8	max. 0,5
Saponification value (mg KOH g^{-1})	196,04	-	-
Chlorella protothecoides	-	-	-
Density (15 °C, kgm^{-3})	882	875-900	860-900
Viscosity (40 °C, mm^2 s^{-1})	4,43	1,9-6	3,5-5,0
Cetane number (min)	-	min.47	min.51
Iodine value (g I2 100 g^{-1})	112,2	-	max.120
CFPP	-	-	-
Pour point (°C)	-	-15 to 10	-
Cloud point (°C)	-	-3 to 12	-
DU (%)	-	-	-
LCSF (wt %)	-	-	-
Acid value (mg KOH g^{-1})	0,29	max. 0,8	max. 0,5
Saponification value (mg KOH g^{-1})	-	-	-

Table **6** shows some quality properties for biodiesel.

Table 6. Quality properties of biodiesel [3].

Carbon Number	Methyl Esters	Weight (%)	Type
C11:0	Undecanoic acid	1,187	SFA
C10:0	Capric acid	1,281	SFA
C13:0	Tedecanoic acid	1,571	SFA
C14:0	Myristic acid	3,541	SFA
C15:1	Cis-10-pentadecanoic	31,11	MUFA
C16:0	Palmitic	1,056	SFA
C17:0	Heptadecanoic acid	29,44	SFA
C17:1	Cis-10-Heptadecanoic	1,705	MUFA
C18:1	Elaidic	7,515	MUFA
C18:1	Oleic	3,744	MUFA
C18:2	Linoeladic acid	2,679	PUFA
C20:0	Arachidic	0,415	SFA
C20:1	Cis-11Eicosenoic	1,084	MUFA
C18:3	Linolenic	0,834	PUFA
C20:2	Cis-11,14 Eicosadienoic	0,772	PUFA
C22:0	Behenic	2,07	SFA
C23:0	Tricosanoic	0,353	SFA
C20:2	Cis-13,16 Docosadienoic	5,223	PUFA
C20:5	Cis-5,8,11,14,17-Eicosapentaneoic	0,201	PUFA
C22:6	Cis-4,7,10,13,16,19-Decosahexaenoic	2,361	PUFA
SFA			42,023
MUFA			45,899
PUFA			12,074
Total			99,996

There are many types of algae used in biofuel production. Table 7 shows the lipid content, lipid efficiencies, and volumetric and areal yields of some of the algae used in biofuel production[3]. The most widely used algae types in biodiesel production are Chlorella, Dunaliella, Isochrysis, Nannochloris, Nannochloropsis Oculata, Neochloris, Nitzschia, Phaeodactylum, and Porphyridium spp. Their fat content varies between 20-50%. As seen in Table 7, Chlorella protothecoides algae we use in this study have the highest lipid content (14.6-57.8% dry weight), the highest lipid yield (1214mg/L/day), and the highest biomass volume (2-7.70 mg/L/day).

Table 7. Selected properties of various algae types [5].

Algae Type	Lipid content (dry, % wt.)	Lipid Efficiency (mg/L/day)	Biomass Volumetric Efficiency (mg/L/day)	Biomass Area Efficiency (mg/m²/day)
Ankistrodesmus sp.	24-21	-	-	11,5-17,4
Botryococcus braunii	25-50	-	0,02	3,0
Chaetoceros muelleri	33,6	21,8	0,07	-
Chaetoceros calcitrans	14,6-16,4	17,6	0,04	-
Chlorella emersonii	25-63	10.3-50	0,036-0,041	0,91-0,97
Chlorella protothecoides	14,6-57,8	1214	2-7,7	-
Chlorella sorokiniana	19-22	44,7	0,23-1,47	-
Chlorella vulgaris	5-58	11,2-40	0,02-0,2	0,57-0,95
Chlorella sp.	10-48	42,1	0,02-0,2	1,61-16,4
Chlorealla pyrenoidosa	2	-	2,90-3,64	72,5-130
Chlorella	18-57	18,7	-	3,5-13,9
Chlorococcum sp.	19,3	53,7	0,28	-
Crypthecodinium cohnii	20-51,1	-	10	-
Duanaliella salina	6-25	116	0,22-0,34	1,6-3,5/20-38
Dunaliealla primolecta	23,1	-	0,09	14
Dunaliella tertiolecta	16,7-71	-	0,12	-
Dunaliella sp.	17,5-67	33,5	-	-
Ellipsoidion sp.	27,4	47,3	0,17	-
Euglena gracilis	14-20	-	7,7	-
Haematococcus pluvialis	25	-	0,05-0,06	10,2-36,4
Isochrysis galbana	7-40	-	0,32-1,6	-
Isochrysis sp.	7,1-33	37,8	0,08-0,17	-
Monodus subterraneus	16	30,4	0,19	-
Monallanthus salina	20-22	-	0,08	12
Nannochloris sp.	20-56	60,9-76,5	0,17-0,51	-
Nannochloropsis oculate.	22,7-29,7	84-142	0,37-0,48	-
Nannochloropsis sp.	12-53	37,6-90	0,17-1,43	1,9-5,3
Neochloris oleoabundans	29-65	90-134	-	-

(Table 7) cont.....

Algae Types	Properties (Experimental)	Standards (ASTM D6751)	Standards (EN14214)
Iodine value (g I2 100 g^{-1})	60,26	-	max.120
CFPP	-5.0863	-	-
Pour point (°C)	-11.6373	-15 to 10	-
Cloud point (°C)	-4.43681	-3 to 12	-
DU (%)	67,04	-	-
LCSF (wt %)	3.62565	-	-
Acid value (mg KOH g^{-1})	0,46	max. 0,8	max. 0,5
Saponification value (mg KOH g^{-1})	164,184	-	-
Chlorella sp.	-	-	-
Density (15 °C, kgm^{-3})	-	875-900	860-900
Viscosity (40 °C, mm^2 s^{-1})	-	1,9-6	3,5-5,0
Cetane number (min)	51,94	min.47	min.51
Iodine value (g I2 100 g^{-1})	98,65	-	max.120
CFPP	-15,30	-	-
Pour point (°C)	-	-15 to 10	-
Cloud point (°C)	-	-3 to 12	-
DU (%)	-	-	-
LCSF (wt %)	0,38	-	-
Acid value (mg KOH g^{-1})	-	max. 0,8	max. 0,5
Saponification value (mg KOH g^{-1})	196,04	-	-
Chlorella protothecoides	-	-	-
Density (15 °C, kgm^{-3})	882	875-900	860-900
Viscosity (40 °C, mm^2 s^{-1})	4,43	1,9-6	3,5-5,0
Cetane number (min)	-	min.47	min.51
Iodine value (g I2 100 g^{-1})	112,2	-	max.120
CFPP	-	-	-
Pour point (°C)	-	-15 to 10	-
Cloud point (°C)	-	-3 to 12	-
DU (%)	-	-	-
LCSF (wt %)	-	-	-
Acid value (mg KOH g^{-1})	0,29	max. 0,8	max. 0,5
Saponification value (mg KOH g^{-1})	-	-	-

present in TAG are of commercial interest because they can be used as a resource for the production of transportation fuels, bulk chemicals, nutraceuticals (ω-3 fatty acids), and food commodities 3-6. Microalgae can grow in sea water-based cultivation media, can have a much higher areal productivity than terrestrial plants, and can be cultivated in photobioreactors at locations that are unsuitable for agriculture, possibly even offshore. For these reasons, microalgae are often considered a promising alternative to terrestrial plants for the production of biodiesel and other bulk products [4 - 7]. Potentially no agricultural land or freshwater (when cultivated in closed photobioreactors or when marine microalgae are used) is needed for their production. Therefore, biofuels derived from microalgae are considered 3rd-generation biofuels. The total cellular content of fatty acids (% of dry weight), the lipid class composition, as well as the fatty acid length and degree of saturation are highly variable between microalgae species. Furthermore, these properties vary with cultivation conditions such as nutrient availability, temperature, pH, and light intensity. For example, when exposed to nitrogen starvation, microalgae can accumulate large quantities of TAG. Under optimal growth conditions, TAG typically constitutes less than 2% of dry weight, but when exposed to nitrogen starvation TAG content can increase to up to 40% of the microalgal dry weight. Microalgae mainly produce fatty acids with chain lengths of 16 and 18 carbon atoms, but some species can make fatty acids of up to 24 carbon atoms in length. Both saturated, as well as highly unsaturated fatty acids, are produced by microalgae. The latter include fatty acids with nutritional benefits (ω-3 fatty acids) like C20:5 (eicosapentaenoic acid; EPA) and C22:6 (docosahexaenoic acid; DHA) for which no vegetable alternatives exist [7, 8, 12 - 15]. The (distribution of) fatty acid chain length and degree of saturation also determines the properties and quality of algae-derived biofuels and edible oils [12, 16]. To develop commercial applications of microalgal-derived fatty acids, reliable analytical methods for quantification of fatty acid content and composition are needed. As also pointed out by Ryckebosch *et al.* [17], analysis of fatty acids in microalgae distinguishes itself from other substrates (*e.g.* vegetable oil, food products, animal tissues, *etc.*) because 1) microalgae are single cells surrounded by rigid cell walls, complicating lipid extraction; 2) microalgae contain a wide variety of lipid classes and the lipid class distribution is highly variable [17]. These different lipid classes have a wide variety of chemical structures and properties, such as polarity. Also, lipid classes other than acyl lipids are produced; 3) microalgae contain a wide variety of fatty acids, ranging from 12 to 24 carbon atoms in length and containing both saturated as well as highly unsaturated fatty acids. Therefore, methods developed to analyze fatty acids in substrates other than microalgae might not be suitable to analyze fatty acids in microalgae. As reviewed by Ryckebosch *et al.* [17], the main difference between commonly used lipid extraction procedures is in the solvent systems that are used.

Because of the large variety of lipid classes present in microalgae, each varying in polarity, the extracted lipid quantity will vary with solvents used [18 - 20]. This leads to inconsistencies in the lipid content and composition presented in the literature [17, 18]. Depending on the solvent system used, methods based on solvent extraction without cell disruption are not available to use, for example, bead beating or sonication, might not extract all lipids because of the rigid structure of some microalgae species [17]. In the case of incomplete lipid extraction, the extraction efficiency of the different lipid classes can vary [21]. This can also influence the measured fatty acid composition because the fatty acid composition is variable among lipid classes.

The fatty acid profiles of the algae are presented in Table **8**. The highest lipid content was observed in *P. tricornitum* followed by *I. galbana* and *P. subcapitata*, while *Tribonema* sp. had the lowest lipid content. The fatty acid profiles varied between the classes. In Cyanophyceae, *Chroococcus* sp. had C16:0 and C18:2 x6 as the dominating fatty acids, with lower amounts of C16:1 and C18:3 x3 (Table **9**) [22].

Table 8. Typical algae oil composition [23].

Properties	Percentage
Fatty acid compositions	%
(C22:1)	2,89
(C22:0)	6,89
(C20:1)	0,63
(C20:0)	1,27
(C18:2)	1,48
(C18:1)	52,18
(C16:0)	30,63
Undefinied	4,03
Physicochemical characteristics	Percentage
Kinetic viscosity (at 40 °C) (cSt)	22,14
Acid value (mg KOH/goil)	0,021
Flash point (°C)	184
Density (g/cm³)	0,9205
Iodine value (g of I_2/100 g)	1,56
Cloud point (°C)	26
Saponification value (mg KOH/g oil)	131,7
Gross calorific value (Cal/gm)	9670

Comparisons between diesel and biodiesel fuels are shown in Table **10**.

Table 9. Fatty acid profile of the microalgae tested. Data are given as mg.g-1 of dry weight.

Algae Type	Fatty Acid Content
Chroococcus sp.	-
Saturated	-
C12:0	2,1
C14:0	0,1
C16:0	21,3
C18:0	0,3
C20:0	0,2
C24:0	-
Sum	24,0
Monosaturated	-
C16:1	1,1
C18:1	0,4
C20:1	-
C22:1	0,1
Polyunsaturated	-
C18:2	10,7
C18:3	1,0
C18:4	-
C20:2	0,1
C20:4	0,1
C20:5	-
C22:5	-
C22:6	-
Sum	11,9
Synechococcus sp.	-
C12:0	0,7
C14:0	5,6
C16:0	3,4
C18:0	-
C20:0	-
C24:0	-

(Table 9) cont.....

Algae Type	Fatty Acid Content
Sum	9,7
C16:1	10,8
C18:1	-
C20:1	-
C22:1	-
Sum	10,8
Polyunsaturated	-
C18:2	-
C18,3	-
C18:4	-
C20:2	-
C20:4	-
C20:5	-
C22:5	-
C22:6	-
Isochrysis galbana	-
Saturated	-
C12:0	-
C14:0	8,9
C16:0	11,5
C18:0	-
C20:0	-
C24:0	-
Sum	20,4
Monosaturated	-
C16:1	10,8
C18:1	13,1
C20:1	-
C22:1	0,6
Sum	17,0
Polyunsaturated	-
C18:2	7,0
C18:3	3,8
C18:4	12,5

(Table 9) cont.....

Algae Type	Fatty Acid Content
C20:2	-
C20:4	-
C20:5	0,8
C22:5	-
C22:6	15,8
Sum	39,9
Pavlova Sp.	-
Saturated	-
C12:0	-
C14:0	7,5
C16:0	13,4
C18:0	0,4
C20:0	-
C24:0	-
Sum	21,3
Monosaturated	-
C16:1	12,8
C18:1	2,9
C20:1	-
C22:1	0,8
Sum	16,5
Polyunsaturated	-
C18:2	2,1
C18:3	1,8
C18:4	4,3
C20:2	-
C20:4	-
C20:5	18,0
C22:5	-
C22:6	13,2
Sum	39,8
Phaeodactylum tricornutum	-
Saturated	-
C12:0	-

(Table 9) cont.....

Algae Type	Fatty Acid Content
C14:0	8,8
C16:0	16,6
C18:0	0,6
C20:0	-
C24:0	1,6
Sum	27,6
Monosaturated	-
C16:1	26,0
C18:1	1,8
C20:1	0,3
C22:1	-
Sum	28,1
Polyunsaturated	-
C18:2	1,5
C18:3	0,3
C18:4	3,3
C20:2	-
C20:4	2,2
C20:5	28,4
C22:5	1,3
C22:6	0,2
Sum	37,2
Porphyridium cruentum	-
Saturated	-
C12:0	-
C14:0	-
C16:0	5,9
C18:0	0,2
C20:0	-
C24:0	-
Sum	6,1
Monosaturated	-
C16:1	-
C18:1	0,1

(Table 9) cont.....

Algae Type	Fatty Acid Content
C20:1	-
C22:1	-
Sum	0,1
Polyunsaturated	-
C18:2	2,1
C18:3	-
C18:4	-
C20:2	0,3
C20:4	6,0
C20:5	6,1
C22:5	-
C22:6	-
Sum	14,5
Rhodomonas baltica	-
Saturated	-
C12:0	2,0
C14:0	4,1
C16:0	6,0
C18:0	0,8
C20:0	0,1
C24:0	4,0
Sum	17
Monosaturated	-
C16:1	0,4
C18:1	3,4
C20:1	0,1
C22:1	0,1
Sum	4,0
Polyunsaturated	-
C18:2	11,7
C18:3	12,0
C18:4	5,1
C20:2	0,1
C20:4	0,2

(Table 9) cont.....

Algae Type	Fatty Acid Content
C20:5	4,4
C22:5	0,2
C22:6	-
Sum	33,7
Oocystis sp.	-
Saturated	-
C12:0	-
C14:0	0,2
C16:0	3,8
C18:0	-
C20:0	-
C24:0	0,1
Sum	4,1
Monosaturated	-
C16:1	1,5
C18:1	3,9
C20:1	-
C22:1	-
Sum	5,4
Polyunsaturated	-
C18:2	6,4
C18:3	8,1
C18:4	0,7
C20:2	-
C20:4	0,5
C20:5	1,1
C22:5	-
C22:6	-
Sum	16,8
Pseudokirchneriella subcapitata	-
Saturated	-
C12:0	-
C14:0	0,1
C16:0	16,2

(Table 9) cont.....

Algae Type	Fatty Acid Content
C18:0	1,3
C20:0	0,2
C24:0	0,7
Sum	18,5
Monosaturated	-
C16:1	1,0
C18:1	31,1
C20:1	0,9
C22:1	0,8
Sum	33,8
Polyunsaturated	-
C18:2	5,1
C18:3	11,4
C18:4	3,0
C20:2	-
C20:4	-
C20:5	-
C22:5	2,1
C22:6	0,1
Sum	21,7
Tribonema sp.	-
Saturated	-
C12:0	-
C14:0	1,1
C16:0	2,5
C18:0	0,1
C20:0	-
C24:0	-
Sum	3,7
Monosaturated	-
C16:1	5,1
C18:1	0,2
C20:1	-
C22:1	-

(Table 9) cont.....

Algae Type	Fatty Acid Content
Sum	5,3
Polyunsaturated	-
C18:2	0,2
C18:3	-
C18:4	0,1
C20:2	-
C20:4	-
C20:5	3,2
C22:5	-
C22:6	-
Sum	3,5
Nannochloropsis oceanica	-
Saturated	-
C12:0	1,2
C14:0	16,9
C16:0	17,2
C18:0	1,8
C20:0	-
C24:0	-
Sum	37,1
Monosaturated	-
C16:1	18,2
C18:1	4,1
C20:1	0,5
C22:1	-
Sum	22,8
Polyunsaturated	-
C18:2	9,7
C18:3	0,5
C18:4	-
C20:2	0,5
C20:4	3,7
C20:5	23,4
C22:5	-

(Table 9) cont.....

Algae Type	Fatty Acid Content
C22:6	-
Sum	37,8

Dashes indicate FA not detected [a]ω-3 fatty acids [b]ω-6 fatty acids [c]ω-9 fatty acids

ANALYSIS OF FUEL PROPERTIES OF ALGAE OIL BIODIESEL

A typical algae oil composition is given in Table **3**.

Yadav *et al* studied using biodiesel in diesel engines in 2018. They used Table **10** for defining fuel and used Table **11** engine specifications. To determine using biodiesel fuel in diesel engines they used Table **12** for exhaust gas accuracy and range [23].

Table 10. Comparison between biodiesel and diesel fuels [23].

Composition Details	Fuel Type
%50 algal biodiesel and 50% DF on a volume basis	AB50
%30 algal biodiesel and 70% DF on a volume basis	AB30
%25 algal biodiesel and 75% DF on a volume basis	AB25
%20 algal biodiesel and 80% DF on a volume basis	AB20
15% algal biodiesel and 85% DF on a volume basis	AB15
10% algal biodiesel and %90 DF on a volume basis	AB10
5% algal biodiesel and 95% DF on a volume basis	AB5
Pure diesel fuel (DF)	DF

Table 11. Specification of the engine [23].

Properties	Engine Characteristic
Kirloskar	Make
Four-stroke, single-cylinder, 3,5 kW, 110 mm	Engine type
1500 rpm	Speed
87,5 mm	Bore
12-18	CR range
Crank angle sensor	Resolution 1 Deg. speed 5500 rpm with TDC pulse
Dynamometer	Eddy current type, water-cooled with loading unit

Table 12. Exhaust gas accuracy and range [23].

Instrument	Accuracy	Unit and Range
Load indicator (LI)	± 2.0%	0-50 kg
Speed sensor (SS)	± 1.0%	0-9999 rpm
Temperature sensor (PS)	± 0.15%	0-300 °C
Crake angle encoder (CAE)	± 0.5%	0-100 bar
Exhaust gas analyzer	-	-
CO_2	± 0.51%	0-20.00%
CO	± 0.006%	0-15.00%
O_2	± 0.1%	0-25.00%
HC	± 12 ppm	0-3000 ppm
NO_x	± 3 ppm	0-5000 ppm

They defined the physicochemical properties of algal biodiesel blends in Table **13** [23].

Table 13. Algal biodiesel blends essential physicochemical properties [23].

Fuel Types	ASTM Standard	Parameters	Values
DF	D1448-1972	Density at 15 °C (g/cm³)	0,828
AB05	D1448-1972	Density at 15 °C (g/cm³)	0,830
AB10	D1448-1972	Density at 15 °C (g/cm³)	0,831
AB15	D1448-1972	Density at 15 °C (g/cm³)	0,833
AB20	D1448-1972	Density at 15 °C (g/cm³)	0,835
AB30	D1448-1972	Density at 15 °C (g/cm³)	0,836
AB30	D1448-1972	Density at 15 °C (g/cm³)	0,838
AB50	D1448-1972	Density at 15 °C (g/cm³)	0,839
DF	D4868	Calorific value (MJ/kg)	42,50
AB05	D4868	Calorific value (MJ/kg)	42,33
AB10	D4868	Calorific value (MJ/kg)	42,13
AB15	D4868	Calorific value (MJ/kg)	42,01
AB20	D4868	Calorific value (MJ/kg)	41,9
AB30	D4868	Calorific value (MJ/kg)	41,66
AB30	D4868	Calorific value (MJ/kg)	41,35
AB50	D4868	Calorific value (MJ/kg)	41,00
DF	D93	Flash point (°C)	67

(Table 13) cont.....

Fuel Types	ASTM Standard	Parameters	Values
AB05	D93	Flash point (°C)	69
AB10	D93	Flash point (°C)	73
AB15	D93	Flash point (°C)	77
AB20	D93	Flash point (°C)	79
AB30	D93	Flash point (°C)	84
AB30	D93	Flash point (°C)	87
AB50	D93	Flash point (°C)	92
DF	D93	Flash point (°C)	74
AB05	D93	Flash point (°C)	82
AB10	D93	Flash point (°C)	86
AB15	D93	Flash point (°C)	92
AB20	D93	Flash point (°C)	96
AB30	D93	Flash point (°C)	100
AB30	D93	Flash point (°C)	104
AB50	D93	Flash point (°C)	110

After determining the fatty acid distribution and other properties of algae oil, the conversion of algae oil to biodiesel with the most appropriate method should be addressed. In studies, it is known that the ideal method for converting algae oil to biodiesel is transesterification [21 - 23]. In this study, the methods and devices used in the characterization of algae oil and in determining the properties of the methyl ester fuel produced, the working principles, and the sensitivity of the devices are shown in Table **14**. After large-scale biodiesel production was carried out in Batman University Vocational School Refinery and Petro-Chemistry Technology Laboratory, the fatty acid distribution of biodiesel was determined at Dicle University Science and Technology Application and Research Center. The fatty acid composition of biodiesel is given in Table **15**.

Table 14. Methods and devices used in oil characterization and determination of fuel properties.

Properties	Unit	Device	Test Method	Precision
Ester content	% (m/m)	Agilent 6890 GC	EN 14103	0,09
Density (15 °C)	kg/m³	DMA38	EN ISO 12185	0,1
Viscosity (40 °C)	mm²/s	HVM 472	EN ISO 3104	0,001
Flash point	°C	HFP 339	EN ISO 3679	0,5
Sulfur content	mg/kg	Flash 2000	EN ISO 20846	0,1

(Table 14) cont.....

Properties	Unit	Device	Test Method	Precision
Polycyclic aromatic hydrocarbon	% mass	HPLC-RID SHIMADZU	EN ISO 5165	-
Cold filter Plugging Point	°C	FPP 5Gs	EN 116	0,1
Lower Heating Point	kj/kg	IKA C5003	TS 1740	1,0
Lubricity (60 °C)	μm	HFRR	EN ISO 12156	-
Water content	mg/kg	KEM Karl-Fischer MKC-501	EN ISO 12937	0,01
Acid value	mg KOH/g	Titration	EN 14104	-
Fatty Acid Distribution	% (m/m)	GSMS-TQ8030	EN ISO 5508	-
Methanol content	% (m/m)	Agilent 6890 GC	EN 14110	-
Monogliserid content	% (m/m)	Agilent 6890 GC	EN 14105	0,03
Digliserid content	% (m/m)	Agilent 6890 GC	EN 14105	0,02
Trigliserid content	% (m/m)	Agilent 6890 GC	EN 14105	0,005
Free Glycerol	% (m/m)	Agilent 6890 GC	EN 14105	0,0006
Total Glycerol	% (m/m)	Agilent 6890 GC	EN 14105	0,014
-	% (m/m)	Agilent 6890 GC	-	-
-	% (m/m)	Agilent 6890 GC	-	-

The ester content, total and free glycerol content, methanol content, and mono-, di- and triglyceride contents of the produced biodiesel fuel were determined at METU Petroleum Research Center. Other features were made in the TÜPRAŞ Batman Refinery fuel analysis laboratory and Batman University Vocational School Refinery and Petro-Chemistry Technology program laboratory.

Table 15. The experimental fatty acid composition was conducted at Batman University.

Fatty Acid	C:D	Close Formula	Mass Weight	Weight %
Palmitic	C16:0	$C_{16}H_{32}O_2$	256,43	4,904
Palmitoleic	C16:1	$C_{16}H_{30}O_2$	254,41	0,228
Stearic	C18:0	$C_{18}H_{36}O_2$	284,48	1,716
Oleic	C18:1c	$C_{18}H_{34}O_2$	282,46	63,958
Linoleic	C18:2c	$C_{18}H_{32}O_2$	280,44	20,506
Linolenic	C18:3	$C_{18}H_{30}O_2$	278,43	7,321
Arachidic	C20:0	$C_{20}H_{40}O_2$	312,54	0,706
Cis-11-Eicosenoic	C20:1	$C_{20}H_{38}O_2$	310,51	0,664

CONCLUSION

The general properties of biodiesel produced from algae are summarized below.

1. Biodiesel is promising alternative diesel fuel, obtained from renewable resources such as vegetable and animal oils, containing oxygen and especially in terms of reducing polluting exhaust emissions. Biodiesel is a methyl or ethyl ester-type fuel containing medium-length C16-C18 fatty acid chains.
2. It does not contain toxic wastes, dissolves quickly in nature like sugar, and reduces the need for fertilization thanks to its nitrogen retention feature. The negative effects on the ozone layer are 50% less than diesel fuel.
3. Biodiesel is an environmentally friendly, biodegradable fuel with anti-toxic effects, which can be obtained from renewable raw materials quickly and easily.
4. At the same time, it is a fuel that can be easily stored, transported, and used with its low sulfur content and high flash point. It also has excellent lubricating properties.
5. Biodiesel and its blends with diesel fuel can be used in diesel engines without any technical changes or with minor modifications. This feature improves the quality of petroleum-derived diesel.
6. For example, it reduces the emission values of environmentally harmful gases formed as a result of combustion, increases the degree of lubrication in the engine, and dissolves deposits that reduce engine power.
7. Since biodiesel is generally obtained from agricultural plants, it does not increase the greenhouse effect, as it accelerates the carbon cycle by converting CO_2 with photosynthesis within the biological carbon cycle.
8. In other words, biodiesel can be considered a natural sink for CO_2 emissions.

CONSENT FOR PUBLICATION

Not applicable.

CONFLICT OF INTEREST

The authors declare no conflict of interest, financial or otherwise.

ACKNOWLEDGEMENT

Declared none.

REFERENCES

[1] Arun J, Gopinath KP, Sivaramakrishnan R, SundarRajan P, Malolan R, Pugazhendhi A. Technical insights into the production of green fuel from CO_2 sequestered algal biomass: A conceptual review on

green energy. Sci Total Environ 2021; 755(Pt 2): 142636.
[http://dx.doi.org/10.1016/j.scitotenv.2020.142636] [PMID: 33065504]

[2]   Mata TM, Martins AA, Caetano NS. Microalgae for biodiesel production and other applications: A review. Renew Sustain Energy Rev 2010; 14(1): 217-32.
[http://dx.doi.org/10.1016/j.rser.2009.07.020]

[3]   Casas-Godoy L, Barrera-Martínez I, Ayala-Mendivil N, *et al.* Chapter 4 - Biofuels. In: Galanakis CM, Ed. Biobased Products and Industries. Elsevier 2020; pp. 125-70.
[http://dx.doi.org/10.1016/B978-0-12-818493-6.00004-X]

[4]   Bessières D, Bazile JP, Tanh XNT, García-Cuadra F, Acien FG. Thermophysical behavior of three algal biodiesels over wide ranges of pressure and temperature. Fuel 2018; 233: 497-503.
[http://dx.doi.org/10.1016/j.fuel.2018.06.091]

[5]   Baldev E, Mubarakali D, Saravanakumar K, *et al.* Unveiling algal cultivation using raceway ponds for biodiesel production and its quality assessment. Renew Energy 2018; 123: 486-98.
[http://dx.doi.org/10.1016/j.renene.2018.02.032]

[6]   Knothe G. Dependence of biodiesel fuel properties on the structure of fatty acid alkyl esters. Fuel Process Technol 2005; 86(10): 1059-70.
[http://dx.doi.org/10.1016/j.fuproc.2004.11.002]

[7]   Schlagermann P, Göttlicher G, Dillschneider R, Rosello-Sastre R, Posten C. Composition of Algal Oil and Its Potential as Biofuel. J Combust 2012; 2012: 1-14.
[http://dx.doi.org/10.1155/2012/285185]

[8]   Ramos MJ, Fernández CM, Casas A, Rodríguez L, Pérez Á. Influence of fatty acid composition of raw materials on biodiesel properties. Bioresour Technol 2009; 100(1): 261-8.
[http://dx.doi.org/10.1016/j.biortech.2008.06.039] [PMID: 18693011]

[9]   Hu Z, Tan P, Yan X, Lou D. Life cycle energy, environment and economic assessment of soybean-based biodiesel as an alternative automotive fuel in China. Energy 2008; 33(11): 1654-8.
[http://dx.doi.org/10.1016/j.energy.2008.06.004]

[10]   Breuer G, Lamers PP, Martens DE, Draaisma RB, Wijffels RH. The impact of nitrogen starvation on the dynamics of triacylglycerol accumulation in nine microalgae strains. Bioresour Technol 2012; 124: 217-26.
[http://dx.doi.org/10.1016/j.biortech.2012.08.003] [PMID: 22995162]

[11]   Hu Q, Sommerfeld M, Jarvis E, *et al.* Microalgal triacylglycerols as feedstocks for biofuel production: Perspectives and advances. Plant J 2008; 54(4): 621-39.
[http://dx.doi.org/10.1111/j.1365-313X.2008.03492.x] [PMID: 18476868]

[12]   Chisti Y. Biodiesel from microalgae. Biotechnol Adv 2007; 25(3): 294-306.
[http://dx.doi.org/10.1016/j.biotechadv.2007.02.001] [PMID: 17350212]

[13]   Draaisma RB, Wijffels RH, Slegers PM, Brentner LB, Roy A, Barbosa MJ. Food commodities from microalgae. Curr Opin Biotechnol 2013; 24(2): 169-77.
[http://dx.doi.org/10.1016/j.copbio.2012.09.012] [PMID: 23084075]

[14]   Wijffels RH, Barbosa MJ. An outlook on microalgal biofuels. Science 2010; 329(5993): 796-9.
[http://dx.doi.org/10.1126/science.1189003] [PMID: 20705853]

[15]   Wijffels RH, Barbosa MJ, Eppink MHM. Microalgae for the production of bulk chemicals and biofuels. Biofuels Bioprod Biorefin 2010; 4(3): 287-95.
[http://dx.doi.org/10.1002/bbb.215]

[16]   Laurens LML, Dempster TA, Jones HDT, *et al.* Algal biomass constituent analysis: Method uncertainties and investigation of the underlying measuring chemistries. Anal Chem 2012; 84(4): 1879-87.
[http://dx.doi.org/10.1021/ac202668c] [PMID: 22242663]

[17]   Iverson SJ, Lang SLC, Cooper MH. Comparison of the bligh and dyer and folch methods for total lipid determination in a broad range of marine tissue. Lipids 2001; 36(11): 1283-7.
[http://dx.doi.org/10.1007/s11745-001-0843-0] [PMID: 11795862]

[18]   Grima EM, Medina AR, Giménez AG, Sánchez Pérez JA, Camacho FG, García Sánchez JL. Comparison between extraction of lipids and fatty acids from microalgal biomass. J Am Oil Chem Soc 1994; 71(9): 955-9.
[http://dx.doi.org/10.1007/BF02542261]

[19]   Lee JY, Yoo C, Jun SY, Ahn CY, Oh HM. Comparison of several methods for effective lipid extraction from microalgae. Bioresour Technol 2010; 101(1) (Suppl. 1): S75-7.
[http://dx.doi.org/10.1016/j.biortech.2009.03.058] [PMID: 19386486]

[20]   Guckert JB, Cooksey KE, Jackson LL. Lipid sovent systems are not equivalent for analysis of lipid classes in the microeukaryotic green alga, Chlorella. J Microbiol Methods 1988; 8(3): 139-49.
[http://dx.doi.org/10.1016/0167-7012(88)90015-2]

[21]   Patil V, Källqvist T, Olsen E, Vogt G, Gislerød HR. Fatty acid composition of 12 microalgae for possible use in aquaculture feed. Aquacult Int 2007; 15(1): 1-9.
[http://dx.doi.org/10.1007/s10499-006-9060-3]

[22]   Thao N, Nguyen T, Thanh B. Biodiesel production from microalgae by extraction – transesterification method. Waste Technology. 2013; p. 1.

[23]   Yadav M, Chavan SB, Singh R, Bux F, Sharma YC. Experimental study on emissions of algal biodiesel and its blends on a diesel engine. J Taiwan Inst Chem Eng 2019; 96: 160-8.
[http://dx.doi.org/10.1016/j.jtice.2018.10.022]

CHAPTER 6

Microbial Fuel Cells (MFCs) Technology

Mesut Yılmazoğlu[1,*]

¹ Chemical Engineering Department, Engineering Faculty, Yalova University, 77100 Yalova, Turkey

Abstract: The purpose of this book chapter is to provide general information regarding microbial fuel cell (MFC) systems, an important type of fuel cell of environmentally friendly energy conversion systems as an alternative to fossil fuel technologies. Besides, it is one of the main motivations of this study to include the academic literature on microbial fuel cells, which is a very popular field of study in recent years. In this context, the history, principles, and different approaches of MFCs are discussed. After that, the materials (anode, cathode, membrane, *etc.*) that make up the system are examined. Finally, different types of microbial fuel cells that can be varied by material design are discussed and presented.

Keywords: Energy, MFC materials, Microbial fuel cell, Single-chamber MFC, Two-chamber MFC.

INTRODUCTION

The fuel cell is a clean, high-efficiency energy technology that does not remove waste materials that cause environmental pollution. Fuel cells are devices that continuously convert the chemical energy of a fuel (hydrogen-rich gas mixture from hydrogen) and an oxidizer (oxygen of air) into electricity and heat form. In other words, it is an electrochemical structure (galvanic cell) that converts the free energy of a chemical reaction into electrical energy. The fuel cell has a high fuel conversion efficiency ranging from 40% to 70%, depending on the thermal value of the fuel [1].

Fuel cells are similar to accumulators or batteries. Both convert chemical energy directly into electricity. It is stored in the accumulator before the use of chemical energy, while the fuel cell can generate electricity as long as energy is supplied from external sources. The principle of operation of fuel cells is the chemical reaction, which is the opposite of water electrolysis. In a fuel cell, gaseous fuels

* **Correspondence author Mesut Yılmazoğlu:** Chemical Engineering Department, Engineering Faculty, Yalova University, 77100 Yalova, Turkey; Tel: +902268155399; E-mail: mesut.yilmazoglu@yalova.edu.tr

Hüseyin Karaca and Cemil Koyunoğlu (Eds.)

(hydrogen) are sent by the anode and oxidizing gases (oxygen) by the cathode. Oxidation reactions at the anode occur at the cathode and reduction reactions occur at the cathode. As a result of the total reaction, water and heat are produced as a product [2]. In the electrolysis reaction, when a direct current is applied to the water, it decomposes in proportional volumes to oxygen and hydrogen. Since electric energy is decomposed into water components when applied, it is taken logically if the process is arranged in the opposite direction, that is when water and heat are obtained as a result of the reaction of oxygen and hydrogen [1 - 3].

Instead of pure hydrogen, hydrocarbons can be used in the fuel cell. However, it is not preferred because it reduces efficiency. The general display of a typical fuel cell is shown in Fig. (**1**).

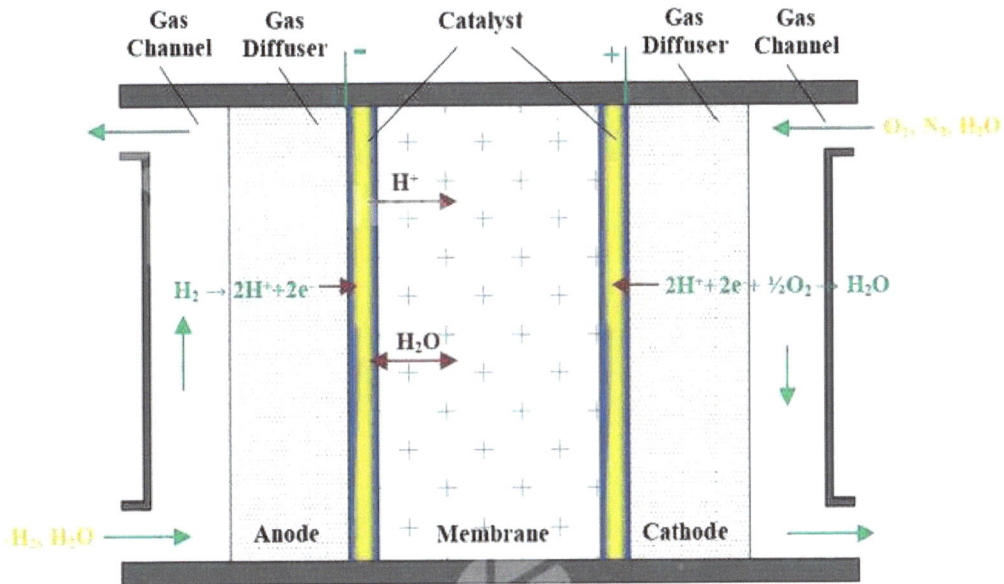

Fig. (1). A typical PEM (Polymer electrolyte membrane) fuel cell [4, 5].

Considering that fossil fuels are consumed day by day, it is necessary to find and use alternative energy sources. Therefore, electrical energy production from renewable energy sources is the most preferred method. Recently, the most emphasized alternative energy source is Microbial Fuel Cells (MFCs). The principle of the MFC, in its purest form, is to convert chemical energy contained in organic waste into electrical energy with the help of microorganisms.

Microbial Fuel Cells (MFCs)

Electric energy production from microorganisms was not a new idea but was first introduced by Botanical professor MC Potter at Durham University in 1911. In this study, he stated that electricity production is the result of the activities of microorganisms, and the electrical effect is affected by temperature, food source, and the number of active microorganisms available. In the study, it was reported between the maximum of 0.3-0.5 V electricity generation. Assuming that people will meet their energy needs from such systems in the future, he established a single microbial fuel cell but did not have a broad knowledge of bacterial metabolism [6].

No significant work has been carried out until the early 1980s. During this time, with the realization that electricity is a significant power source, Young *et al.* (1966), using intensive studies on MFC, used microorganisms for the production and removal of electrochemical products. As a result, they made three types of biochemical fuel cells [7]. In the late 1980s, Benotto proposed the reduction and oxidation process that takes place in microorganisms, which is still valid today with many studies [8].

In 2004, the relationship between electricity generation and wastewater treatment was demonstrated using MFCs, and it was shown that domestic wastewater could be treated at the application scale when generating electrical energy. A first lab-scale MFC prototype was developed by Bruce Logan and showed that when bacteria are fed with microbial nutrients, such as glucose acetate, or organic compounds in wastewater, energy is produced [9].

MFC technology is a promising sustainable energy source that can use organic materials, and generate electrical energy, recently through microorganisms [10, 11]. One of the reasons why MFC systems are regarded as renewable energy sources is that they are carbon-neutral, that is, they release only stable carbon into the atmosphere as a result of oxidation of organic materials [12]. MFC can also be considered an electrochemical hybrid system because they are a system that combines microbial and electrochemical processes [10]. In short, MFCs are bioelectrochemical reactors that convert chemical energy in organic compounds into electrical energy through catabolic reactions of microorganisms under anaerobic conditions. In MFC, dissolved organic materials are converted into electrical energy by bioelectrochemical means, that is, by the catalytic reaction of microorganisms [13].

In a study, Varol and Uğurlu (2016) investigated the use of *Spirulina platensis*, a blue-green alga, as a renewable energy source in discrete and semi-discrete systems. In intermittent studies, by reducing 89-93% of volatile solids, a solids concentration of 0.6-5% was achieved under initial conditions. The similar biogas production was between 210 and 260 mL / g UK (volatile solid). In continuous studies, a two-phase anaerobic fermentation system was used. In the first part, biogas production from waste sludge is examined under the total hydraulic retention time of 14 days. As a result, the system has reached 2620 mL of biogas per day (525 mL of biogas / g UK day). It developed biogas production as a result of the fermentation of waste sewage sludge with *S. platensis*. As a result of this process, 2880 mL/day of total biogas production (640 mL biogas/day UK) was provided [14].

In another study, a new design of a photosynthetic microbial fuel cell was made. This system is a membrane-free system with a gold slot anode and a cathode of graphite-carbon fabric. *Spirulina platensis* was developed and allowed to form a fine biofilm on the anode. As a result of this study, it was found that the chlorophyll content of the biofilm depends on the electrical performance of the microbial fuel cell. The maximum open-circuit voltage of the microbial fuel cell was found to be 0.49 V. When the microbial fuel cell was connected to an external resistance of 1 kOhm, and the maximum power density was achieved as ten mW.m^2. This designed microbial fuel cell provides higher open-circuit voltage and power density compared to other microbial fuel cells reported in the literature [15].

Spirulina platensis was attached to the anode of a microbial fuel cell without a membrane and media to produce electricity. An increase in open-circuit voltage has been observed with light intensity and optimum biomass field intensity [16]. The highest open-circuit voltage observation for the microbial fuel cell was measured in the dark at 0.39 V, and a density of biomass area was determined on the anode surface of 1.2 g.cm^2. Also, it has been observed that the microbial fuel cell with a biomass field density of 0.75 g.cm^2 produces 1.64 mW.m^2 electrical energy in the dark and is superior to 0.132 mW.m^2 produced in light. As a result, it has been found that the microbial fuel cell can be used to generate electrical energy under both day and night conditions [16].

In a study carried out by Taşkan (2016), the electricity production performance of the sludge was investigated by using sediment-type microbial fuel cells. As a result of the study, it has been observed that the electricity production capacity of the sludge continues for 34 days and after this time the amount of electricity produced in the system starts to decrease. This showed that an electrochemically active biofilm structure was formed on the graphite electrode surface. With the adaptation of the electrogenic microorganisms to the environment, the consumption of organic substances in the sludge has increased, and electricity production has increased. On the 34th day, the highest voltage was obtained as

158 ± 0.7 mV. The reduction in the amount of electricity produced is thought to occur as a result of the microorganisms on the electrode surface consuming organic matter in the sludge [17].

Fang *et al.* Investigated the availability of nitrate as an oxidant in the cathode section, without ventilation, and denitrification bacteria, for sustainable power generation in a microbial fuel cell. In the study, a maximum carbon power of 7.2 mW/m^2 was obtained in two compartments Nafion-MFC where flat carbon paper as anode and cathode electrode and raw domestic wastewater in the anode section as a substrate and nitrate solution in the cathode section, and daily nitrate removal, 0.57 mg (NO_3 - N) /L.day. When using a Pt catalyst on the cathode electrode, 16 times more power (117.7 mW / m^2) was obtained, while nitrate removal was 2 mg (NO_3 - N) /L. These results revealed that nitrate could be successfully used as an oxidant to produce power without ventilation and also to remove nitrate from water. However, it has been reported that it is necessary to control the process to prevent the reduction of nitrate to ammonia and to reduce it to N_2 gas-only [18].

MFC components and materials are presented in Table **1**.

Table 1. Components and materials used in MFC.

Component	Materials
Anode Electrode	Graphite, graphite/carbon fabric, graphite/carbon paper, Pt, reticulated glassy carbon
Cathode Electrode	Graphite, graphite fabric, graphite paper, carbon fabric, Pt, reticulated glassy carbon
Anode Compartment	Glass, polycarbonate, plexiglass
Cathode Compartment	Glass, polycarbonate, plexiglass
Proton Conductive Membrane	Nafion, Ultrex, salt bridge, electrolyte only (without membrane)
Electrode Catalyst	Pt, MnO_2, Fe (III), polyamines, immobilized electron carriers on the anode.

ANODE MATERIALS

The type and amount of microorganisms in the anode compartment seriously affect the power density obtained from MFC and affects the material and structure of the anode, microbial development, electron transfer [19].

For an active anode material following characteristics should be achieved [20].

• It should have high electrical conductivity and low resistance.
• It should be biocompatible with the microorganism; it should not react with the

liquid in the reactor and should not damage the microorganisms in the anode compartment.

- It should be resistant to chemical stability and corrosion.
- It should not allow microorganisms to enter and clog.
- It should have a large surface area and be mechanically resistant.

Also, the microorganism must be able to adhere to the material and achieve a good electrical connection. It should be known how the electrode material affects the electron transfer of the microorganism.

Carbon was generally used as the anode material in MFC studies. The use of carbon-based electrodes in these systems is quite common because they have high conductivity and provide a surface where microorganisms can grow and adhere. Carbon paper is a slightly brittle material, but it is easy to connect. Copper is also a conductive material, but it will have a toxic effect over time, so it should not be preferred. It is more appropriate to use stainless steel or titanium instead of copper. Apart from these, graphite rods, graphite felt, graphite plate, graphite plate, graphite granule, graphite fiber, and graphite brush can be listed as metal and metal coating materials [20].

CATHODE MATERIALS

The cathode material is another factor that limits power generation in MFCs and dramatically affects performance. The choice of cathode material dramatically affects performance and increases the variety of experiments. The development of catalysts using a chemical solution in the cathode section is very challenging in MFC research. The choice of cathode material varies greatly depending on the application. It is challenging to regulate the chemical reaction occurring at the cathode [21].

In the cathode section, a three-phase reaction is formed: solid (cathode electrode), gas (electrons, protons, and all of the oxygen), and liquid (catholyte). The catalyst must be on the conductive surface; however, for the electrons and protons in different phases to reach the same point, the catalyst must be exposed to both water and air. The most suitable electron acceptor in MFCs is oxygen due to its high oxidation potential, accessibility, low cost, and sustainability. To increase the oxygen reduction rate, platinum (Pt) catalyst is generally used instead of dissolved oxygen. Using Pt as a cathode electrode significantly increases cathode cost [22]. To decrease the amount of Pt that increases the cost in MFCs, it should be used as low as 0.1 mg.cm^{-2}. In the long run, Pt needs to be thoroughly investigated and lower cost catalysts found. Recently, it has been proposed to use noble metal catalysts such as pyrolyzed iron (II) phthalocyanine [23].

The cost of cathode materials constitutes 47% of the cost of air cathode MFCs. Many researchers think that the use of high specific area carbon materials will be more appropriate than the use of catalysts in the cathode compartment. Deng *et al.* (2010) used the activated carbon fiber felt electrode as an air cathode electrode in upstream MFC, and a power density of 315 mW/m^2 was obtained. He compared these values with platinum-coated carbon paper and achieved a power density of 124 mW/m^2 (0.2 mg pt/cm^2) with platinum-coated carbon paper [24].

MEMBRANE MATERIALS

The membrane is the material that physically separates the anode and cathode compartments in MFCs and provides proton transfer from anode to cathode to produce electric current [25, 26]. Membranes must have a permeable structure to enable proton transition from anode to cathode [26]. The feature of the membrane is essential in this separation process [27]. By using the membrane, it is ensured that the substances in the cathode and anode compartments do not mix [28]. If the membrane is a proton or cation-selective membrane, this membrane only allows the passage of positively charged elements [29]. Another feature of the membrane is that it prevents oxygen diffusion from the cathode to the anode and prevents the passage of other cations in the cathode solution to the anode [30]. In MFC systems, microorganisms in the anode section thrive in anaerobic environments.

The electrons formed as a result of the destruction of organic matter in the microorganism are transferred to the anode electrode and sent to the cathode section through a circuit, while protons emerge simultaneously in the environment as a result of the destruction [31]. Protons are transmitted to the cathode section by a selectively permeable membrane. While the membrane allows the passage of cations, it also prevents the passage of oxygen, which is the anion in the cathode section, into the anode section, which creates a dangerous environment for anaerobic microorganisms [32]. In the absence of membranes, oxygen and substrate diffusion increases, but due to this increase, yield and bio-electrocatalytic activities of anode microorganisms decrease [33]. The disadvantages of the membranes are that they are costly and reduce system performance. The high cost limits the use of materials for large-scale applications. Also, increased substrate diffusion causes blockage of the membrane surface. The low rate of proton transfer in the membrane affects the reaction rate at the anode and cathode. Accumulation of protons at the anode also slows down and stops microbial activity. Cation exchange membranes, anion exchange membranes, and ultrafiltration membranes are generally used as separators between the electrodes [31, 32]. Frequently used cation exchange membranes; Nafion 117 and Ultrex CMI-7000. However, it is not an obligation to use the membrane in MFC systems.

According to the use of a membrane, MFC systems are divided into two with and without membranes. In membrane-free systems, the transport of protons takes place with water [34, 35].

A typical MFC system with a proton conductive polymeric membrane is presented in Fig. (**2**) [36].

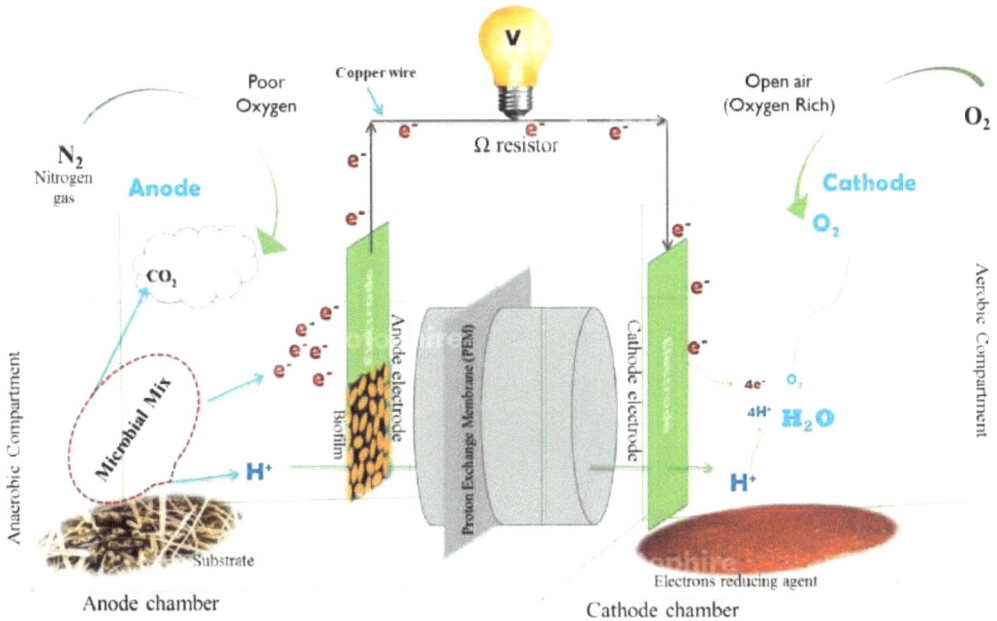

Fig. (2). A typical MFC system and polymer electrolyte membrane [36].

MFC TYPES

Many different types of MFC systems have been developed for MFC studies to date. These; two-chamber MFC, single-chamber MFC, tubular single-chamber MFC, and multi-reactor pitched MFC. Commonly used types are H-type two-chamber and a single-chamber air cathode.

TWO-CHAMBER MFC

The two-compartment MFC consists of the anode compartment, the cathode compartment, and the proton exchanger membrane separating these two compartments. The H-shaped MFC system has the most widely used and inexpensive design. In this system, the two glass vials are usually linked by a tube containing a cation membrane. In H type MFC, the membrane is attached in the middle of the tube connecting the bottle (Fig. **3a**). Cubic MFC, which has a cube-shaped design, is one of the two-chamber MFC reactors. It is operated intermittently. The most important difference from H type MFC is the size of the membrane used. While the membrane located between the tube in the H type MFC is tiny compared to the anode and cathode sections, the membrane used in the cubic MFC, and the anode and cathode sections are of equal size. This feature of the membrane provides a smoother transition compared to H-type MFC in the passage of protons formed in the anode section. However, in cubic type MFC, oxygen diffusing the coulombic effect is more than diffusing into the anode section [37, 38] (Fig. **3b**).

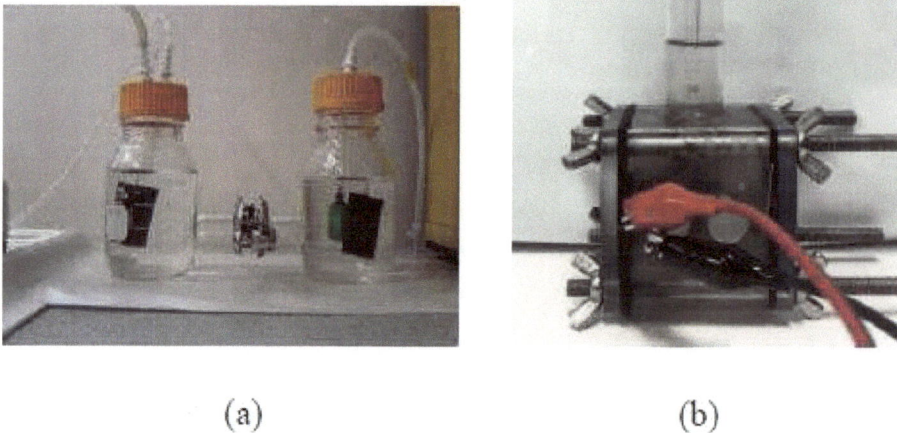

(a)   (b)

Fig. (3). Typical H type MFC (**a**) and cubic type MFCs (**b**).

While the organic matter oxidizes in anaerobic conditions in the anode compartment, the formed electrons pass from the anode to external resistance. Protons passing through the membrane into the cathode compartment combine with oxygen in the cathode compartment to form water. The point to be noted in two-chamber MFCs is that the selected membrane does not allow the passage of the proton and does not allow the passage of the substrate and the electron acceptor (usually O_2) at the cathode. The most crucial disadvantage of two-chamber MFC is that there is a risk of O_2 passing into the anode chamber during the continuous ventilation of the cathode chamber [37 - 39].

SINGLE-CHAMBER MFC

The first study with single-chamber MFC was sold in 1989 by Sell *et al.*. Simpler and more efficient single-chamber MFCs have been developed by removing the cathode chamber to reduce weaknesses such as high volume and extra cost for ventilation in two-chamber MFCs. The cathode electrode is in direct contact with the air, and the protons produced pass from the anode solution to the porous air cathode. The reactor has a simple design, and since there is no cathode compartment, no additional costs are required for ventilation[37-39]. Different reactor designs are available in single-chamber MYHs, either anode electrode adjacent to the membrane or separate from the membrane. These systems are simpler in design and do not require extra cost for ventilation [37, 38].

A typical single-chamber MFC is presented in Fig. (**4**) [39].

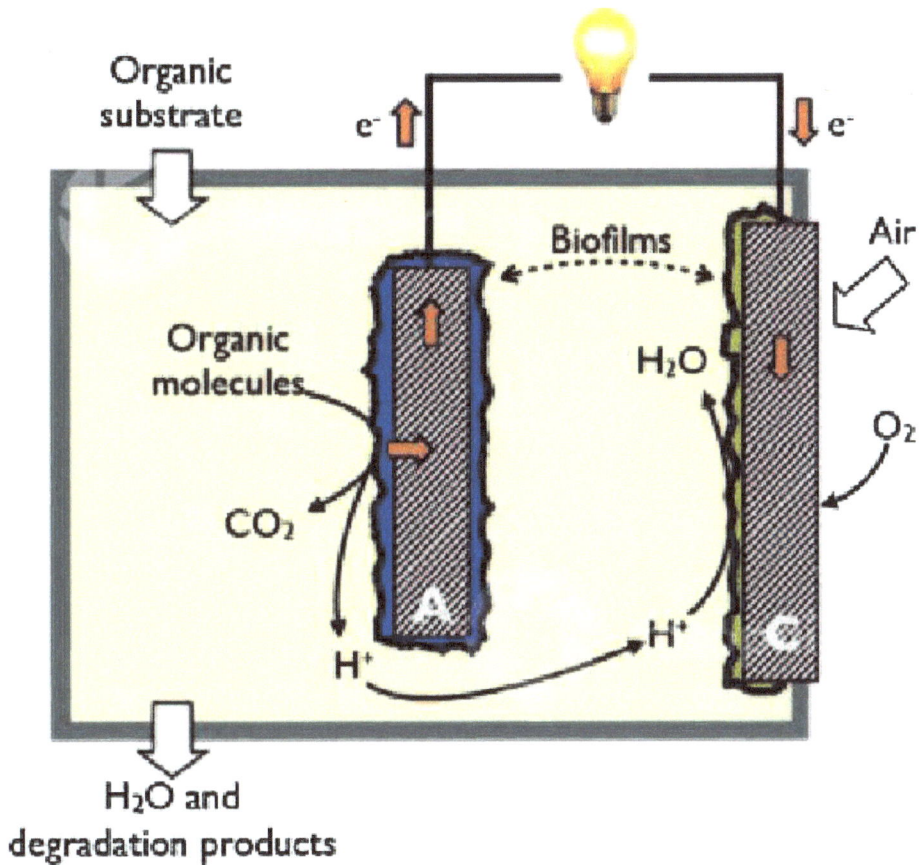

Fig. (4). A typical single-chamber MFC [39].

A typical view of the tubular reactor, designed by Rabaey [40], with the cathode inside and the anode chamber inside, is given in the literature. In the study, the maximum power densities obtained when using acetate, glucose, and hospital wastewater as substrates were recorded as 90 W/m^3, 66 W/m^3 and 48 W/m^3, respectively. Aelterman used curtained MFC in his studies. The reactor consists of adjacent fuel cells. With this design, it was aimed to increase the cell potential, and the maximum power density obtained in the study was recorded as 258 W/m^3. He used an upstream MFC reactor modified from the upstream mud bed anaerobic reactor (UASB) in their work. The maximum power density obtained in the study is 170 mW/m^2 [41]. When the proton exchange membrane (PDM) is not used in a single chamber MFC, power generation can be further increased, and installation costs can be reduced [42]. In a study conducted with a single-chamber air cathode MFC designed by Liu and Logan[9], it was reported that while the PDM was lifted, the maximum power density increased to 494 ± 21 mW/m^2 when PDM was lifted. However, the Coulomb yield was only 9-12% in membrane-free systems, while it was nearly 50% when PDM was used. This indicates that there is a significant amount of oxygen diffusion from the cathode to the anode chamber when the membrane is not used [9, 43].

On the other hand, although the diffusion of oxygen into the anode chamber increases when PDM is not used, aerobic biofilm formation on the inner side of the cathode facing the anode removes the oxygen to the anode chamber and prevents the loss of anaerobic conditions in the anode chamber. Not using PDM also reduces the cost of materials needed for the design of an MFC [37, 38]. The use of a single-chamber MFC has many advantages over a two-chamber MFC. These systems are it increases mass transfer to the cathode, reduce operating costs as water is not required to be ventilated, there is a general reduction in reactor volume, and their design is simple [43].

CONCLUSION

With industrial developments, the energy need cannot be met with traditional fossil resources and the energy production-consumption gap is gradually increasing. Fossil-based energy resources will be depleted in the future, and it will be imperative to find new environmentally friendly technologies for sustainable wastewater treatment and energy production. Microbial fuel cells (MFCs) are bioelectrochemical energy systems that convert organic matter into electrical energy through the metabolic activity of microorganisms. In this system, wastewater treatment and electricity generation can be done at the same time. In this study, a general introduction to MFC technology was made, system components and membrane materials were discussed in detail. The main types of MFC systems are presented. MFC systems are a technology that is developing day by day. In order to find out whether the foreseen advantages of MFCs can be achieved in practice, it is necessary to examine the challenges and feasibility of these technologies in situ and determine how they will be redesigned for sustainable wastewater treatment and energy generation in the future. In order to obtain the desired potential from MFC technologies, optimization of qualities such as anode-cathode electrode materials, membrane electrolyte materials, and investment costs is required.

CONSENT FOR PUBLICATION

Not applicable.

CONFLICT OF INTEREST

The author declares no conflict of interest, financial or otherwise.

ACKNOWLEDGEMENT

Declared none.

REFERENCES

[1]   Shah RK. Introduction to fuel cells. In: Basu S, Ed. Recent Trends in Fuel Cell Science and Technology. New York, NY: Springer New York 2007; pp. 1-9.
        [http://dx.doi.org/10.1007/978-0-387-68815-2_1]

[2]   Adem Yılmaz S Ü. Yakıt Pili Teknolojisi. Technological Applied Sciences (NWSATAS) 2017; 12(4): 185-92.

[3]   Hickner MA, Ghassemi H, Kim YS, Einsla BR, McGrath JE. Alternative polymer systems for proton exchange membranes (PEMs). Chem Rev 2004; 104(10): 4587-612.
        [http://dx.doi.org/10.1021/cr020711a] [PMID: 15669163]

[4]   Larminie J, Dicks A. Introduction. Fuel Cell Systems Explained John Wiley & Sons Ltd 2003; 1-24.

[5]   Taner T. Alternative energy of the future: A technical note of pem fuel cell water management. Journal of Fundamentals of Renewable Energy and Applications 2015; 5: 1-4.

[6]   Potter MC, Waller AD. Electrical effects accompanying the decomposition of organic compounds. Proc R Soc Lond, B 1911; 84(571): 260-76.
        [http://dx.doi.org/10.1098/rspb.1911.0073]

[7]   Young TG, Hadjipetrou L, Lilly MD. The theoretical aspects of biochemical fuel cells. Biotechnol Bioeng 1966; 8(4): 581-93.
        [http://dx.doi.org/10.1002/bit.260080410]

[8]   Thurston CF, Bennetto HP, Delaney GM, Mason JR, Roller S D, Stirling JL. In Glucose Metabolism in a Microbial Fuel Cell Stoichiometry of Product Formation in a Thionine-mediated Proteus vulgaris Fuel Cell and its Relation to Coulombic Yields. 1985.

[9]   Liu H, Logan BE. Electricity generation using an air-cathode single chamber microbial fuel cell in the presence and absence of a proton exchange membrane. Environ Sci Technol 2004; 38(14): 4040-6.
        [http://dx.doi.org/10.1021/es0499344] [PMID: 15298217]

[10]   Nancharaiah YV, Venkata Mohan S, Lens PNL. Metals removal and recovery in bioelectrochemical systems: A review. Bioresour Technol 2015; 195: 102-14.

[http://dx.doi.org/10.1016/j.biortech.2015.06.058] [PMID: 26116446]

[11]   Lovley DR. Microbial fuel cells: Novel microbial physiologies and engineering approaches. Curr Opin Biotechnol 2006; 17(3): 327-32.
[http://dx.doi.org/10.1016/j.copbio.2006.04.006] [PMID: 16679010]

[12]   Clauwaert P, Rabaey K, Aelterman P, *et al.* Biological denitrification in microbial fuel cells. Environ Sci Technol 2007; 41(9): 3354-60.
[http://dx.doi.org/10.1021/es062580r] [PMID: 17539549]

[13]   Palanisamy G, Jung HY, Sadhasivam T, Kurkuri MD, Kim SC, Roh SH. A comprehensive review on microbial fuel cell technologies: Processes, utilization, and advanced developments in electrodes and membranes. J Clean Prod 2019; 221: 598-621.
[http://dx.doi.org/10.1016/j.jclepro.2019.02.172]

[14]   Varol A, Ugurlu A. Biogas production from microalgae (Spirulina platensis) in a two stage anaerobic system. Waste Biomass Valoriz 2016; 7(1): 193-200.
[http://dx.doi.org/10.1007/s12649-015-9442-8]

[15]   Lin CC, Wei CH, Chen CI, Shieh CJ, Liu YC. Characteristics of the photosynthesis microbial fuel cell with a Spirulina platensis biofilm. Bioresour Technol 2013; 135: 640-3.
[http://dx.doi.org/10.1016/j.biortech.2012.09.138] [PMID: 23186678]

[16]   Fu CC, Su CH, Hung TC, Hsieh CH, Suryani D, Wu WT. Effects of biomass weight and light intensity on the performance of photosynthetic microbial fuel cells with Spirulina platensis. Bioresour Technol 2009; 100(18): 4183-6.
[http://dx.doi.org/10.1016/j.biortech.2009.03.059] [PMID: 19386488]

[17]   Taşkan E. Sediment Tipi Mikrobiyal Yakıt Hücresi Kullanılarak Arıtma Çamurlarından Elektrik Üretimi. Fırat Üniversitesi Mühendislik Bilimleri Dergisi 2016; 28(1): 15-21.

[18]   Zang Y. Energy recovery from waste streams with microbial fuel cell (MFC)-based Technologies. PhD Thesis. DTU Miljø.

[19]   Santoro C, Arbizzani C, Erable B, Ieropoulos I. Microbial fuel cells: From fundamentals to applications. A review. J Power Sources 2017; 356: 225-44.
[http://dx.doi.org/10.1016/j.jpowsour.2017.03.109] [PMID: 28717261]

[20]   Wei J, Liang P, Huang X. Recent progress in electrodes for microbial fuel cells. Bioresour Technol 2011; 102(20): 9335-44.
[http://dx.doi.org/10.1016/j.biortech.2011.07.019] [PMID: 21855328]

[21]   Cheng S, Liu H, Logan BE. Increased performance of single-chamber microbial fuel cells using an improved cathode structure. Electrochem Commun 2006; 8(3): 489-94.
[http://dx.doi.org/10.1016/j.elecom.2006.01.010]

[22]   HaoYu E, Cheng S, Scott K, Logan B. Microbial fuel cell performance with non-Pt cathode catalysts. J Power Sources 2007; 171(2): 275-81.
[http://dx.doi.org/10.1016/j.jpowsour.2007.07.010]

[23]   Harnisch F, Savastenko NA, Zhao F, Steffen H, Brüser V, Schröder U. Comparative study on the performance of pyrolyzed and plasma-treated iron(II) phthalocyanine-based catalysts for oxygen reduction in pH neutral electrolyte solutions. J Power Sources 2009; 193(1): 86-92.
[http://dx.doi.org/10.1016/j.jpowsour.2008.12.049]

[24]   Deng Q, Li X, Zuo J, Ling A, Logan BE. Power generation using an activated carbon fiber felt cathode in an upflow microbial fuel cell. J Power Sources 2010; 195(4): 1130-5.
[http://dx.doi.org/10.1016/j.jpowsour.2009.08.092]

[25]   Chae KJ, Choi M, Ajayi FF, Park W, Chang IS, Kim IS. Mass Transport through a Proton Exchange Membrane (Nafion) in Microbial Fuel Cells. Energy Fuels 2008; 22(1): 169-76.
[http://dx.doi.org/10.1021/ef700308u]

[26]   Rismani-Yazdi H, Carver SM, Christy AD, Tuovinen OH. Cathodic limitations in microbial fuel cells: An overview. J Power Sources 2008; 180(2): 683-94.
[http://dx.doi.org/10.1016/j.jpowsour.2008.02.074]

[27]   Torres CI, Lee HS, Rittmann BE. Carbonate species as OH- carriers for decreasing the pH gradient between cathode and anode in biological fuel cells. Environ Sci Technol 2008; 42(23): 8773-7.
[http://dx.doi.org/10.1021/es8019353] [PMID: 19192796]

[28]   Steele BCH, Heinzel A. Materials for fuel-cell technologies. Materials for Sustainable Energy. 224-31.
[http://dx.doi.org/10.1142/9789814317665_0031]

[29]   Peighambardoust SJ, Rowshanzamir S, Amjadi M. Review of the proton exchange membranes for fuel cell applications. Int J Hydrogen Energy 2010; 35(17): 9349-84.
[http://dx.doi.org/10.1016/j.ijhydene.2010.05.017]

[30]   Rinaldi A, Mecheri B, Garavaglia V, Licoccia S, Di Nardo P, Traversa E. Engineering materials and biology to boost performance of microbial fuel cells: A critical review. Energy Environ Sci 2008; 1(4): 417-29.
[http://dx.doi.org/10.1039/b806498a]

[31]   Patil SA, Hägerhäll C, Gorton L. Electron transfer mechanisms between microorganisms and electrodes in bioelectrochemical systems. Bioanal Rev 2012; 4(2-4): 159-92.
[http://dx.doi.org/10.1007/s12566-012-0033-x]

[32]   Chang CT, Wang CT, Wang CH, Chong WT. Effect of oxygen gradient on the organic degradation and power performance of single sediment microbial fuel cells. Energy Procedia 2017; 105: 654-61.
[http://dx.doi.org/10.1016/j.egypro.2017.03.371]

[33]   Bajracharya S, Vanbroekhoven K, Buisman CJN, Pant D, Strik DPBTB. Application of gas diffusion biocathode in microbial electrosynthesis from carbon dioxide. Environ Sci Pollut Res Int 2016; 23(22): 22292-308.
[http://dx.doi.org/10.1007/s11356-016-7196-x] [PMID: 27436381]

[34]   Goswami R, Mishra VK. A review of design, operational conditions and applications of microbial fuel cells. Biofuels 2018; 9(2): 203-20.
[http://dx.doi.org/10.1080/17597269.2017.1302682]

[35]   Harnisch F, Schröder U, Scholz F. The suitability of monopolar and bipolar ion exchange membranes as separators for biological fuel cells. Environ Sci Technol 2008; 42(5): 1740-6.
[http://dx.doi.org/10.1021/es702224a] [PMID: 18441829]

[36]   Flimban S, Hassan S, Rahman M, Oh S-E. The effect of Nafion membrane fouling on the power generation of a microbial fuel cell. Int J Hydrogen Energy 2018.

[37]   Sørensen B, Spazzafumo G. Hydrogen and Fuel Cells: Emerging Technologies and Applications. Elsevier Science 2018.

[38]   Zhang X, Li X, Zhao X, Li Y. Factors affecting the efficiency of a bioelectrochemical system: A review. RSC Advances 2019; 9(34): 19748-61.
[http://dx.doi.org/10.1039/C9RA03605A] [PMID: 35519388]

[39]   Zhang X, Li X, Zhao X, *et al.* Bioelectric field accelerates the conversion of carbon and nitrogen in soil bioelectrochemical systems. J Hazard Mater 2020; 388: 121790.
[http://dx.doi.org/10.1016/j.jhazmat.2019.121790] [PMID: 31818651]

[40]   Rabaey K, Clauwaert P, Aelterman P, Verstraete W. Tubular microbial fuel cells for efficient electricity generation. Environ Sci Technol 2005; 39(20): 8077-82.
[http://dx.doi.org/10.1021/es050986i] [PMID: 16295878]

[41]   Aelterman P, Rabaey K, Pham HT, Boon N, Verstraete W. Continuous electricity generation at high voltages and currents using stacked microbial fuel cells. Environ Sci Technol 2006; 40(10): 3388-94.
[http://dx.doi.org/10.1021/es0525511] [PMID: 16749711]

[42]   Raju CH AI, M.B. , Rao PJ. Studies on development of microbial fuel cell for waste water treatment using aceto bacteraceti (Vinegar). International journal for research & development in technology 8(5): 58-69.

[43]   Arenas EG, Rodriguez Palacio MC, Juantorena AU, Fernando SEL, Sebastian PJ. Microalgae as a potential source for biodiesel production: Techniques, methods, and other challenges. Int J Energy Res 2017; 41(6): 761-89.
[http://dx.doi.org/10.1002/er.3663]

CHAPTER 7

Algae Cultivation in Different Systems

Fevzi Yaşar[1,*]

[1] *Chemistry and Chemical Process Technology Department, Vocational School of Technical Sciences, Batman University, 72100 Batman, Turkey*

Abstract: Since algae are simple organisms that contain chlorophyll, they can be found anywhere on earth where they can use light for photosynthesis. Although pool-type open systems are generally used, closed photobioreactors are also used in the cultivation of algae. The low investment and operating costs of the outdoor pools made the system preferable in the industry. However, the difficulty of controlling the production conditions and the risk of contamination appears as the disadvantages of the system. It is necessary to compare fundamental aspects such as effective use of light in large-scale culture systems, temperature, hydrodynamic balance in algae culture, and maintaining the continuity of the culture. The ideal growth of each algae species takes place in culture media with its specific conditions. For example, Spirulina grows best at high pH and bicarbonate concentration, Chlorella in nutrient-rich media, and Dunaliella salina at very high salinity.

Keywords: Fuel from Algae oil, Photobioreactor, Open-closed pond systems.

INTRODUCTION

Microalgae absorb solar energy under phototrophic growth conditions. Microalgae absorb carbon dioxide from nutrients in aquatic life and the air. The most important factor affecting microalgae's production efficiency is reactor selection. Outdoor pools are systems that require large areas and are used in industrial activities due to the low capital and investment cost. Temperature control in these systems is difficult due to seasonal and daily fluctuations. Fluctuations in temperature, poor light intensity, dissolved oxygen concentration, and the effect of pH can limit the growth parameters of microalgae. With low biomass productivity (0.1-1.5 g/L), the formation of unwanted species and operational weakness make system control difficult.

* **Correspondence author Fevzi Yaşar:** Chemistry and Chemical Process Technology Department, Vocational School of Technical Sciences, Batman University, 72100 Batman, Turkey; E-mail: fevzi.yasar@batman.edu.tr

Hüseyin Karaca and Cemil Koyunoğlu (Eds.)

Open systems consist of different structures such as large and shallow circular pools, algae pools. Circular and agitated ponds revolving around the central shaft are the earliest known algae growing tanks. These systems are designed to be 10.000 m^2 due to the difficulty of constructing the larger mixer.

The most common system used for microalgae production is algae ponds. These systems are designed in the form of oval-shaped and closed-loop circulation channels, in the range of 0.2-0.5 meters, to be mixed with an impeller to ensure homogeneity. Algae ponds can be made of fiberglass, concrete, or membranes.

Chlorella and spirulina are the most commonly grown microalgae species in open systems. When considering microalgae cultivation in open systems, many parameters such as land cost, the biology of microalgae, nutrients and water in the system, final product to be formed and local climatic conditions should be considered [1, 2].

OPEN POND SYSTEMS

Production in the open system is completely open to nature, depending on the climatic conditions of the region where the product is made. Open systems are shallow open-channel raceway pool systems where the water depth is not less than 15cm and not more than 25cm; circulation is provided by paddle mixers [1, 2]. In the production of the open-air system, factors such as temperature, lighting, evaporation, and water loss cannot be intervened. Nutrients such as nitrogen, phosphorus, and carbon can be partially controlled. The amount of carbon can be determined by measuring the acidity of water. Due to disadvantages such as being impure (mixed with other species), being directly affected by external factors, and being unhygienic, the waste of Spirulina, which is produced in the open system, increases, and the productivity decreases. These negativities directly affect the production costs [1, 3]. In addition, labor costs in the production in open systems are quite high. The main reason for the use of open systems is that the initial establishment costs are quite low, the seasonal cycle can be caught, the production can be started quickly, and the product can be purchased early. In addition, such production systems are preferred and made in hot regions where seasonal values do not change much, water is abundant, production will not be affected much [4, 5]. The quality standard of the product obtained as a result of the products made in open systems varies according to the seasonal conditions and differences of the day [1, 6, 7]. This brings together more than one quality classification. As a result of this classification, the usage area of the product changes. Therefore, the value decreases depending on the quality class of the product [8]. The yield of the products made in the open system is not certain and varies in a wide range [9]. This creates uncertainty in the production. In order to eliminate these negativities and reduce risks, the number of production pools and systems in open systems is increased (Fig. **1**). Parallel to this increase, personnel, energy, land, and consumption expenses also increase [10, 11]. Another important factor in Spirulina production in open systems is that the ponds are open to poultry, reptiles, amphibians, rodents, and other regional wildlife species and visitors. The biggest problem that cannot be dealt with in open-type production systems is flies and aquatic insects, and their destruction and removal are carried out with pesticides. The use of pesticides also creates a toxic effect. It is not as easy as it seems to eliminate this toxicity. The use of such chemicals and drugs causes significant losses in products to be used as human food but also poses some risks. At this point, the point to be considered is that the product to be offered for human consumption should first of all be purified from bacteria that cause disease and transmitted to animals and possibly toxic substances that have been used. This purification process brings additional workmanship and chemical applications to the product. In these applications, it causes various concerns in terms of consumability, as well as spoiling the quality, organic structure, and naturalness of the product [7, 12 - 15].

Fig. (1). Algae production in pools and ponds.

CLOSED POND SYSTEMS

Although the algae production area is less in closed production systems, the production amount is slightly higher since all conditions can be controlled and intervened. In the closed production system, all the criteria of *Spirulina* can be followed and kept under control. Closed production systems have higher investment costs than open production systems. This is a factor that should be taken into account when investing. Temperature and lighting factors, which are important in Spirulina production in closed systems, can be changed independently of seasonal conditions. The most important issue in closed systems is that every point and parameter of the system can be controlled completely [16 - 20]. The system must be isolated from the natural environment and external factors, otherwise, it has no functionality. All production parameters of closed systems Spirulina should be examined, considering the smallest detail, and the investment should be made fully during the establishment phase, without missing any point and avoiding any expense. After the closed systems are fully installed and start production, the product quality is produced in a single standard. The quality and standard of the product is determined in proportion to the knowledge, skills, and experience of the personnel who produced it. What is obtained in closed systems is pure within the framework of a correct structuring and production protocol. When the necessary conditions are met, the risk of transmission of other species is eliminated. The environment is hygienic as it is insulated (this depends on the production protocol). Depending on the applied production protocol, the quality remains constant and standard. Standard yield is obtained as long as the conditions are kept under control in the closed system. In closed system production, production can be directed by intervening in small volumes. Since the system is isolated from the natural environment, it is also away from the contamination of disease-causing bacteria originating from regional wildlife such as poultry, reptiles, rodents, and amphibians. In addition, in closed systems, as in open systems, the risk of flies and aquatic insects does not exist if precautions are taken. Accordingly, no chemical intervention is required. Closed systems do not disturb the naturalness of production [16 - 20].

VARIOUS PHOTOBIOREACTORS FOR ALGAL CULTIVATION

Vertical Column Photobioreactors

In addition to the general tubular photobioreactor design, the design that received the most attention was the British "Biotechna Grassser A.P. Ltd" patented by the "BIOCOIL" system. This system is a cylindrically wound vertical tube system.

The cylindrical section manages to provide maximum light by occupying minimum space. This is an important advantage over horizontally placed tubular photobioreactors. The system is made of flexible transparent pipes with an inner diameter of 1.6cm–5cm. Circulation is provided with the help of a pump to be selected according to the algae type, and temperature control is provided by adding a heating unit in which a certain length tube is used as a heat exchanger. *Tetraselmis spp.*, *Isochrysis galbama*, *Phaeodactylum tricornutum*, *Chaetoceros spp.*, and *Spirulina* species were grown in a pilot diameter up to 700 lt in this system, semi-continuously for 4 months. In addition, the system has the flexibility to work with pure cultures with the necessary modifications. It is a design suitable for scaling up and the system can be controlled automatically and measures can be taken to increase efficiency and reduce costs. Vertical photobioreactors can be categorized into bubble column and airlift reactors (Fig. **2**) based on their liquid flow patterns inside the photobioreactors [21 - 23].

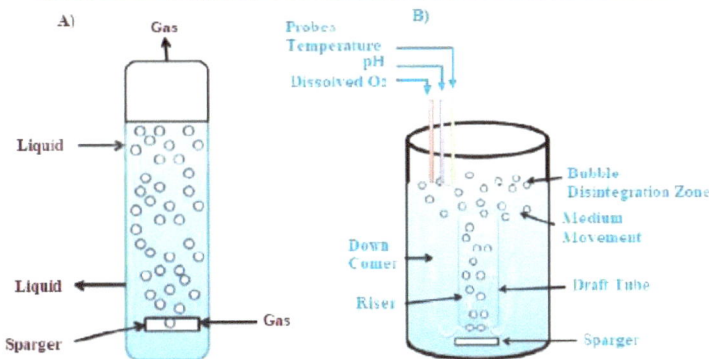

Fig. (2). Schematic diagram of vertical bioreactor: **A)** Bubble column reactor **B)** Air-lift column reactor.

TUBULAR PHOTOBIOREACTOR

The tubular system consists of parallel reactors designed in the form of tubes for large-scale microalgae production. The system has tubular photoreactors of varying scales and is being developed to adjust the exposure of the photoreactors to the desired amount of sunlight. The pumping is done by mechanical means, by applying a small amount of force to the organisms, utilizing an airlift (airlift) pump. Tubular systems can be produced with a capacity of 20-5000 lt and in larger volumes for larger production requirements. The system is designed for long-term operation. The culture can work for more than 6 months without collapsing. The tubular continuous production system is more efficient than other production methods. The system is convenient in terms of meeting both the labor and excessive capacity production needs. Since the system is a closed-loop production system, the control of the species becomes easier, and more products can be made depending on the effective control of light and other parameters affecting growth. While automation is brought to production in the tubular system, the initial investment costs are very high. The system allows long-term production without culture collapse, unlike pond productions In this system, the microalgae are cultivated in transparent tubes (Figs. **3 - 5**) [24 - 29].

In the production of *Spirulina platensis*, it was observed that especially the efficiency of use increased in this type of cylindrical photobioreactor, which was formed from 1.6 cm inner diameter tubes as 60m light receiver and 10m heat exchanger. Because CO_2 can be given to the system against the flow direction. In addition, the utilization of light was more efficient than a horizontal tubular system of the same length. Instead of a cylindrical shape, a conical design was also tried. In this way, the difficulty of making maximum use of the light, especially when the sun is at its highest point, has been overcome in the systems installed in the outdoor environment. Thanks to the conical shape, the vertical helical tubular photobioreactor efficiently utilized light even at its peak. In the light of these studies, scale enlargement was carried out. As the length, which is the problem of tubular systems in scaling up, increases, the accumulated gases are removed from the system and the need for more surface area arises. In such photobioreactors, the process of forming the system by combining small parts, instead of making the system from a single long piece, yielded successful results. Each cone-shaped unit is arranged side by side like a beehive and does not cut off each other's light, ensuring the easy installation of the system and reducing the time required for system maintenance. In addition to these advantages, the problems caused by the pump that will be required to circulate a large-scale system, especially in sensitive types, have been overcome by using a few smaller pumps [30, 31].

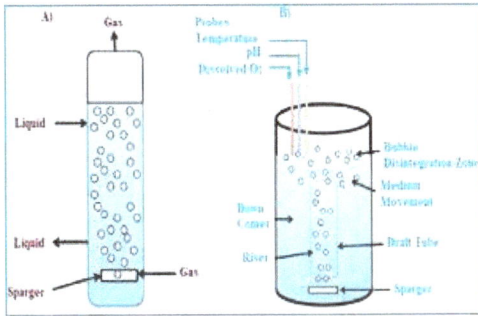

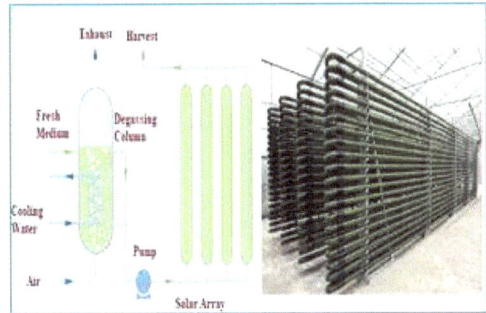

Fig. (3). A typical tubular reactor system.

Fig. (4). A horizontal tubular photobioreactor [30].

Helical Tubular Photobioreactors

Helical tubular photobioreactors are suitable for the production of small-scale algae. Artificial lighting applied to photobioreactors is more effective but more expensive compared to natural lighting. Although the length of the tubes varies depending on the biomass concentration, light intensity, flow rate, and oxygen concentration, they should not exceed 80 meters. While production continues in photobioreactors, carbon dioxide is consumed by algae. As a result of this consumption, the pH value rises. During production, carbon dioxide must be given to the environment so that the pH value does not rise too much. Keeping the temperature constant in photobioreactors positively affects the success of the production. Photobioreactors can be relatively expensive compared to open system pools. But growing axenic monocultures has advantages such as minimizing or preventing contamination. In addition, biocultural conditions such as pH, light intensity, CO_2, and temperature are better controlled. It also has the possibilities of preventing water evaporation in the reactor, low losses for CO_2, high cell concentration possibilities, and production with complex biopharmaceuticals. [32 - 37].

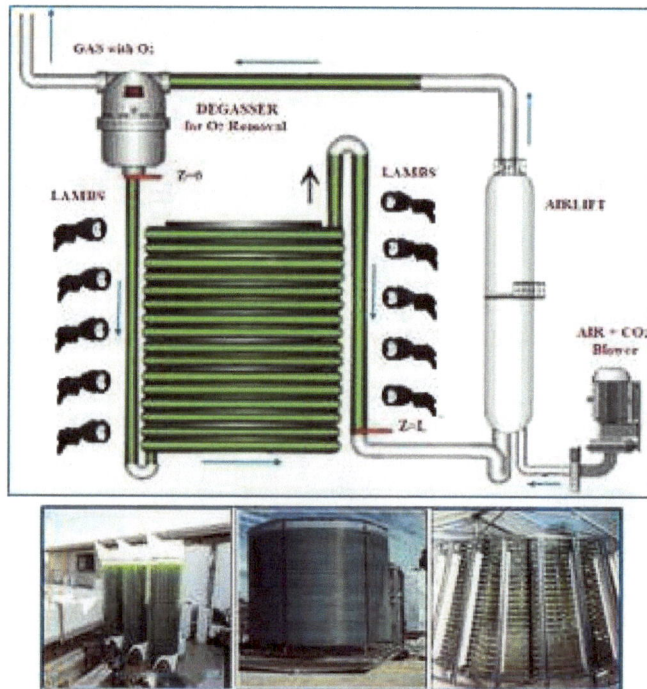

Fig. (5). Helical tubular photobioreactors.

Bubble Column Photobioreactor (BC-PBR)

If the bubble-column photobioreactors are up to 0.19 m in diameter, the final biomass condensation and growth rate are similar to those of the tubular photobioreactor. In bubble-column photobioreactors, when air is supplied from below, flow occurs as eddies in the column. To avoid this, the bubble columns are either equipped with traction tubes or made as split cylinders. In the draft tube photobioreactors, an aspirating tube is placed in the middle and mixing occurs between the ascending and descending regions on the walls of the draft tube. In split photobioreactor types, on the other hand, it is positioned as the rising region on one half of the cylinder and the descending region on the other half [38 - 42].

The cool-white, fluorescent light was vertically positioned at a distance of 12 mm from BC-PBR (Fig. **6**) [43].

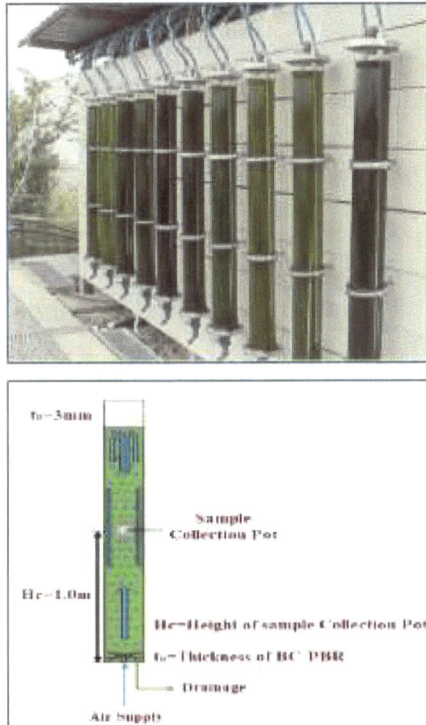

Fig. (6). Bubble column photobioreactors.

Flat Plate Photobioreactor

Flat-panel photobioreactors feature important advantages for the mass production of photoautotrophic microorganisms and may become a standard reactor type for the mass production of several algal species [44]. Flat-plate reactors are characterized by a high surface-to-volume ratio, which leads to the best photosynthetic efficiencies observed for any PBR [45]. Among the photobioreactors used for microalgae cultivation, the flat plate photobioreactor (FPPBR) is widely reported for outdoor cultivation with its characteristics of large illumination surface area, short light path, easy cleanup, and low oxygen buildup [46]. Artificially illuminated flat-plate reactors are often vertical, with the light source incident on the reactor from one side. Outdoor flat-plate reactors are typically tilted at an angle that allows optimal exposure to solar irradiation [47].

The region immediately adjacent to the illuminated reactor surface is the aphotic zone where light saturation, and consequently the photoinhibition of algal growth and H_2 production processes, repeatedly occurs. Also, the light energy available to algal cells decreases exponentially away from this photic zone it has been estimated that fora have fully grown culture of *C. reinhardtii*, effective light penetration is limited to a depth of 0.8 mm [48]. It is therefore important to minimize these light gradients and control the light-dark cycles of the algal cells utilizing an effective agitation system [49]. Flat-plate reactors are subject to relatively low mass transfer rates because the space between panels, known as the light path, is restricted and thus reduces the clearance efficiency of the dissolved O_2 produced by photosynthesis [29]. Good O_2 diffusion rates through the reactor are required to achieve optimal algal biomass growth. Flat-plate reactors provide operational flexibility as they may be run in both batch and continuous modes [50]. The height and width of flat-plate reactors are the two dimensions available for scale-up, but only up to a practical limit of 2-3m [51]. Typical limitations of flat-plate reactors are the scaling-up requirements (the need for many compartments and support materials), the difficulty in controlling culture temperature, the possibility of algal cell clustering on the reactor wall, and the incompatibility with certain algal strains (Fig. **7**) [29].

Fig. (7). Flat-panel photobioreactors.

Heterotrophic Cultivation

Harmful algal blooms, which can lower oxygen levels and produce potent and environmentally persistent biotoxins, have been acknowledged as a severe threat to freshwater habitats and thus algal treatment and disposal are currently receiving considerable attention [52, 53]. Algae can be autotrophic and heterotrophic and are capable of generating substantial biomass that contains proteins, polysaccharides, and lipids [54]. Therefore, algae are a potential alternative biomass resource for the production of sustainable bioenergy and biomaterials, such as biomethane and short-chain fatty acids (SCFAs), through anaerobic fermentation [55 - 57]. Notably, anaerobic fermentation of algae is conducive to SCFAs accumulation, which is accomplished with a shorter fermentation time and a higher yield compared to biomethane production. SCFAs produced from anaerobic fermentation of organic waste have been reported to be utilized as carbon sources for the biological removal of nitrogen and phosphorus and feedstocks for bioenergy generation [58, 59]. However, anaerobic fermentation of algae for SCFAs production has faced challenges due to the rigid structure of algae, which is composed of abundant hard hydrolyzable organic matters such as cellulose [60, 61]. Accordingly, interest in developing new approaches to enhance SCFAs accumulation during anaerobic fermentation of algae has gradually increased. Algae anaerobic fermentation is influenced by pretreatment strategies, biomass characteristics, and the operating parameters of the fermentation process [62, 63]. Physical and chemical pretreatment methods including thermal treatment, microwave, sonication, and chemical additions positively affect algae anaerobic fermentation for methane production, which was beneficial to the destruction of the cell wall and organic matter solubilization [64 - 66]. Some researchers and companies are pursuing an alternative approach to growing algae using sunlight: growing them in the dark on sugars, in a so-called "heterotrophic" fermentation. The algae convert the sugars to oil and biomass, which can be converted into biofuels, chemicals, nutritional products, cosmetics, *etc.* [3]. Biodiesel manufacturers continue to hone the closed container and closed-pond processes, with some developing a variation known as fermentation. This technique cultivates algae that "eats" sugar in closed containers to spur growth. Fermentation is attractive to growers because it provides complete control over the environment (Fig. **8**) [67].

Mixotrophic Cultivation

A hybrid system of enclosed photobioreactor (PBR) and open raceway pond (ORP) has been suggested as a cost-effective solution for maintaining the predominant growth of target algal strains for biomass applications [68, 69]. While the PBR requires generally an order of magnitude higher costs for setup than ORP, its effectiveness in maintaining algal monoculture enables the operation of PBR as a continuous source of the inoculum of desirable algal species throughout the cultivation of ORP. For example, Huntley *et al.* demonstrated large-scale hybrid cultivation and concluded that PBRs were economical to provide a continuous and consistent inoculum for short-term cultures that pre-vents biological system crashes compared to longer-term open pond cultures [70, 71]. Also, the design of the hybrid system has been proposed and successfully demonstrated for continuous outdoor cultivation of green alga, Haematococcus Pluvialis for several years [72, 73]. Moreover, Benemann and Oswald operated a hybrid system with high aerial productivity of 70.4 g m−2day−1and 35% algal lipid yield [69, 74]. Although extensive studies have been carried out on open and closed cultivation methods, limited work has been carried out on two-stage hybrid cultivation systems. Two-phase hybrid cultivation systems have been proposed as an advantageous microalgae cultivation system, as they can essentially separate biomass growth from the lipid accumulation phase. Recently, a life cycle analysis demonstrated a considerably reduced environmental impact when comparing various open and closed cultivation systems with hybrid cultivation. A recent study on large-scale hybrid cultivation concluded that PBRs were economical to provide a continuous and consistent inoculum for short-period batch open pond cultures that prevented biological system crashes compared to longer-term open pond cultures (Fig. **9**) [71].

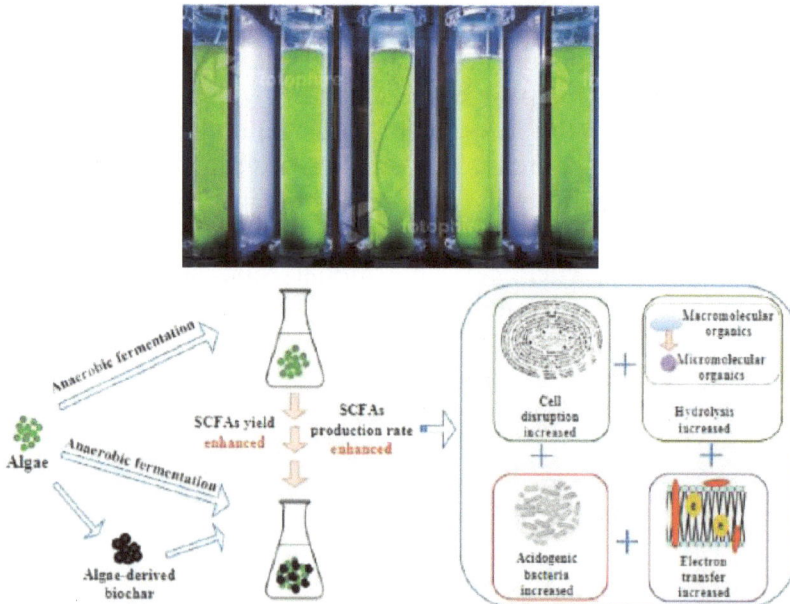

Fig. (8). Fermentation Systems.

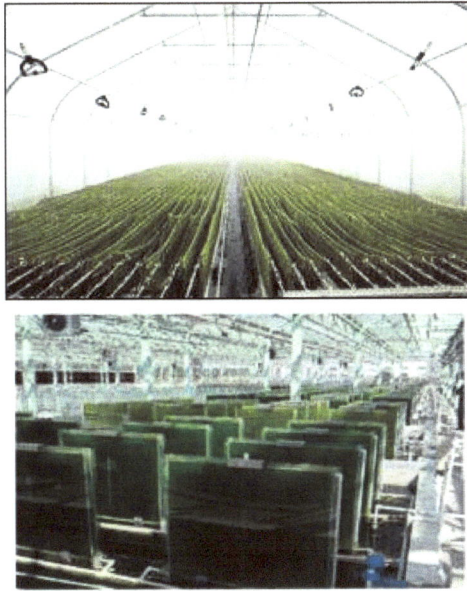

Fig. (9). Hybrid systems.

Integrated Systems

Integrated aquaculture has been proposed as an environmentally friendly way of recycling wastes, especially those produced through the cultivation of high trophic level species, which require the supply of exogenous energy (food). The cultivation of filter-feeders and seaweeds around fish culture cages has been tested for waste recycling. However, success has not been total, partly because the number of filter-feeders and seaweed needed to remove a significant proportion of the wastes produced from intensive large-scale cultivation systems is very large [75]. The systems-level impacts of integrating algae cultivation with wastewater treatment were found to significantly reduce environmental impact. Sensitivity analysis showed that algal productivity most significantly affected fuel selling price, emphasizing the importance of optimizing biomass productivity [76]. Muller-Feuga *et al.* [77] successfully investigated microalgae production with an integrated model that combined the simple monochromatic radiation Beer-Lambert law, the mean bulk flow, and the photosynthetic growth model considering photoinhibition, photo limitation, and the L/D cycles, in a batch and continuous tubular PBR, respectively. With this method, in addition to the removal and distribution of unwanted and toxic substances, the wastewater treatment capabilities of algae that absorb CO_2 and nutrients are strengthened. Integrated systems are the derivation of the pool concept in which the integration of anaerobic and aerobic biological processes in wastewater ponds is represented. Algae have much more productivity and high protein synthesis than other agricultural products. Based on these arguments, it is necessary to reduce the production costs of vital foodstuffs, improve, and facilitate algae production. Besides, high-efficiency algae can be used in wastewater treatment processes. [18, 78].

In the excretion process, as the name indicates algae excrete certain chemicals into the culture in which they are grown. Instead of biomass which stores the oil content, these chemicals are used to synthesize fuels. Here cultivation over large areas and harvesting are not needed and thus excretion method reduces the operating cost. All the six methods discussed above have their advantages and disadvantages. Hybrid systems which comprise of two methods eliminate each other's limitations but multiply the cost. While the comparison cost-effective method is found to be an excretion process, however, its productivity is less under the current situations. These situations include lack of a proper site, scarcity of water, transportation difficulties, *etc*; the site is said to be perfect if problems of water scarcity and transportation have been solved [79, 80]. Most of the companies that grow algae for biofuel production are working on algae modification by secretion. This method is aimed to release chemicals useful for the environment or culture where algae are grown, instead of storing oils in biomass. This approach has the advantage of avoiding the need to produce and process an initial or small amount of algae biomass that will continuously produce oil, which significantly increases the total cost (Fig. **10**).

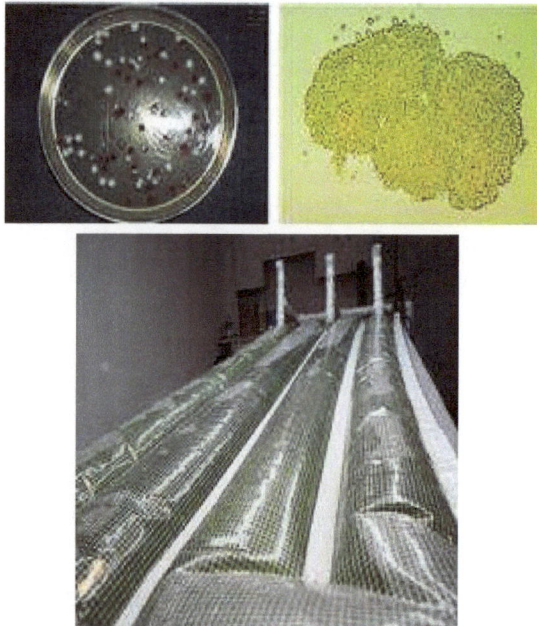

Fig. (10). Excretion processes.

CONCLUSION

The results obtained for the choice of algae pool systems are summarized below.

1. Since algae are simple organisms containing chlorophyll, they can be found anywhere on earth where they can use light for photosynthesis.

2. Although pond-type open systems are generally used, closed photobioreactors are also used for growing algae.

3. The low investment and operating costs of outdoor pools have made the system preferable in the industry. However, the difficulty of controlling the production conditions and the risk of contamination is the disadvantages of the system.

4. It is necessary to compare basic issues such as effective use of light, temperature, hydrodynamic balance in algae culture, and maintaining the culture in large-scale culture systems.

5. The ideal development of each algae species takes place in cultural environments where specific conditions are provided. For example, Spirulina grows best in high pH and bicarbonate density, Chlorella in nutrient-rich media, Dunaliella salina in very high salinity.

CONSENT FOR PUBLICATION

Not applicable.

CONFLICT OF INTEREST

The author declares no conflict of interest, financial or otherwise.

ACKNOWLEDGEMENT

Declared none.

REFERENCES

[1]   Chiaramonti D, Prussi M, Casini D, *et al*. Review of energy balance in raceway ponds for microalgae cultivation: Re-thinking a traditional system is possible. Appl Energy 2013; 102: 101-11. [http://dx.doi.org/10.1016/j.apenergy.2012.07.040]

[2]   J.A.V.. An Open Pond System for Microalgal Cultivation. Biofuels from Algae. 2014; pp. 1-22.

[3]   Industrial Biorefineries & White Biotechnology. http://allaboutalgae.com/open-pond/2015; 35-90.

[4]   Zittelli GC, Biondi N, Rodolfi L, Tredici MR. Photosynthesis in microalgae. In: Richmond A, Hu C, Eds. Handbook of Microalgal Culture: Applied Phycology and Biotechnology. 2nd ed. Oxford: Wiley Blackwell 2013; pp. 225-66. [http://dx.doi.org/10.1002/9781118567166.ch13]

[5]   Acién Fernández FG, Fernández Sevilla JM, Molina Grima E. Photobioreactors for the production of microalgae. Rev Environ Sci Biotechnol 2013; 12(2): 131-51.
[http://dx.doi.org/10.1007/s11157-012-9307-6]

[6]   Chisti Y. Raceways-based production of algal crude oil. In: Posten C, Walter C, Eds. Microalgal Biotechnology, de Gruyter. Berlin 2013; pp. 195-216.

[7]   J.A.V Costa, Freitas BCB, Santos TD, Mitchell BG, Morais MG. Open pond systems for microalgal culture. book: Biofuels from Algae 2019.

[8]   Slegers PM, Lösing MB, Wijffels RH, van Straten G, van Boxtel AJB. Scenario evaluation of open pond microalgae production. Algal Res 2013; 2(4): 358-68.
[http://dx.doi.org/10.1016/j.algal.2013.05.001]

[9]   Kumar K, Mishra SK, Shrivastav A, Park MS, Yang JW. Recent trends in the mass cultivation of algae in raceway ponds. Renew Sustain Energy Rev 2015; 51: 875-85.
[http://dx.doi.org/10.1016/j.rser.2015.06.033]

[10]   Mendoza JL, Granados MR, de Godos I, *et al.* Fluid-dynamic characterization of real-scale raceway reactors for microalgae production. Biomass Bioenergy 2013; 54: 267-75.
[http://dx.doi.org/10.1016/j.biombioc.2013.03.017]

[11]   Tasca AL, Bacci di Capaci R, Tognotti L, Puccini M. Biomethane from short rotation forestry and microalgal open ponds: system modeling and life cycle assessment. Bioresour Technol 2019; 273: 468-77.
[http://dx.doi.org/10.1016/j.biortech.2018.11.038] [PMID: 30469137]

[12]   Sutherland DL, Turnbull MH, Craggs RJ. Increased pond depth improves algal productivity and nutrient removal in wastewater treatment high rate algal ponds. Water Res 2014; 53: 271-81.
[http://dx.doi.org/10.1016/j.watres.2014.01.025] [PMID: 24530547]

[13]   Zhao B, Su Y. Process effect of microalgal-carbon dioxide fixation and biomass production: A review. Renew Sustain Energy Rev 2014; 31: 121-32.
[http://dx.doi.org/10.1016/j.rser.2013.11.054]

[14]   Bitog JP, Lee I-B, Lee C-G, *et al.* Application of computational fluid dynamics for modeling and designing photobioreactors for microalgae production: A review. Comput Electron Agric 2011; 76(2): 131-47.
[http://dx.doi.org/10.1016/j.compag.2011.01.015]

[15]   Apel AC, Weuster-Botz D. Engineering solutions for open microalgae mass cultivation and realistic indoor simulation of outdoor environments. Bioprocess Biosyst Eng 2015; 38(6): 995-1008.
[http://dx.doi.org/10.1007/s00449-015-1363-1] [PMID: 25627468]

[16]   Egbo MK, Okoani AO, Okoh IE. Photobioreactors for microalgae cultivation – An Overview International Journal of Scientific & Engineering Research. 2018; 9(11): 2229-5518.

[17]   Singh NK, Dhar DW. Microalgae as second generation biofuel. A review. Agron Sustain Dev 2011; 31(4): 605-29.
[http://dx.doi.org/10.1007/s13593-011-0018-0]

[18]   Yaşar F. Biodiesel production from algae oil and its use in a diesel engine as alternative fuel [Ph.D Thesis].[Batman]:Batman University; 2016.

[19]   Fernández I, Acién FG, Fernández JM, Guzmán JL, Magán JJ, Berenguel M. Dynamic model of microalgal production in tubular photobioreactors. Bioresour Technol 2012; 126: 172-81.
[http://dx.doi.org/10.1016/j.biortech.2012.08.087] [PMID: 23073105]

[20]   Zhang Q, Wu X, Xue S, Liang K, Cong W. Study of hydrodynamic characteristics in tubular photobioreactors. Bioprocess Biosyst Eng 2013; 36(2): 143-50.
[http://dx.doi.org/10.1007/s00449-012-0769-2] [PMID: 22729676]

[21]   Sharma M, Thukral N, Soni NK, Maji S. Microalgae as Future Fuel: Real Opportunities and

Challenges. Journal of Thermodynamics & Catalysis 2015; 6(1): 1-11.

[22]   Pandey A, Lee DJ, Chisti Y, Soccol CR. Biofuels from algae. Elsevier 2014.

[23]   Loubière K, Olivo E, Bougaran G, Pruvost J, *et al.* A new photobioreactor for continuous microalgal production in hatcheries based in external-loop airlift and swirling flow. 2009; 132-47.

[24]   Acien G, Grima EM, Reis A, *et al.* Photobioreactors for the production of microalgae. Microalgae-Based Biofuels and Bioproducts 2017.
[http://dx.doi.org/10.1016/B978-0-08-101023-5.00001-7]

[25]   Díaz JP, Inostroza C, Acién Fernández FG. Fibonacci-type tubular photobioreactor for the production of microalgae. Process Biochem 2019; 86: 1-8.
[http://dx.doi.org/10.1016/j.procbio.2019.08.008]

[26]   Wolkers H, Barbosa MJ, Kleinegris D, Bosma R, Wijffels RH. Microalgae: the green gold of the future?. Wageningen: Wageningen UR 2011.

[27]   Qiu F. Integration of algae in architecture. 2014; 1-63.

[28]   Norsker NH, Barbosa MJ, Vermuë MH, Wijffels RH. Microalgal production — A close look at the economics. Biotechnol Adv 2011; 29(1): 24-7.
[http://dx.doi.org/10.1016/j.biotechadv.2010.08.005] [PMID: 20728528]

[29]   Sierra E, Acién FG, Fernández JM, García JL, González C, Molina E. Characterization of a flat plate photobioreactor for the production of microalgae. Chem Eng J 2008; 138(1-3): 136-47.
[http://dx.doi.org/10.1016/j.cej.2007.06.004]

[30]   Zhang X. Microalgae removal of CO2 from flue gas. IEA Clean Coal Centre 2015; pp. 1-95.

[31]   Singh RN, Sharma S. Development of suitable photobioreactor for algae production – A review. Renew Sustain Energy Rev 2012; 16(4): 2347-53.
[http://dx.doi.org/10.1016/j.rser.2012.01.026]

[32]   Travieso L, Hall DO, Rao KK, Benítez F, Sánchez E, Borja R. A helical tubular photobioreactor producing Spirulina in a semicontinuous mode. Int Biodeterior Biodegradation 2001; 47(3): 151-5.
[http://dx.doi.org/10.1016/S0964-8305(01)00043-9]

[33]   Morita M, Watanabe Y, Okawa T, Saiki H. Photosynthetic productivity of conical helical tubular photobioreactors incorporatingChlorellasp. under various culture medium flow conditions. Biotechnol Bioeng 2001; 74(2): 136-44.
[http://dx.doi.org/10.1002/bit.1103] [PMID: 11370002]

[34]   Tredici MR, Zittelli GC. Efficiency of sunlight utilization: Tubular *versus* flat photobioreactors. Biotechnol Bioeng 1998; 57(2): 187-97.
[http://dx.doi.org/10.1002/(SICI)1097-0290(19980120)57:2<187::AID-BIT7>3.0.CO;2-J] [PMID: 10099193]

[35]   Watanabe Y, de la Noüe J, Hall DO. Photosynthetic performance of a helical tubular photobioreactor incorporating the cyanobacterium *spirulina platensis*. Biotechnol Bioeng 1995; 47(2): 261-9.
[http://dx.doi.org/10.1002/bit.260470218] [PMID: 18623400]

[36]   Tredici MR, Chini Zittelli G, Rodolfi L. Photobioreactors. In: Flickinger MC, Anderson S, Eds. Encyclopedia of industrial biotechnology: bioprocess, bioseparation, and cell technology. Hoboken: Wiley 2010; Vol. 6: pp. 3821-38.
[http://dx.doi.org/10.1002/9780470054581.eib479]

[37]   Borowitzka MA. Commercial production of microalgae: ponds, tanks, tubes and fermenters. J Biotechnol 1999; 70(1-3): 313-21.
[http://dx.doi.org/10.1016/S0168-1656(99)00083-8]

[38]   Vo HNP, Bui XT, Nguyen TT, *et al.* Effects of nutrient ratios and carbon dioxide bio-sequestration on biomass growth of Chlorella sp. in bubble column photobioreactor. J Environ Manage 2018; 219: 1-8.
[http://dx.doi.org/10.1016/j.jenvman.2018.04.109] [PMID: 29715637]

[39]   Chen CY, Yeh KL, Aisyah R, Lee DJ, Chang JS. Cultivation, photobioreactor design and harvesting of microalgae for biodiesel production: A critical review. Bioresour Technol 2011; 102(1): 71-81.
[http://dx.doi.org/10.1016/j.biortech.2010.06.159] [PMID: 20674344]

[40]   Mayers JJ, Flynn KJ, Shields RJ. Influence of the N:P supply ratio on biomass productivity and time-resolved changes in elemental and bulk biochemical composition of Nannochloropsis sp. Bioresour Technol 2014; 169: 588-95.
[http://dx.doi.org/10.1016/j.biortech.2014.07.048] [PMID: 25103036]

[41]   Jacob-Lopes E, Cacia Ferreira Lacerda LM, Franco TT. Biomass production and carbon dioxide fixation by Aphanothece microscopica Nägeli in a bubble column photobioreactor. Biochem Eng J 2008; 40(1): 27-34.
[http://dx.doi.org/10.1016/j.bej.2007.11.013]

[42]   Sardans J, Rivas-Ubach A, Peñuelas J. The C:N:P stoichiometry of organisms and ecosystems in a changing world: A review and perspectives. Perspect Plant Ecol Evol Syst 2012; 14(1): 33-47.
[http://dx.doi.org/10.1016/j.ppees.2011.08.002]

[43]   Khoo CG, Lam MK, Lee KT. Pilot-scale semi-continuous cultivation of microalgae Chlorella vulgaris in bubble column photobioreactor (BC-PBR): Hydrodynamics and gas–liquid mass transfer study. Algal Res 2016; 15: 65-76.
[http://dx.doi.org/10.1016/j.algal.2016.02.001]

[44]   Tamburic B, Zemichael FW, Crudge P, Maitland GC, Hellgardt K. Design of a novel flat-plate photobioreactor system for green algal hydrogen production. Int J Hydrogen Energy 2011; 36(11): 6578-91.
[http://dx.doi.org/10.1016/j.ijhydene.2011.02.091]

[45]   Akkerman I, Janssen M, Rocha J, Wijffels RH. Photobiological hydrogen production: photochemical efficiency and bioreactor design. Int J Hydrogen Energy 2002; 27(11-12): 1195-208.
[http://dx.doi.org/10.1016/S0360-3199(02)00071-X]

[46]   Zhang QH, Wu X, Xue SZ, Wang ZH, Yan CH, Cong W. Hydrodynamic characteristics and microalgae cultivation in a novel flat-plate photobioreactor. Biotechnol Prog 2013; 29(1): 127-34.
[http://dx.doi.org/10.1002/btpr.1641] [PMID: 23011867]

[47]   Carvalho AP, Meireles LA, Malcata FX. Microalgal reactors: a review of enclosed system designs and performances. Biotechnol Prog 2006; 22(6): 1490-506.
[http://dx.doi.org/10.1002/bp060065r] [PMID: 17137294]

[48]   Janssen M, Tramper J, Mur LR, Wijffels RH. Enclosed outdoor photobioreactors: Light regime, photosynthetic efficiency, scale-up, and future prospects. Biotechnol Bioeng 2003; 81(2): 193-210.
[http://dx.doi.org/10.1002/bit.10468] [PMID: 12451556]

[49]   Hankamer B, Lehr F, Rupprecht J, Mussgnug JH, Posten C, Kruse O. Photosynthetic biomass and H$_2$ production by green algae: from bioengineering to bioreactor scale-up. Physiol Plant 2007; 131(1): 10-21.
[http://dx.doi.org/10.1111/j.1399-3054.2007.00924.x] [PMID: 18251920]

[50]   O P. Photobioreactors: production systems for phototrophic microorganisms. Appl Microbiol Biotechnol 2001; 57(3): 287-93.
[http://dx.doi.org/10.1007/s002530100702] [PMID: 11759675]

[51]   Posten C. Design principles of photo-bioreactors for cultivation of microalgae. Eng Life Sci 2009; 9(3): 165-77.
[http://dx.doi.org/10.1002/elsc.200900003]

[52]   Ghimire A, Kumar G, Sivagurunathan P, et al. Bio-hythane production from microalgae biomass: Key challenges and potential opportunities for algal bio-refineries. Bioresour Technol 2017; 241: 525-36.
[http://dx.doi.org/10.1016/j.biortech.2017.05.156] [PMID: 28601770]

[53]   Sun R, Sun P, Zhang J, Esquivel-Elizondo S, Wu Y. Microorganisms-based methods for harmful algal

blooms control: A review. Bioresour Technol 2018; 248(Pt B): 12-20.
[http://dx.doi.org/10.1016/j.biortech.2017.07.175] [PMID: 28801171]

[54]   Vo Hoang Nhat P, Ngo HH, Guo WS, *et al.* Can algae-based technologies be an affordable green
       process for biofuel production and wastewater remediation? Bioresour Technol 2018; 256: 491-501.
       [http://dx.doi.org/10.1016/j.biortech.2018.02.031] [PMID: 29472123]

[55]   Kim SH, Mudhoo A, Pugazhendhi A, *et al.* A perspective on galactose-based fermentative hydrogen
       production from macroalgal biomass: Trends and opportunities. Bioresour Technol 2019; 280: 447-58.
       [http://dx.doi.org/10.1016/j.biortech.2019.02.050] [PMID: 30777703]

[56]   Colling Klein B, Bonomi A, Maciel Filho R. Integration of microalgae production with industrial
       biofuel facilities: A critical review. Renew Sustain Energy Rev 2018; 82: 1376-92.
       [http://dx.doi.org/10.1016/j.rser.2017.04.063]

[57]   Saratale RG, Kuppam C, Mudhoo A, *et al.* Bioelectrochemical systems using microalgae – A concise
       research update. Chemosphere 2017; 177: 35-43.
       [http://dx.doi.org/10.1016/j.chemosphere.2017.02.132] [PMID: 28284115]

[58]   Duan X, Wang X, Xie J, Feng L, Yan Y, Zhou Q. Effect of nonylphenol on volatile fatty acids
       accumulation during anaerobic fermentation of waste activated sludge. Water Res 2016; 105: 209-17.
       [http://dx.doi.org/10.1016/j.watres.2016.08.062] [PMID: 27619497]

[59]   Pham TN, Nam WJ, Jeon YJ, Yoon HH. Volatile fatty acids production from marine macroalgae by
       anaerobic fermentation. Bioresour Technol 2012; 124: 500-3.
       [http://dx.doi.org/10.1016/j.biortech.2012.08.081] [PMID: 23021950]

[60]   Velazquez-Lucio J, Rodríguez-Jasso RM, Colla LM, *et al.* Microalgal biomass pretreatment for
       bioethanol production: a review. Biofuel Research Journal 2018; 5(1): 780-91.
       [http://dx.doi.org/10.18331/BRJ2018.5.1.5]

[61]   Kendir E, Ugurlu A. A comprehensive review on pretreatment of microalgae for biogas production. Int
       J Energy Res 2018; 42(12): 3711-31.
       [http://dx.doi.org/10.1002/er.4100]

[62]   Passos F, Uggetti E, Carrère H, Ferrer I. Pretreatment of microalgae to improve biogas production: A
       review. Bioresour Technol 2014; 172: 403-12.
       [http://dx.doi.org/10.1016/j.biortech.2014.08.114] [PMID: 25257071]

[63]   Rodriguez C, Alaswad A, Mooney J, Prescott T, Olabi AG. Pre-treatment techniques used for
       anaerobic digestion of algae. Fuel Process Technol 2015; 138: 765-79.
       [http://dx.doi.org/10.1016/j.fuproc.2015.06.027]

[64]   Klassen V, Blifernez-Klassen O, Wobbe L, Schlüter A, Kruse O, Mussgnug JH. Efficiency and
       biotechnological aspects of biogas production from microalgal substrates. J Biotechnol 2016; 234: 7-
       26.
       [http://dx.doi.org/10.1016/j.jbiotec.2016.07.015] [PMID: 27449486]

[65]   Ometto F, Quiroga G, Pšenička P, Whitton R, Jefferson B, Villa R. Impacts of microalgae pre-
       treatments for improved anaerobic digestion: Thermal treatment, thermal hydrolysis, ultrasound and
       enzymatic hydrolysis. Water Res 2014; 65: 350-61.
       [http://dx.doi.org/10.1016/j.watres.2014.07.040] [PMID: 25150520]

[66]   Wang M, Lee E, Dilbeck MP, Liebelt M, Zhang Q, Ergas SJ. Thermal pretreatment of microalgae for
       biomethane production: experimental studies, kinetics and energy analysis. J Chem Technol
       Biotechnol 2017; 92(2): 399-407.
       [http://dx.doi.org/10.1002/jctb.5018]

[67]   https://www.thoughtco.com/making-biodiesel-from-algae-85138

[68]   Yun J-H, Cho D-H, Lee S, *et al.* Hybrid operation of photobioreactor and wastewater-fed open
       raceway ponds enhances the dominance of target algal species and algal biomass production. Algal
       Res 2018; 29: 319-29.

[http://dx.doi.org/10.1016/j.algal.2017.11.037]

[69]   Schenk PM, Thomas-Hall SR, Stephens E, *et al.* Second generation biofuels: high-efficiency microalgae forbiodiesel production. BioEnergy Res 2008; 1(1): 20-43.
[http://dx.doi.org/10.1007/s12155-008-9008-8]

[70]   Huntley ME, Johnson ZI, Brown SL, *et al.* Algal Res 2018; 29: 319-29.
[http://dx.doi.org/10.1016/j.algal.2017.11.037]

[71]   Narala RR, Garg S, Sharma KK, *et al.* Comparison of microalgae cultivation in photobioreactor, open raceway pond, anda two-stage hybrid system. Front Energy Res 2016; 4(29): 1-10.

[72]   Huntley ME, Redalje DG. CO2mitigation and renewable oil from photosyntheticmicrobes: a new appraisal. Mitig Adapt Strategies Glob Change 2007; 12(4): 573-608.
[http://dx.doi.org/10.1007/s11027-006-7304-1]

[73]   Adesanya VO, Cadena E, Scott SA, Smith AG. Life cycle assessment on microalgal biodiesel production using a hybrid cultivation system. Bioresour Technol 2014; 163: 343-55.
[http://dx.doi.org/10.1016/j.biortech.2014.04.051] [PMID: 24852435]

[74]   Benemann JR, Oswald WJ. Systems and Economic Analysis of Microalgae Pondsfor Conversion of CO2to Biomass. 1996.

[75]   Troell M., Kautsky N., Integrated algal farming: a review. Cah Biol Mar 2001; 42: 83-90.

[76]   Barlow J, Sims RC, Quinn JC. Techno-economic and life-cycle assessment of an attached growth algal biorefinery. Bioresour Technol 2016; 220: 360-8.
[http://dx.doi.org/10.1016/j.biortech.2016.08.091] [PMID: 27595701]

[77]   Muller-Feuga A, Le Guédes R, Pruvost J. Benefits and limitations of modeling for optimization of Porphyridium cruentum cultures in an annular photobioreactor. J Biotechnol 2003; 103(2): 153-63.
[http://dx.doi.org/10.1016/S0168-1656(03)00100-7] [PMID: 12814874]

[78]   Cowan AK, Render DS. Integrated algae pondıng system technical description. Institute for Environmental Biotechnology, Rhodes University 2012.

[79]   Mehrabadi A, Craggs R, Farid MM. Wastewater treatment high rate algal ponds (WWT HRAP) for low-cost biofuel production. Bioresour Technol 2015; 184: 202-14.
[http://dx.doi.org/10.1016/j.biortech.2014.11.004] [PMID: 25465780]

[80]   Tachoth V, Rose A. Site Selection for Large-Scale Algae Cultivation towards Biodiesel Production International Journal of Renewable Energy Research. 2016; 6.

CHAPTER 8

Use of Microbial Fuel Cells (MFCs) in Food Industry Wastewater Treatment

Mesut Yılmazoğlu[1,*]

[1] *Chemical Engineering Department, Engineering Faculty, Yalova University, 77100 Yalova, Turkey*

Abstract: Since MFC degrades simple carbohydrates *i.e.* glucose, acetate, and butyrate, and countless organic substances such as pig wastewater, domestic wastewater, and manure sludge waste, the biochemical energy generated by the catalytic reactions of microorganisms and converts the waste produced into energy. It promises a sustainable wastewater treatment to balance the operating cost. This chapter is a review of the advantages of microbial fuel cell treatment of food industry wastewater, which creates high organic pollution in the industrial field.

Keywords: Food industry wastewater, Industrial wastewater treatment, Microbial fuel cell.

INTRODUCTION

Rivers, groundwater strata, and groundwater are particularly contaminated by humanitarian actions such as energy and food production. This pollution occurs when large amounts of foreign and harmful substances or substances that cannot be removed by natural cleaning are mixed into water, and this mixing makes the water unusable. More than 10 million children die before their 5th birthday each year. Four million of these children die before turning a month old and 2 million from diarrhea. This current situation necessitates the purification of water from pathogens for its sustainable use. Today, 300-500 million tons of heavy metals, solvents, toxic waste, and other waste materials accumulate every year due to industrial activities. Others break down to form non-toxic compounds. Industries based on organic raw materials such as the food production industry are the sectors that contribute the most to the organic pollution problem. The food industry in developing countries produces more than 50% of organic water pollutants. The water supply, sewerage, and wastewater systems require energy

---

[*] **Correspondence author Mesut Yılmazoğlu:** Chemical Engineering Department, Engineering Faculty, Yalova University, 77100 Yalova, Turkey; Tel: +902268155399; E-mail: mesut.yilmazoglu@yalova.edu.tr

Hüseyin Karaca and Cemil Koyunoğlu (Eds.)

for the extraction, distribution, collection, treatment, and disposal of water. Traditionally, water supply systems rely on surface water resources, large distribution systems, and the mixing of treated wastewater. Today, additional energy input is needed to meet the increasing water demand and to protect human and environmental health [1].

It is known from the literature and managerial experiences that 25-40% of the operating costs in a conventional wastewater treatment plant result from energy consumption. This value varies between approximately 0.2-2.1 kW-hour per m^3 of treated wastewater. Typically, in a conventional wastewater treatment plant, the main contributors to energy consumption are mixed fluid aeration (55-70%), primary and secondary settling by sludge pumping (15.6%), and sludge dewatering (7%) [2].

Today, by using microbial fuel cells for the treatment of wastewater with limited use for biogas production with high concentrations of nitrogen and sulfur, the wastewater treatment will be carried out under anaerobic conditions, while electrical energy will be obtained directly. The system, which can be operated with high loading rates under anaerobic conditions, will be treated without ventilation, which corresponds to almost half of the energy needed for the treatment, as mentioned above, and electrical energy will be obtained directly from the chemical energy stored in organic matter. As there is only CO_2 output as a result of wastewater treatment, there is no need for an additional gas treatment unit [3].

COMPONENTS OF MFC

Many different types of MFC reactors have been used in studies carried out so far.; These are generally two-compartment, H-type MFC reactors, U-tube MFC reactors, tubeless MFC reactors, sediment-type MFC reactors, and single-compartment MFC. The H-type MFC reactor consists of two flasks, usually in the form of an H, combined with a tube containing a membrane that separates the anode and cathode from each other. Separators are usually in the form of a cation exchange membrane or salt bridge such as Nafion or Ultrex. The point that should be taken into consideration in designs is that the membrane allows protons to pass while not allowing the passage of food or the electron acceptor in the cathode compartment. Accordingly, MFC; consists of anode, cathode, and selectively permeable proton exchange membrane [4].

Anode Cell

An anode material should be: biocompatible, suitable mechanical strength, low internal resistance, high conductivity, high surface area, porosity, non-corrosive, non-clogging by bacteria, cheap, easily manufactured, chemically inert, and scalable in larger sizes. The most important of these features is that it must have electrical conductivity, unlike other biofilm reactors. Also, the bacteria must be able to attach to the material and provide a good electrical connection. There are many anode materials used in MFCs. Materials such as plain graphite, graphite plate, carbon paper felt or foam, mesh vitreous carbon (RVC), graphite granules, graphite rods, graphite fibers, graphite fiber brushes, and anode due to their stability, high conductivity, and high specific surface area in a microbial graft mixture are being used [5, 6].

The use of carbon-based electrodes has long been proven and enabled the production of carbon in a large number of analytical and industrial applications due to its high efficiency in heterogeneous electron transfer kinetics [5, 6].

Cathode Cell

The cathode material is another factor that limits power generation in MFCs and significantly affects performance. Different cathode materials were used in the studies according to the reactor and wastewater type. In addition to the use of graphite-based materials, it can also be used as an anode electrode [9]. In the cathode section; A three-phase reaction occurs: solid (cathode electrode), gas (electrons, protons, and all of the oxygen), and liquid (catholyte). Ferricyanide $[Fe(CN)_6]$ has a very popular use as an experimental electron acceptor in MFC systems due to its good performance. Oxygen is the most suitable electron acceptor for MFC due to its high oxidation potential, low cost, sustainability, and no chemical waste generation [3, 7 - 9].

PROTON SELECTIVE MEMBRANE

MFCs are divided into two according to the presence of membrane in MFCs. The presence of the membrane is not mandatory [3].

Proton conductive membranes are MFC components that physically separate the anode and cathode compartments and allow the passage of protons to the cathode to generate electrical current, and ensure that the mixtures in the cathode and anode compartments do not mix. These membranes must be permeable to allow proton passage from the anode to cathode. The most widely used membrane type in proton selective systems is the Nafion membrane [3].

ELECTRICTY FORMATION IN MFCS

Obtaining electrical energy from an MFC is possible only if the reaction taking place is thermodynamically suitable. The total cell electromotive force of the reaction is expressed by the potential difference between the anode and cathode. The most important factor determining these potential values is the nature of the chemical reaction that takes place [7]. For example, if feeding is made with glucose,

The reaction at the anode:

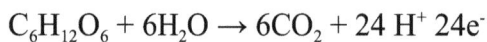

$$C_6H_{12}O_6 + 6H_2O \rightarrow 6CO_2 + 24\,H^+\,24e^-$$

Reaction at the cathode:

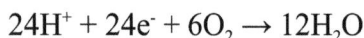

$$24H^+ + 24e^- + 6O_2 \rightarrow 12H_2O$$

When feeding with the anode,

The reaction at the anode:

$$CH_3COO^- + 2H_2O \rightarrow 2CO_2 + 7H^+ + 8e^-$$

The reaction at the cathode:

$$O_2 + 4H^+ \rightarrow 4e^- + 2H_2O$$

As can be understood from the equations written above, the electricity generation through electron flow will vary according to the type of feeding.

In short, as the substrate turns into carbon dioxide and water, an electric current also occurs [3].

FOOD INDUSTRY WASTEWATER CHARACTERIZATION

Baker's yeast industry is one of the largest fermentation industries. The annual amount of bread yeast produced in the world is approximately 2 million tons. Approximately 15-20% of this is produced in our country. Pollution caused by wastewater resulting from baker's yeast production adversely affects the environment and especially threatens human health. Baker's wastewater is an important source of pollution because it has high chemical oxygen content (COD) and contains dark-colored and non-biodegradable organic impurities. (Pala and Erden, 2005). In this industry, it is very difficult to break down the high molecular weight melonoid type compounds found in the dark brown color, sugar beet molasses used as raw material[17]. Discharge limits of yeast wastewater are given in Table 1. Yeast industry wastewater can be reduced to these limits and discharged [8 - 16].

Table 1. Yeast wastewater discharge limits [17].

Parameter	Unit	Composite Sample for 2 hours	Composite Sample for 24 hours
Chemical oxygen demand (COD)	mg/L	1200	1000
Suspended solid (SS)	mg/L	200	100
Oil and grease	mg/L	60	30
pH		6-9	6-7

Whey is as weak as milk. Quickly acidifies. In the meantime, lactose, which increases the value of whey, turns into lactic acid to a large extent depending on the activity of microorganisms. The characterization of whey is given in Table **2** [19].

Table 2. Characterization of Whey [19].

Parameter	Whey
COD (mg/I)	18400-70000
ROD (mg/I)	8300-40000
SS (mg/I)	7000
Oil-Gres (mg/I)	4095
TKN (mg/I)	280-1100
PO4-P (mg/I)	110-135
Detergen (mg/I)	-
pH	5-9,5

Here, the values are the 24-hour composite sample discharge limits: Biological oxygen demand (BOD) should be reduced to 40 mg/l, chemical oxygen demand (COD) to 160 mg/l, oil and grease to 30 mg/l, and pH 6-9.

To ensure the sustainable use of water, food wastewater can be used as irrigation water after the necessary treatment is provided as given in Table **3** below [15].

Food Industry Wastewater Treatment

Wastewater is discharged to a water source at the endpoint, whether it is treated or not. It is usually given to a surface water source such as a sea, river, or lake or rarely leaked to the ground. If wastewater is discharged without treatment, it causes eutrophication, which is called overfeeding, the transformation of surface waters into an anaerobic environment, fish deaths, and odor spreading. This situation makes it difficult and even prevents the supply of drinking water from water sources. Besides, wastewater discharge to water resources causes an increase in infectious diseases such as cholera, typhoid, and jaundice. The purpose of the treatment of wastewater is to prevent pollution of surface water resources, groundwater, and ground, and to protect public health [10].

Table 3. Suitability of food wastewater for agricultural irrigation [15].

1	Can be used as irrigation water if there is suitable land nearby	Beer, Malt, Wine, Potatoes, Vegetables, Canned Food, Marmalade, Fruit canned, milk, potato starch, and factories, *etc.*
2	Suitable for use as irrigation water under certain conditions	Sugar, Rice, and Grain starch, Slaughterhouse, Meat combined plants, Margarine, Paper, Fish Meal, Canned Fish, Mining, *etc.*
3	Not suitable for use as irrigation water	Lacquer and paint factories, Soap Factory, Inorganic chemical factories, Pharmaceutical, Metal, Mineral oil factories, Viscose rayon factory, *etc.*

James Lovelock Gaia hypothesis, the earth as a self-regulating entity comes alive. The basis of this arrangement includes chemical elements such as carbon, nitrogen, and sulfur in the sea atmosphere and millions of tons of mobility in earth conditions. The conversion between solid and gaseous elements is mostly carried out by microorganisms such as archaea and bacteria [11]. In this context, water polluted by various human activities is purified as a result of the decomposition activities of microorganisms using many different methods. With its high organic matter content, *Salmonella* spp. *Listeria monocytogenes, Escherichia coli, Shigella* spp., *Vibrio* spp. And Brucella disease agent *Brucella* spp [12]. Food wastewater containing many pathogenic microorganisms is necessary to be treated. This wastewater is treated using many methods. In the study of Çetinkaya (2018), dairy industry wastewater with high biodegradable properties was used. Biogas production from waste and wastewater plays an important role in providing renewable energy around the world. Biochemical Methane Potential (BMP) tests are considered innovative technologies in determining the methane yield of wastes, as they are easier to perform than continuous experiments[13]. Pulido (2016) conducted a review on the importance of membranes in olive oil wastewater treatment. In light of the literature review obtained, he stated that olive oil wastewater can be effectively treated with the membrane, and nano filter (NF) membranes are more effective than other membranes[14]. In the study by Özan and Açıkgöz, it was realized that the boiled meat wastewater, diluted 1/4 and 1/10 of the food industry, was operated in a laboratory/pilot-scale membrane bioreactor (MBR) system [16].

Use of MFCs in Food Industry Wastewater Treatment

Food wastewater, which constitutes about half of organic wastewater, is also treated using microbiological methods. As a conventional method, the activated sludge method is widely used despite organic contamination. However, many factors restrict the use of the activated sludge method such as the large space requirement required, long waiting times, excessive sludge formation, and the considerable energy requirement of the ventilation process applied during microorganism growth [8].

MFC, which is an alternative method for energy recovery while treating food tack waters with high biodegradability, is a system in which microorganisms are used as catalysts. As given in Fig. (**1**), the MFC System is integrated with the food wastewater treatment system: Wastewater is sent to the anode section utilizing pumps. When the electrons are fed from the bottom, while the electrons pass to the cathode section by the salt bridge, the clear water is collected from the upper part of the reactor. While the anode section is an anaerobic environment, the air is supplied to the cathode section, which should be aerobic[8]. Nitrogen gas is given to the anode section to be anaerobic. Wastewater is continuously fed to ensure continuous energy harvesting from the system. Mixed culture anaerobic microorganisms using alternative electron acceptors such as nitrate, sulfate, and CO_2 are used in the anode compartment instead of heterotroph and terminal oxygen as the electron acceptor [4, 12].

MFCs are routinely operated as closed system devices in which the anodic chamber is kept under anaerobic conditions. This is necessary to facilitate the growth of obligate anaerobic bacteria capable of electron transfer, such as *G. sulfurreducens*. Bacterial fuel sources that can play a role in electron transfer until now are wastewater, marine sediment, freshwater sediment, soil, and activated sludge. Bacteria in such systems can produce electrons by oxidizing substrates isolated in the anodic chamber. These electrons flow from the anode to the cathode through an external circuit, resulting in electricity generation due to the difference in potential due to electron flow. Protons produced at the anode can pass through the solution and the membrane. Nafion membranes are the most frequently used separators in MFC due to their desired properties such as high conductivity to cations, selective permeability to protons, low internal resistance, and capability [5].

On the other hand, water is released as a result of the combination of hydrogen ions transferred to the cathode section with MFC technology and oxygen. Thus, the clean water released as a result of the treatment provided by this technology can be evaluated as irrigation water or process cooling water, which highlights another advantageous aspect. Thanks to the established system; It is aimed to reduce the cost of wastewater treatment for enterprises by ensuring the efficient use of food wastes. In the study by Erenler and Ülke (2018), molasses wastewater with high sugar, pectin, and protein content has been observed to have higher energy output compared to olive black water and whey [8].

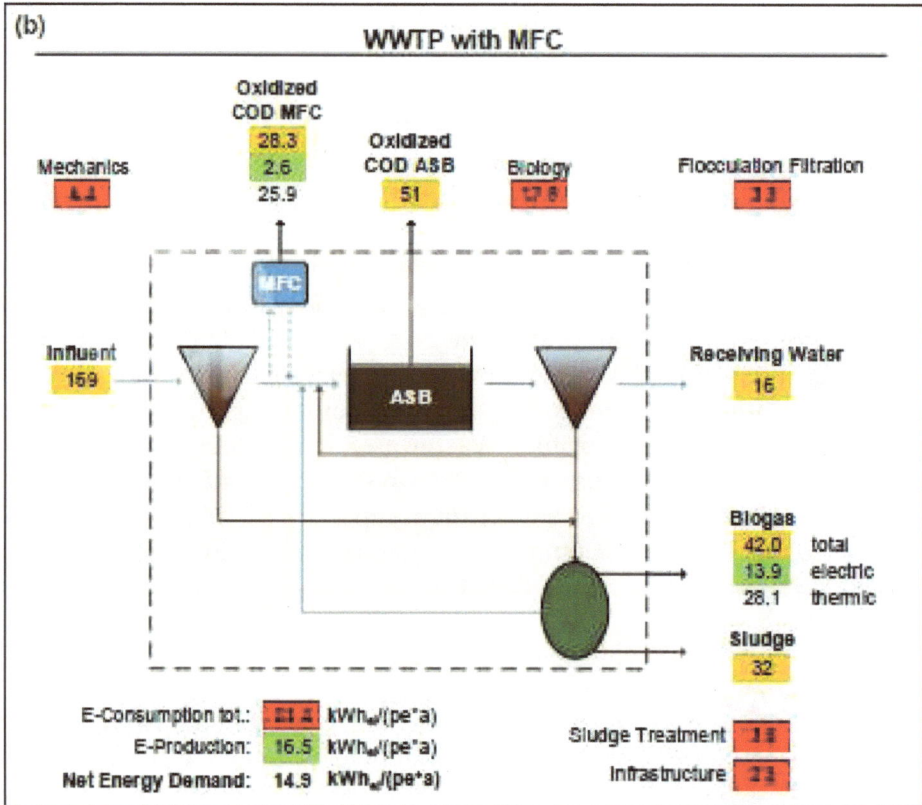

Fig. (1). MFC with wastewater system [17].

MFCs can produce hydrogen from the fermentation of sugars in wastewater, which can be used as a fuel source in other renewable energy technologies. He studied hydrogen production in food processing wastewater along with electricity generation. Model bacterial strains currently used in MFCs are Shewanella spp. and Geobacter spp., which are iron-reducing species. These bacteria can degrade organic substances for the nutrient cycle, for example, iron oxides found in both soil and sediments [5].

Most of the modern MFC technologies developed so far are used as a fuel source that operates based on energy recovery through the biological decomposition of wastewater, organic-rich waste. The ability to generate electricity from wastewater can play an important role in renewable energy generation. It has been reported that in 2012, 5% of the total energy consumption of the USA was used to facilitate water and wastewater treatment plants. However, for the proper treatment of wastewater (according to relevant regulations and standards), the issues surrounding the quality of MFC treated wastewater have not yet been adequately addressed. Treatment of wastewater with only MFCs may not be a viable option due to solid wastewater quality requirements [5].

Therefore, it may be necessary to add other steps such as membrane technology and traditional treatment technology (after the MFC stage), and integrated MFCs. Other steps may need to be added, such as MFCs. MFCs offer a significant advantage over other renewable energy sources as they can be applied to wastewater treatment. Another advantage of MFC technologies is that they are less dependent on other renewable energy technologies (eg solar and wind) than geographic location and seasonal change [5].

Min and Logan developed a flat sheet MFC in which electrodes are joined into a single strip, usually separated by a PDM, to duplicate parameters used in conventional hydrogen fuel cells. This allows the electrodes to be kept close to increase the proton conduction between the two electrodes. However, in MFCs, PDMs such as Nafion are often used as oxygen permeable, which can have a detrimental effect on obligate anaerobes when used as preferred bacterias in the anodic chamber. Other conditions that can cause voltage reversal and thus damage power outputs include lack of oxygen at the cathode, impedance differences, catalyst deficiency, and insufficient fuel. An example of this is when a cell is shorted to indicate that the voltage is reversed; The researchers used diodes connected in parallel to a hydrogen fuel cell due to their low ohmic resistances, so when one or more of the cells fails, MYH can automatically short-circuit. Avoiding fuel starvation, *i.e.* providing sufficient substrate at the anode and oxygen at the cathode (for air cathodes) has been shown to reduce the likelihood of voltage reversal. It has also been shown that operating the MFC configuration at low current densities also prevents voltage reversal [5].

CONSEQUENCE-RESTRICTING FACTORS

It has been reported that the limiting factors of MCBs are high associated costs (especially due to electrode materials and the use of PDM), low energy outputs, and limited lifetimes; Currently, the main factors limiting the power output of MFC technologies are the electron transfer rate for the anode and the electrochemical properties of the material. Besides, it has been shown that PDM is the main source of internal resistance (Rin) of MFCs. In light of these limiting factors, MFCs are currently not able to achieve theoretical power outputs and therefore it is not yet possible to apply this technology to the industry [18].

The overall efficiency and performance of a VQA can be affected by a wide variety of factors. Other performances limiting factors include biofouling (leading) while trying to improve MFC's performance for industrial and social applications. Electrode surface blockage and eventually a reduction in surface area), catalyst inactivity (if present), and excessive biofilm growth - possibly leading to the production of non-conductive debris [18].

According to Sun *et al.*, it turned out that when the dominant bacterium in an MFC layout was *Geobacter anoededucens*, a bilayer biofilm developed over time with an inner dead nucleus and a living cell layer. The results show that the outer layer is responsible for the current generation and that the dead inner layer continues as an electrically conductive matrix. It can be predicted that this ongoing electrochemical activity may be due to the mechanism of electron transfer, for example, Geobacter spp. is well known for its electrochemical activity due to nanowires, even still having a viable connection to the electrode surface. Other characterization factors that can have a detrimental effect on both power outputs and the efficiency of an MFC are the crossing of electro-catalysts (if any) and organic compounds or electron acceptors from anode to cathode (or vice versa). The passage of electron acceptors from the cathodic chamber into the anode has previously been shown to disrupt biofilm formation and lead to biofilm inactivation, which can significantly reduce MFC performance due to the flow of internal currents and the generation of mixed potentials (*i.e.* a short-circuit system).

CONCLUSION

Microbial fuel cells (MFCs) are current biochemical energy conversion systems that perform simultaneous electricity generation and wastewater treatment. MFC systems can be used in all sectors where industrial wastewater is present. In this section, research on the use of MFC systems in the food industry containing complex organic waste is presented.

Food wastewater, which constitutes about half of organic wastewater, is also treated using microbiological methods. As a conventional method, the activated sludge method is widely used despite organic contamination. However, many factors restrict the use of the activated sludge method such as the large space requirement required, long waiting times, excessive sludge formation, and the considerable energy requirement of the ventilation process applied during microorganism growth.

Most of the modern MFC technologies developed so far are used as a fuel source that operates based on energy recovery through the biological decomposition of wastewater, organic-rich waste. The ability to generate electricity from wastewater can play an important role in renewable energy generation. However, for the proper treatment of wastewater, the issues surrounding the quality of MFC treated wastewater have not yet been adequately addressed. Treatment of wastewater with only MFCs may not be a viable option due to wastewater quality requirements. For this reason, MFC systems integrated into traditional treatment systems and membrane systems may be more effective.

CONSENT FOR PUBLICATION

Not applicable.

CONFLICT OF INTEREST

The author declares no conflict of interest, financial or otherwise.

ACKNOWLEDGEMENT

Declared none.

REFERENCES

[1]   Gautier C, Petrol Su ve İklim. Ankara, Tübitak Popüler Bilim Kİtapları. 2014.

[2]   Türkmenler H. Atık su arıtma tesislerinde enerji verimliliği. Politeknik dergisi 2017; 20(2): 495-502.

[3]   Demir Ö, Gümüş E. Mikrobiyal yakıt hücreleri ile çamur arıtımı ve elektrik üretimi. sinop üniversitesi fen bilimleri dergisi 2016; 1(2): 81-9. Ö. Demir, E. Gümüş, 2016. Mikrobiyal yakıt hücreleri ile çamur arıtımı ve elektrik üretimi, sinop üniversitesi fen bilimleri dergisi, 1(2):81-89.

[4]   Kılıç A, Uysal Y, Çınar O. Laboratuvar ölçekli bir mikrobiyal atıksudan elektrik üretimi. pamukkale üniversitesi mühendislik bilimleri dergisi 2011; 17(1): 43-9.

[5]   Slatea AJ, Whiteheada KA, Brownsona DAC, Banks CE. Microbial fuel cells: An overview of current technology, renewable and sustainable energy reviews 2019; 101: 60-81.

[6]   Özcan E. Mikrobiyal yakıt hücrelerinde membrane ve işletme şartlarındaki değişimin güç üretimine etkisi. Hacettepe üniversitesi, Fen bilimleri enstitüsü, Yüksek lisans tezi, Ankara 2013; 170.

[7]   Aydın İ, Çalışıyor A, Üstün Ö. Düşük güçlü sistemler için alternative enerji kaynağı: Mikrobiyal yakıt hücreleri. İstanbul Teknik Üniversitesi 2015.

[8]   Erenler AŞ, Ülke EN. Mikrobiyal yakıt hücre teknolojisini kullanarak gıda endüstrisi atıklarından elektrik enerjisi üreitim. Batman üniversites yaşam bilimleri dergisi 2018; 8(2): 22-36.

[9]   Köroğlu EO. Mikrobiyal yakıt hücrelerinde evsel atıksulardan elektrik üretimi. Yıldız Teknik üniversitesi, Fen bilimleri enstitüsü, Yüksek lisans tezi 2013; 103.

[10]   Samsunlu A. Atıksuların arıtılması, Güncelleştirilmiş baskı, İstanbul, Birsen yayınevi. 2011.

[11]   Barker S, Griffiths C, Nicklin J. Mikrobiyoloji, 4 baskı. Ankara: Nobel Akadneik Yayıncılık 2013.

[12]   Sağlam D, Şeker E. Gıda kaynaklı bakteriyel patojenler, Kocatepe. Vet Jounery 2016; 9(2): 105-13.

[13] Çetinkaya AY. Süt Endüstrisi Atıksuyunun Biyometanizasyon potansiyelinin incelenmesi. Politeknik dergisi 2018; 21(2): 457-60.

[14] Bektüre SI. Nanofiltrasyon yöntemi ile deri ve zeytinyağı atıksularından renk giderimi, Eskişehir osmangazi üniversitesi. Fen bilimleri enstitüsü, Yüksek lisans tezi 2017; 54s..

[15] Aşık BB, Katkat AV. Gıda Sanayii arıtma tesisi atık suyunun sulama suyu olarak kullanım olanağı. Uludağ üniversitesi ziarat fakültesi dergisi 2005; 19(2): 23-31.

[16] Özan K, Açıkgöz Ç. Haşlama et atıksuyunun membrane biyoreaktör sisteminde arıtımı. Bibad 2015; 8(2): 7-11.

[17] Ünal T. Ekmek mayası endüstrisi separasyon projesi atıksularından ozon ve ozon (hidrojen peroksit oksidasyonu ile renk giderimi, İstanbul Teknik Üniversitesi, Fen bilimleri enstitüsü. Yüksek lisans tezi 2011; 95a

[18] Sun G, Thygesen A, Ale MT, Mensah M, Poulsen FW, Meyer AS. The significance of the initiation process parameters and reactor design for maximizing the efficiency of microbial fuel cells. Appl Microbiol Biotechnol 2014; 98: 2415.

[19] Altunışık İ, İleri R, Artır R. Küçük ve orta ölçekli süt endüstrisi atıksularının önarıtılmasında bentonite ve sepiyolitin kullanılabilirliği. SAU Fen bilimleri dergisi 2002; 6(2): 65-76.

CHAPTER 9

Development of Algae Oil Production With Ecological Engineering

Cemil Koyunoğlu[1,2,*] and Hüseyin Karaca[3]

[1] *Energy Systems Engineering Department, Faculty of Engineering, Yalova University, Yalova, Turkey*

[2] *Fuel Oil Analysis Laboratory, Yalova University Central Research Laboratory, Yalova University, Yalova, Turkey*

[3] *Chemical Engineering Department, Faculty of Engineering, Inonu University, Malatya, Turkey*

Abstract: It is well known that the use of algae-based fuels is not widespread as they do not meet the necessary fuel standards. The amount of oil products obtained from traditional biomass sources exceeds that obtained from algae. However, ecology can provide more assistance in increasing the current efficiency of oil production from algae. Depending on the species, effective results can be obtained through the production of fatty acids and biofuels. The amount of fuel produced by a single species should be weighted against the amount of oil produced by a combination of species. The presence of genetic diversity, particularly in two types of algae mixtures, is unavoidable in order to improve the quality of fuel produced. The lipid obtained under genetically diverse population is highlighted in this study.

Keywords: Biodiesel quality, Ecological science, Fatty acid, Genetic diversity, Mixed algae.

INTRODUCTION

Humans, like all living things, require energy. As a result, humans have changed their lifestyles for years in search of energy resources, waged wars, and even changed their industries based on energy production. The most significant advantage of algae over traditional biomass sources is that they can be grown ubiquitously and it is relatively easy to separate the fatty acids produced from the biomass. The fatty acid methyl ester is a well-known product during the esterification process of fatty acids extracted from algae biomass. The different

* **Correspondence author Cemil Koyunoğlu:** Energy Systems Engineering Department, Faculty of Engineering, Yalova University, Yalova, Turkey; Tel: +90 226 8155378; E-mail: cemil.koyunoglu@yalova.edu.tr

Hüseyin Karaca and Cemil Koyunoğlu (Eds.)

types of acid compositions and fatty acids determine the quality of algae biomass. Innovative technologies, such as the hydrothermal liquefaction method, can be used to determine the conversion efficiency of wet algae biomass to crude oil [1].

The fact that the biomass with a fat-to-acid ratio is generally low in the whole mass is a significant bottleneck. The primary reason for this is that living organisms consume energy and store some products while producing fatty acids.

Using genetic engineering technologies, more than one algal strain can improve both the quality and quantity of biofuel produced. Such genetic manipulations can also reduce the increasing cost and constraints in fatty acid production.

FINDINGS

A scenario for oil production by mixed algae is depicted in Fig. (**1**) [2], showing the life cycle greenhouse gas emissions for biofuels produced from algae cultures grown with treated wastewater. Carbon emissions are due to the use of electricity when the water is pumped out of the system after the algae have grown. The high concentration of algae biomass is an important outcome of this life cycle study. In other words, mixed algae harvest more efficiently. It has been determined that an environment with minimal chemical inputs during algae cultivation without the use of electricity can only be produced under conditions such as precipitation and clumping. However, the cultivation of algae in a large precipitation tank will cost additional infrastructure. However, the carbon emission effect of this setup on the cost was not considered such as the cost of dewatering (*i.e.*, 50.2 g CO_2 eq/MJ). Drying was thought to be accomplished through centrifugation and heating. The steam of coal plants can be used for heating. The sunlight can also be used to dry the algae. However, considering the intensity of the sunlight during the time of processing, such drying should be weighed against time and space. Improving the tools such as pumps, and centrifuges has a significant impact on system cost [2]. When three different algae species were combined and supplemented with 5% $MgCl_2$, $ZnCl_2$ and NaH_2PO_3 for fuel production the oil production increased by 6.3%, 16.2%, and 0.71%, respectively [3].

The importance of algae mixtures for the production of high oil content was also emphasized in other studies. Bio-oil produced by mixing 15 different algae species can be converted into biodiesel at 60% efficiency and while 96% of nutrient removal was also achieved [4]. In addition, another study showed the increased yield of methane by mixed algae in household waste using biogas technology [5]. A study using CO_2 conversion of *Chlorella vulgaris* and *Scenedesmus obliquus* species to be 14.9% and 13.85% when CO_2 feeding rates were 140.91 and 129.82 mg/L per day, respectively. The same study found that the mixture of *Scenedesmus obliquus* and *Chlorella pyrenoidosa* has the greatest potential (biomass production up to 20%, CO_2 conversion at 10%) for biomass production and CO_2 conversion compared to other species. In conclusion, it is predicted that liquefaction is an ideal technology for the production of fuel from algae [6]. The rate of lipid removal was shown to increase ten-fold at a temperature range of 80–140 °C for 12 minutes using the hydrothermal wave process and mixed algae cultures combined with an alternative electroflotation method produced the best algae-growing environment [7].

Effluent cultivated algae / whole cell pyrolysis

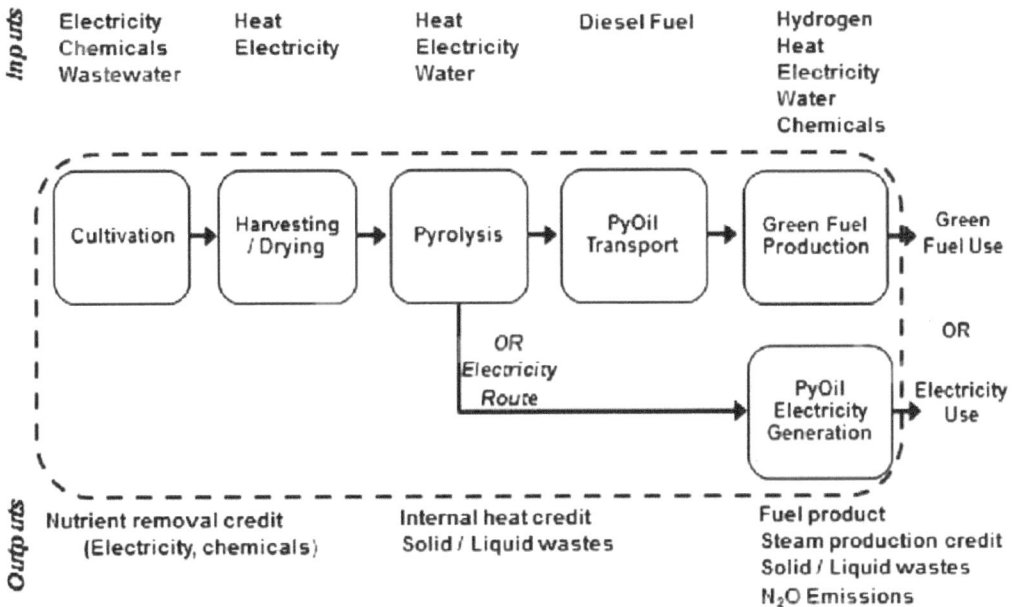

Inputs				
Electricity	Heat	Heat	Diesel Fuel	Hydrogen
Chemicals	Electricity	Electricity		Heat
Wastewater		Water		Electricity
				Water
				Chemicals

```
Cultivation → Harvesting → Pyrolysis → PyOil → Green Fuel → Green
              / Drying                  Transport   Production    Fuel Use

                                   OR                              OR
                              Electricity
                                 Route              PyOil        Electricity
                                              →  Electricity   →    Use
                                                 Generation
```

Outputs		
Nutrient removal credit	Internal heat credit	Fuel product
(Electricity, chemicals)	Solid / Liquid wastes	Steam production credit
		Solid / Liquid wastes
		N_2O Emissions

Fig. (1). The stages envisioned during an algae fuel production [2].

CONCLUSION

The evaluation of algae-based biofuel in terms of environmental engineering is summarized below.

1. C16-C18 methyl esters of algal lipid are easily and quickly decomposed and dissolved.
2. The biofuel is biodegradable and non-toxic as 99.6% of the biodiesel derived from canola was decomposed in about 21 days.
3. The biofuel is environmentally friendly and is a renewable energy source, and it can be produced using local resources.
4. Unlike diesel, biofuel does not contribute to the accumulation of CO_2 in the

atmosphere and, as a result, it does not contribute to the greenhouse gas effect. Because the CO_2 produced by the combustion of biodiesel is re-used by the plants that produce biodiesel.

CONSENT FOR PUBLICATION

Not applicable.

CONFLICT OF INTEREST

The authors declare no conflict of interest, financial or otherwise.

ACKNOWLEDGEMENT

Declared none.

REFERENCES

[1]   Jackrel SL, Narwani A, Bentlage B, *et al.* Ecological engineering helps maximize function in Algal oil production. Appl Environ Microbiol 2018; 84(15): e00953-18.
       [http://dx.doi.org/10.1128/AEM.00953-18] [PMID: 29776927]

[2]   Handler RM, Shonnard DR, Kalnes TN, Lupton FS. Life cycle assessment of algal biofuels: Influence of feedstock cultivation systems and conversion platforms. Algal Res 2014; 4: 105-15.
       [http://dx.doi.org/10.1016/j.algal.2013.12.001]

[3]   Vo Hoang Nhat P, Ngo HH, Guo WS, *et al.* Can algae-based technologies be an affordable green process for biofuel production and wastewater remediation? Bioresour Technol 2018; 256: 491-501.
       [http://dx.doi.org/10.1016/j.biortech.2018.02.031] [PMID: 29472123]

[4]   Kligerman DC, Bouwer EJ. Prospects for biodiesel production from algae-based wastewater treatment in Brazil: A review. Renew Sustain Energy Rev 2015; 52: 1834-46.
       [http://dx.doi.org/10.1016/j.rser.2015.08.030]

[5]   Manninen K, Huttunen S, Seppälä J, Laitinen J, Spilling K. Resource recycling with algal cultivation: Environmental and social perspectives. J Clean Prod 2016; 134: 495-505.
       [http://dx.doi.org/10.1016/j.jclepro.2015.10.097]

[6]   Arun J, Gopinath KP, Sivaramakrishnan R, SundarRajan P, Malolan R, Pugazhendhi A. Technical insights into the production of green fuel from CO_2 sequestered algal biomass: A conceptual review on green energy. Sci Total Environ 2021; 755(Pt 2): 142636.
       [http://dx.doi.org/10.1016/j.scitotenv.2020.142636] [PMID: 33065504]

[7]   Nautiyal P, Subramanian KA, Dastidar MG. Recent advancements in the production of biodiesel from Algae: A Review. Reference Module in Earth Systems and Environmental Sciences 2014.
       [http://dx.doi.org/10.1016/B978-0-12-409548-9.09380-5]

CHAPTER 10

Microalgae Culture

Oya Işık[1,*] and **Leyla Uslu**[1]

[1] *Marine Biology Department, Fisheries Faculty, Cukurova University, 01330 Adana, Turkey*

Abstract: Energy is becoming one of the most expensive production inputs nowadays. Energy reserves are starting to run out and their polluting effects have been seen around the world. Therefore, there is an urgent need for renewable energy sources instead of fossil fuels. One of these energy sources is algae biomass, which is seen as promising for biofuel production.

Keywords: Algae Biomass, Biodiesel, Lipid Production, Microalgal Biotechnology.

INTRODUCTION

Algae are aquatic, using solar energy, oxygen-producing photosynthetic autotrophs that are unicellular, colonial, or are constructed of filaments or composed of simple tissues [1]. Microalgae contain valuable metabolites which can be utilized in biofuels, health supplements, pharmaceuticals, and cosmetics [2]. They also have applications in wastewater treatment and atmospheric CO_2 mitigation. Microalgae produce a wide range of bioproducts, including polysaccharides, lipids, pigments, proteins, vitamins, and bioactive compounds [3]. The interest in microalgae as a renewable and sustainable feedstock for biofuel production has inspired a new focus in biorefinery. Growth enhancement techniques and genetic engineering may be used to improve their potential as a future source of renewable bioproducts. First, the cultivation of microalgae does not need much land as compared to that of *terraneous* plants [4] and biodiesel produced from microalgae will not compromise the production of food and other products derived from crops. Second, microalgae grow extremely rapidly and many algal species are rich in oils [5]. Today, studies conducted in many countries to increase the production of microalgal oil are aimed at obtaining high amounts of oil with biomass efficiency, and success is being achieved in cost

[*] **Correspondence author Oya Işık:** Marine Biology Department, Fisheries Faculty, Cukurova University, 01330 Adana, Turkey; Tel: +90 5322617942; Fax: +90 3223386439; E-mail: oyaisik@cu.edu.tr

Hüseyin Karaca and Cemil Koyunoğlu (Eds.)

analysis studies. In recent years, the use of microalgal biomass to obtain biofuel has become increasingly important [6]. The properties of biodiesel obtained from microalgae seem to be similar to those obtained from oilseed crops and fossil fuels. However, biodiesel from microalgae is considered to be an environmentally friendly and sustainable fuel for the future [7]. Unfortunately, the cost associated with producing biodiesel from microalgae is relatively high; therefore, the question rises as to why are microalgae a source of biomass? Microalgae are considered to be the most efficient biological system using solar energy and inorganic compounds for the production of organic compounds *via* photosynthesis. Algae are plants that have a simple reproductive system, do not have vascular bundles, and the entire biomass can be harvested and utilized. **a.** Algae are considered to be a very efficient biological system for harvesting solar energy for the production of organic compounds *via* the photosynthetic process. **b.** Algae are non-vascular plants, lacking (usually) complex reproductive organs, making the entire biomass available for harvest and use. **c.** Many species of algae can be cultivated to produce commercially- valuable compounds, such as proteins, carbohydrates, lipids, and pigments [8]. d. Algae are microorganisms having a simple cell division cycle, enabling them to complete their cell cycle within a few hours and can be made a genetic selection and strain screening relatively quick and easy. They also undergo much rapid growth and production processes than other crops. **e.** Infertile soils where land cultivation cannot be done can be utilized for microalgae production using seawater or brackish water. **f.** With microalgae production in shallow ponds, protein, and other metabolites can be obtained without concreting the soil, and algae production can be carried out in different models, from simple systems to fully automatic systems, depending on the degree of use of technology [9]. Pawar [10] reports that autotrophic cultivation of microalgae can offer a reliable solution for future energy security in addition to sequestration of carbon dioxide (CO_2). However, the economics of the biodiesel production process mainly depends on the capital as well as operating costs of the photobioreactors (PBRs) and dewatering units which are the major hurdle in the successful implementation of the algal biofuel mission. The laboratory studies have revealed important aspects of microalgae growth kinetics under various conditions and the data can be used as a basis to design an optimal largescale process. The growth of microalgae in the PBR depends on various factors, and the complete process is complex and needs complete monitoring of pH, gas sparging time, dissolved oxygen (O_2) level, biomass concentration, the temperature of the culture media, and so on. The role of nutrient media (composition and micronutrients) also plays a crucial role in determining maximum growth. Microalgae production in ponds, and open system, has both advantages and disadvantages compared to production in closed systems, tubular or panel photo-bioreactors (PBR).

CULTIVATION OF ALGAE IN PONDS

Constructing ponds for microalgae production can be considered cheaper than building closed systems; production costs may vary according to countries and raw material. It is important that the inner surface of the pond is smooth and easy to clean, and that there are no areas where there is no aeration in the pools. Another important issue is that there is no toxic substance released from the inner surface of the pond to the water.

The biggest advantage of these open ponds (Fig. **1**) is their simplicity, resulting in low production costs and low operating costs [11]. With paddlewheels providing the flow, algae are kept suspended in the water and are circulated back to the surface on a regular frequency. The ponds are usually kept shallow because the algae need to be exposed to sunlight, and sunlight can only penetrate the pond water to a limited depth [12].

Fig. (1). Algae ponds in Cukurova University.

3. PHOTOBIOREACTORS

Photobioreactors are closed and light-receiving production systems for microalgae. Photobioreactors have the advantage that cultures are free from contaminants and there is no evaporation loss. It can be in different designs depending on the criteria of the most efficient use of light, taking up less area, and being easy to clean (Figs. **2-5**).

Fig. (2). Tubular photobioreactor in Cukurova University.

Fig. (3). Tubular photobioreactor in Cukurova University.

Fig. (4). Tubular photobioreactor in Cukurova University.

Fig. (5). Algae stock cultures in the laboratory.

Although the construction costs of photobioreactors are high, pH, temperature, light and CO_2 concentration, *etc.*, conditions can be better controlled. CO_2 loss in photobioreactors is minimal. Microalgae cell concentration is higher in photobioreactors. The production of biopharmaceuticals in photobioreactors is successful.

In photobioreactors, the culture is fully illuminated [12]. Different types of photobioreactors have been designed and developed for the production of algae.

There are physical and chemical conditions that must be met in microalgal production. In addition to the appropriate temperature and illumination, a nutrient medium is required, which can vary depending on the species of algae. The better the nutrients are dissolved in the water, the higher the algae cells benefit from these nutrient elements.

To prevent the precipitation of microalgae cells and to ensure that algae cells benefit from nutrients and light equally, the cultures should be mixed/aerated. In the laboratory environment, cultures with volumes usually greater than 2 liters are mixed/aerated with the help of a compressor. Mixing in raceways and ponds is commonly done with a pedal system. Mixing speed is important in raceways ponds. The optimum mixing speed is set as 20-30 cm sec^{-1} (Fig. **6**).

Fig. (6). No disease occurred in plants treated with *Spirulina platensis* (The bottom one of the three photos above).

In the Faculty of Fisheries, Çukurova University, stock cultures are maintained in the laboratory, while there is an outdoor unit with 5 ellipsoid ponds of which dimensions are $5.0 \times 1.0 \times 0.30$ m^3. The ponds are usually kept shallow because the algae need to be exposed to sunlight, and sunlight can only penetrate the pond water to a limited depth. The algae culture ponds are in the greenhouse. There are also tubular and panel photobioreactors in the Algal biotechnology unit outdoor.

The volume of the tubular photobioreactor used in our studies is approximately 300 liters with 150 liters' collection tank. The tubular photobioreactor was horizontally installed and made of transparent acrylic tubes of 2.6 cm in diameter. The flow rate of the cultures is adjusted to 0.3 m s^{-1}. by a circulation pump. The CO_2 gas inlet is provided by the flowmeter. pH and temperature were measured continuously by probes. Outdoor flat-panel PBRs were 10 mm thick transparent glass material, 50,0 cm wide and 50, 0 high with 1, 3, 5, 7, and 10 cm light paths in batch systems. The volumes of the PBR were 2 L in 1 cm, 6 L in 3 cm, 10 L in 5 cm, 15 L in 7 cm, and 21 L in 10 cm. The culture mixture was provided with 2% CO_2 enriched air [13]. Continuous measurements are made by placing pH and temperature probes. Nutrient addition is made with a peristaltic pump.

The volume of stock algae cultures (Fig. **7**) is increased in the laboratory to be used as inoculation before starting production in ponds or PBRs. The inoculation rates can be applied at around 20% or 30% under optimum environmental conditions, depending on the season, the type of algae, and the growth stage. The fact that the starter cultures are in the logarithmic phase ensures a short lag phase. Care should be taken to avoid photoinhibition on the first day of inoculation due to the low cell density and shining sunlight in the culture medium [14].

Sterilization is an extremely important issue in the preservation of stock cultures in a laboratory environment. An autoclave is used for the sterilization of liquids. In the sterilization of glass and metal materials, an oven or dry heat sterilizer is used. Microalgae inoculation in solid and liquid nutrients should be done in a sterile cabinet and bunsen flame.

Fig. (7). Bread enriched with (top) and without (bottom) *Spirulina*.

In recent years, studies are carried out to reduce the production costs in many countries for the transition to industrial-scale applications.

In order to reduce the cost of nutrient input in *Spirulina* cultures, a study on the effect of different proportions of nutrients on dry matter, protein, and chlorophyll-*a* levels was conducted. In the study carried out intending to obtain high biomass productivity to not adversely affect the food chemistry of *Spirulina* and create a different chemical composition of nutrients to reduce the production cost, carbonate 25, 50, 75, 100%, carbonate and microelements, part I and part II of growth media, part I, part II and microelements 25, 50, 75% were applied missing. The Control group closest to the dry matter content 2.36 g L^{-1} with the carbonate and microelements 50% missing in the applied group was detected. In our studies, we have tried to prefer species that are easy to culture, do not cost too much, and whose culture has been proven in the outdoor environment before [15].

One of the valuable metabolites of microalgae cells is their pigments. Microalgal pigments are extensively used in various industries, including the food, nutraceutical, pharmaceutical, aquaculture, and cosmetic industry. Also, it has been used in clinical/research laboratories, which are effective as the label for antibodies and receptors [16]. Antioxidants, anti-inflammatory, neuroprotective, and hepatoprotective properties are also exhibited by phycobiliproteins [16].

The study was carried out to determine the effects of light intensity, temperature, nitrogen deficiency, and aeration on the vegetative growth and form of the cysts of the green alga *Haematococcus pluvialis* in laboratory conditions, vegetative growth of the cultures were found similar at the light intensities of 27 and 48 μmol photon $m^{-2}s^{-1}$ and % 2-10 inoculation amounts. *H. pluvialis* cultures were exposed to the 177 and 379 μmol photon $m^{-2}s^{-1}$ light intensities and N-deficient conditions to form a cyst and accumulate astaxanthin. While 27 and 48 μmol photon $m^{-2}s^{-1}$ light intensities caused the accumulation of astaxanthin at the low cell concentrations, 177 μmol photon $m^{-2}s^{-1}$ irradiance was insufficient to turn red the cells in the nitrogen-deficient medium and high cell densities. The highest astaxanthin content of % 3.605 was obtained from the cultures aerated, at the irradiance of 379 μmol photon $m^{-2}s^{-1}$ and 24°C. In the study, it was observed that 35°C temperature and 177 μmol photon $m^{-2}s^{-1}$ irradiance caused the death of the cells [17].

Growth and production of carotenoid in three *Dunaliella* species (*Dunaliella salina* (Dunal) Teodoresco, *Dunaliella bardawil* Ben-Amotz & Avron, and *Dunaliella* sp.) were investigated using flat-plate photobioreactors in outdoor conditions with two optical paths (3 cm and 5 cm). The experiment was conducted in duplicate and lasted four weeks during which light intensity, temperature, pH, and optical density were checked daily. The pigment production (total carotenoid and chlorophyll *a*) was monitored every two days. To induce additional stress besides temperature and light intensity, two different salt concentrations were used, *i.e.* 6% and 8% NaCl. The highest growth in all treatment groups was noticed for *Dunaliella* sp. followed by *D. bardawil* and *D. salina*. *D. salina* produced a higher content of carotenoid concentrations corresponding to 5 cm/8% and 5 cm/6% groups; 779.102 ± 0.434 μg.mL^{-1} and 694.326 ± 0.098 μg.mL^{-1} were registered at the end of the experiment. The same species had also greater content of β-carotene [18].

Algae can be used as a soil fertilizer. In a study we conducted with the Faculty of Agriculture, biological control of tomato bacterial wilt disease by *Spirulina platensis* was investigated. In the study, the effect of the potential use of *Spirulina platensis* as a biological control agent against bacterial wilt disease of tomato caused by the bacterium *Clavibacter michiganensis subsp. michiganensis*. A preliminary study was conducted by using *Spirulina* sp. purchased from the market and two individual experiments were carried out in greenhouse conditions with freshly harvested *Spirulina platensis* in the Faculty of Fisheries. The effect of *Spirulina platensis* on disease development, plant's dry weight, and total chlorophyll content of leaves was investigated. In the preliminary trial, the soil treated with *Spirulina* sp. suppressed the bacterial wilts of tomato caused by *Clavibacter michiganensis subsp. michiganensis* by 28.12%. In the first and second experiments with *Spirulina platensis*, the disease incidence was inhibited by 88% and 81%, respectively. In Spirulina platensis treated soil, while the dry weight of diseased plants advanced at the ratios of 41%, 88%, and 68%, the amount of the dry matter in healthy plants was increased by 26%, 20%, and 21%. The chlorophyll content in diseased and healthy plants was raised by 20% and 23%, respectively (Fig. **6**). Bio fertilization of the soil by *Spirulina platensis* has increased the plant dry weight, chlorophyll content of the leaves, and suppressed the disease development. This treatment can be considered a new approach to bacterial wilt disease management in traditional and organic tomato farming. *Spirulina* applications with the minimized cost of production studies and applications of greenhouse or field trials will be widespread [19].

In another study, the effect of *Spirulina platensis* on the growth of Begonia (*B. semperflorens*) plant was investigated, and the leaf size was greater in the groups containing *Spirulina* [20].

Algae are also used as food additives. In a study, we conducted on this subject. The study aimed to increase the nutrient content of bread prepared with white flour, using the valuable metabolites included in *Spirulina platensis* (Fig. **7**). In this study, conventional pieces of bread were added with 10% of Spirulina. The nutrient composition, protein, and lipid content were evaluated and microbiological and sensory analyses were conducted on the pieces of bread with microalgal biomass. The addition of microalgal biomass resulted in a protein content increase, ranging from 7.40% to 11.63%. While the Calcium, Magnesium, and Iron contents of bread with *S. platensis* were 721.2, 336.6, 41.12 ppm, conventional bread contained 261.7 ppm Calcium, 196 ppm Magnesium, and 8.72 ppm Iron. Adding *Spirulina* showed a positive effect on the volatile compounds of bread. The results for the sensory assessment of bread enriched with *Spirulina* were considered satisfactory even if some algae flavor in the samples was perceived. In addition, *Spirulina* inhibited mold growth in bread stored at room conditions and was observed to have a positive effect on the inhibition of mold growth. These results show that the use of microalgae can enhance the nutritional quality of bread without a negative impact on its shelf life [21].

In another study, pasta is produced from semolina with the addition of *Spirulina platensis* at three different levels (5, 10, and 15% w/w) to enhance the sensorial and nutritional quality (Fig. **8**). Pasta samples were evaluated for cooking quality (weight increase, cooking loss, volume increase), microbiological quality (total mold and yeast count), color (*L, a, b*), and sensory characteristics. The cooking loss in all pasta samples was similar and below the technologically acceptable limit (<8%). According to color measurements and evaluations of panelists, adding *Spirulina* in pasta provided an appealing green tone. Also, mold and yeast were not detected in the control and pasta enriched with *Spirulina*. Sensory evaluation indicated that pasta enriched with 10% *S. platensis* has a good overall score the same as the control group. In this study, the investigation results of pasta enriched with *Spirulina* indicated satisfactory technological attributes [22].

Fig. (8). Pasta enriched with (left) and without (right) *Spirulina.*

CONCLUSION

We know that three-quarters of the world is covered with oceans and seas, so human beings should be able to make more use of the seas. Due to its valuable metabolites, the advantages of microalgae culture over land plants should be investigated for more economical microalgae production. Algal biomass is valuable biomass for oil production, so the cost of microalgal biomass production should be reduced. The main input that increases the production cost is the nutrients. One of the main elements of the nutrient medium is carbon dioxide. The use of CO_2 in industrial flue gases should be implemented. Additionally, nitrogen, phosphorus, and carbon, which are essential elements of the nutrient medium, efforts should be made to produce micro and trace elements at a lower cost.

CONSENT FOR PUBLICATION

Not applicable.

CONFLICT OF INTEREST

The authors declare no conflict of interest, financial or otherwise.

ACKNOWLEDGEMENT

Declared none.

REFERENCES

[1]   Guiry MD. How many species of algae are there? J Phycol 2012; 48(5): 1057-63.
        [http://dx.doi.org/10.1111/j.1529-8817.2012.01222.x] [PMID: 27011267]

[2]   Das P, Aziz SS, Obbard JP. Two phase microalgae growth in the open system for enhanced lipid
        productivity. Renew Energy 2011; 36(9): 2524-8.
        [http://dx.doi.org/10.1016/j.renene.2011.02.002]

[3]   Brennan L, Owende P. Biofuels from microalgae—a review of technologies for production,
        processing, and extractions of biofuels and co-products. Renew Sustain Energy Rev 2010; 14(2): 557-
        77.
        [http://dx.doi.org/10.1016/j.rser.2009.10.009]

[4]   Chisti Y. Biodiesel from microalgae. Biotechnol Adv 2007; 25(3): 294-306.
        [http://dx.doi.org/10.1016/j.biotechadv.2007.02.001] [PMID: 17350212]

[5]   Huang G, Chen F, Wei D, Zhang X, Chen G. Biodiesel production by microalgal biotechnology. Appl
        Energy 2010; 87(1): 38-46.
        [http://dx.doi.org/10.1016/j.apenergy.2009.06.016]

[6]   Eloka-Eboka AC, Inambao FL. Effects of CO_2 sequestration on lipid and biomass productivity in
        microalgal biomass production. Appl Energy 2017; 195: 1100-11.
        [http://dx.doi.org/10.1016/j.apenergy.2017.03.071]

[7]   Sohi SMH, Eghdami A. Biodiesel production using marine microalgae *Dunaliella salina*. J Biodivers
        Environ Sci 2014; 4: 177-82.

[8]   Cohen Z, Vonshak A, Richmond A. Effect of environmental conditions on the fatty acid composition
        of the red alga *Porphyridium cruentum*: Correlation to growth rate. J Phycol 1988; 24(3): 328-32.
        [http://dx.doi.org/10.1111/j.1529-8817.1988.tb04474.x]

[9]   Vonshak A. Recent advances in microalgal biotechnology. Biotechnol Adv 1990; 8(4): 709-27.
        [http://dx.doi.org/10.1016/0734-9750(90)91993-Q] [PMID: 14543692]

[10]   Pawar S. Effectiveness mapping of open raceway pond and tubular photobioreactors for sustainable
        production of microalgae biofuel. Renew Sustain Energy Rev 2016; 62: 640-53.
        [http://dx.doi.org/10.1016/j.rser.2016.04.074]

[11]   Ugwu CU, Aoyagi H, Uchiyama H. Photobioreactors for mass cultivation of algae. Bioresour Technol
        2008; 99(10): 4021-8.
        [http://dx.doi.org/10.1016/j.biortech.2007.01.046] [PMID: 17379512]

[12]   Singh RN, Sharma S. Development of suitable photobioreactor for algae production-a review. Renew
        Sustain Energy Rev 2012; 16(4): 2347-53.
        [http://dx.doi.org/10.1016/j.rser.2012.01.026]

[13] Qiang H, Guterman H, Richmond A. Physiological characteristics of *Spirulina platensis* (Cyanobacteria) cultured at ultrahigh cell densities. J Phycol 1996; 32(6): 1066-73.
[http://dx.doi.org/10.1111/j.0022-3646.1996.01066.x]

[14] Richmond A, Ed. Handbook of microalgal mass culture. CRC press 1986.

[15] Kendirli K. Spirulina kültürlerinde besin elementlerinin farklı oranlarda kullanımının kuru madde, protein ve klorofil-a düzeyine etkisi. Adana: Yüksek Lisans Tezi 2010.

[16] Begum H, Yusoff FMD, Banerjee S, Khatoon H, Shariff M. Availability and utilization of pigments from microalgae. Crit Rev Food Sci Nutr 2016; 56(13): 2209-22.
[http://dx.doi.org/10.1080/10408398.2013.764841] [PMID: 25674822]

[17] Köksal M, Isik O, Uslu L, Mutlu Y. Isik, sicaklik, besin eksikligi ve havalandirmanin *Haematococcus pluvialis* Flotow'da büyüme ve astaksantin miktarina etkisi. J FisheriesSciences Com 2012; 6(4): 297.

[18] Hamed I, Işık O, Ak Çimen B, Uslu L, Kafkas E, Zarifikhosroshahi M. Influence of stress factors on growth and pigment production in three *Dunaliella* species cultivated outdoors in flat-plate photobioreactors. Plant Biosystems-An International Journal Dealing with all Aspects of Plant Biology 2020; 1-9.

[19] Yiğenoglu CY, Işık O, Aysan Y, Ak B, Uslu L. Biological control of tomato bacterial wilt and canker disease by *Spirulina platensis*. IOBC-WPRS, Working Group "Integrated Control in Protected Crops, Temperate Climate.

[20] Yalçin Mendi NY, Torun AA, Uslu L, Çürük P, Işik O. The usage of spirulina on acclimatization of *in vitro* begonia (*B. semperflorens*) plantlets. Prooceedings of the 23rd International Eucarpia Symposium Section Ornamentals, Colorful Breeding and Genetics, Acta Horticulturae, Leiden, Netherlands.

[21] Ak B, Avsaroglu E, Isik O, *et al*. Nutritional and physicochemical characteristics of bread enriched with microalgae *Spirulina platensis*. Int J Eng Res Appl 2016; 6: 30-8.

[22] Özyurt G, Uslu L, Yuvka I, *et al*. Evaluation of the cooking quality characteristics of pasta enriched with *Spirulina platensis*. J Food Qual 2015; 38(4): 268-72.
[http://dx.doi.org/10.1111/jfq.12142]

CHAPTER 11

Different Species of Algae

Cemil Koyunoğlu[1,*] and **Hüseyin Karaca**[2]

[1] *Energy Systems Engineering Department, Engineering Faculty, Yalova University, Yalova, Turkey*

[2] *Chemical Engineering Department, Engineering Faculty, Inonu University, Malatya, Turkey*

Abstract: The oil yield of many microalgae species varies. Therefore, the selection of algae species is important. Features that are effective in selecting the appropriate species for algae production are growth and productivity, minimum contamination, and easy harvesting. However, the percentages of carbohydrate, fat, and protein in the structure of algae provide preliminary information about the oil yield to be obtained from that algae species.

Keywords: Algae types, Biofuel production, Macro-algae.

INTRODUCTION

In 1969, Whittaker classified blue-green algae into 5 different kingdoms, which is applicable till date. Algae are from the monera kingdom. Cyanophyta species belong to the Eubacteria kingdom in the Procaryota group in the triple classification system. Cyanophyceae species are morphologically divided into four orders. These are:

1: Chroococcales

2: Oscillatoriales

3: Nostocales

4: Stigonematales.

In the following sections, information about the parameters to be considered in determining algae species and the general classification of algae species are given.

---

* **Correspondence author Cemil Koyunoğlu:** Energy Systems Engineering Department, Engineering Faculty, Yalova University, Yalova, Turkey; Tel: +902268155378; E-mail: cemil.koyunoglu@yalova.edu.tr

Hüseyin Karaca and Cemil Koyunoğlu (Eds.)

KEY FEATURES FOR DETERMINING ALG SPECIES

Generally, when determining algae species, physical determinations are made of the places where they are collected. These are the sea, ocean, river, *etc.*, along with the properties of water. Major parameters include pH, temperature, electrical conductivity, light transmittance, dissolved oxygen, and salinity.

Temperature

Different species live at different temperature values. Temperature, seasons, geographic location, water area, depth of water affect the absorbed solar energy and the amount of molten material.

pH

It is vital in terms of the pH value in aquatic environments, which is between 6-10. The variety and density of blue-green algae increase at high pH (low alkaline environment). However, exceptionally, some species live in 4-4.5 pH. So, this value can be taken as a lower limit for blue-green algae. Living in low pH values, blue-green algae can be used as a competitive agent where nitrogen fixation is possible.

Dissolved Oxygen (mg/l)

The dissolved oxygen value in the water, which is directly related to the living creatures in the water, determines the quality of the water and the value of the amount of oxygen used by living things. The amount of oxygen dissolved in water is determined by salinity, water temperature, and biological factors occurring in water.

Electrical Conductivity (µS/cm)

1 cm^3 of the solution at 25 °C is the opposite of its resistance in ohms and its value is determined by solids dissolved in water.

Salinity (%)

Salinity is the amount of salt content dissolved in water. The total density of water content can also be expressed as salinity. pH and density change the salinity.

ALGAE SPECIES

Chroococcus turgidus

Cells can be easily distinguished by the large dark granules they contain (Fig. **1**).

They are found as single cells, or often in groups of 2 to 4 cells, rarely 8 cells. Cells are round or ellipsoid-shaped. The unsheathed length is 8–32 µm, and the sheathed is 13-25 µm. They live planktonically. They float freely in the body of water. Their gelatinous or mucous matrix sheath is colorless. Their colors are blue-green, yellowish, or olive green [1 - 87].

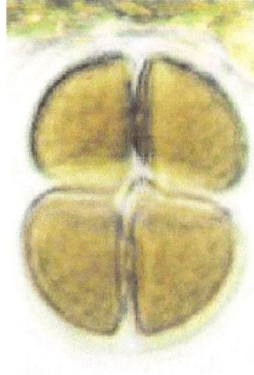

Fig. (1). *Chroococcus turgidus* (scaled with 10 µm) [88].

There are fuel recovery studies on this species [88, 89].

Chroococcus limneticus

It is generally in the form of a plate. Cells are free-floating and can be found in groups of 4-32. Cells are blue-green, olive green, or yellowish. It is 6-12 µm without sheath, 8-14 µm in size if it is sheathed, and is colorless (Fig. **2**) [9, 20, 21, 30, 77, 82, 90 - 131]. There are fuel recovery studies on this species [93].

Fig. (2). *Chroococcus limneticus* (scaled with 10 µm) [132].

Chroococcus minimus

There may be an unspecified envelope around each cell and is not lamellate. The shape of the cells is either spherical or hemispherical after division, and the cells have a homogeneous cell content without aerotops (Fig. **3**). Cells are pale blue-green. It is 1.6-3 µm wide [3, 30, 101, 113, 133 - 138].

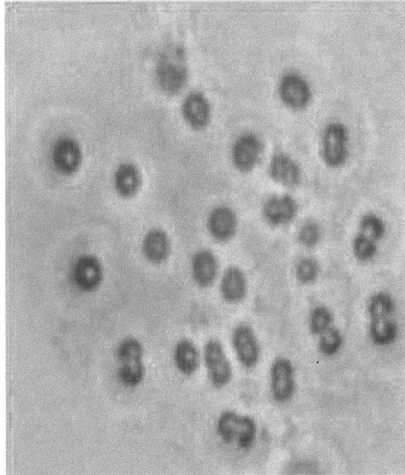

Fig. (3). *Chroococcus minimus* (scaled with 10 µm) [139].

Chroococcus minör

They are cells that are irregularly and more or less clustered and sometimes in groups of 2-4 cells. Colonies are microscopic, irregular, gelatinous, rarely large, misshapen, dirty blue-green or olive-green colors. Mucilage is delicate, broad, flowing, colorless, non-lamellate, and sometimes barely visible. Cells are spherical or hemispherical, irregular during division, 3-4 µm in width, pale blue-green in color [5, 11, 17, 28, 31, 32, 46, 60, 61, 90, 91, 140 - 177] (Fig. **4**). There are fuel recovery studies on this species [5, 141].

Microcystis aeruginosa

Cells are ellipsoid and generally 3-4 (7) µm in size, more or less tightly located, mostly gas vacuoles; it is slightly reddish. Young colony spherical or tall and hardy; the old colony may be networked and separated from each other. Its distribution area is quite wide (Fig. **5**). There are fuel recovery studies on this species [179 - 262].

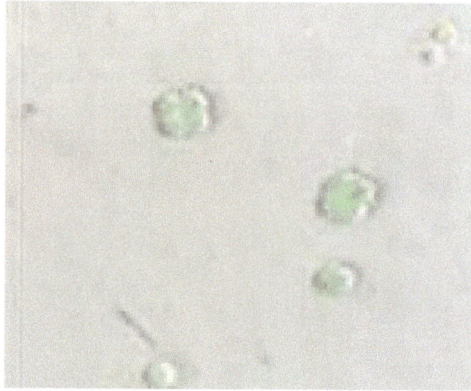

Fig. (4). *Chroococcus minör* (scaled with 10 μm) [178].

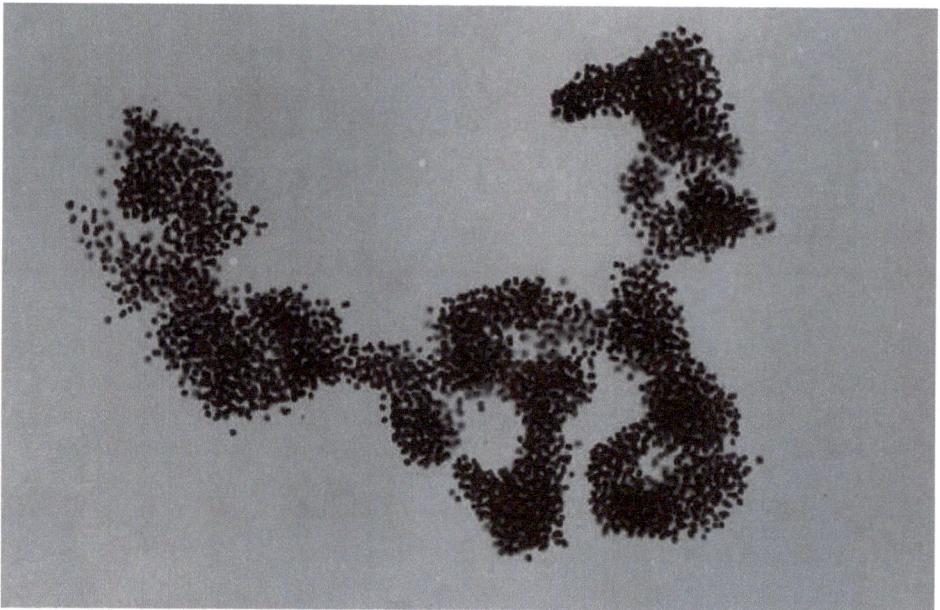

Fig. (5). *Microcystis aeruginosa* (scaled with 10 μm) [263].

Microcystis flos-aquae

Cells can be spherical, 3.5- 4.6 (5.6) μm wide. The mucilage cover is thin, transparent, colorless, and fluid. Mucilage does not exceed the boundary of envelope cell clusters (or at most 1 μm) around colonies of tightly arranged, more or less spherical or irregular colony border, non-lobule, densely clustered cells. Colonies are microscopic, rarely macroscopic (Fig. 6) [1, 115, 143, 200, 231, 264 - 333]. There are fuel recovery studies on this species [292, 295, 334].

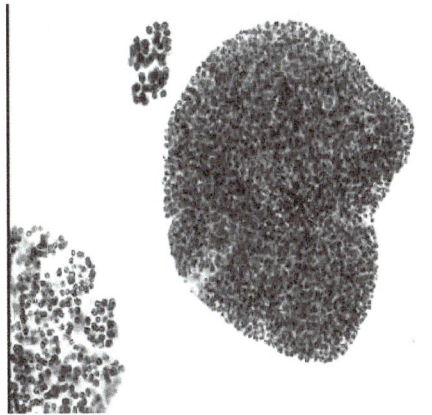

Fig. (6). *Microcystis flos-aquae* (scaled with 10 μm [335].

Gomphosphaeria aponina

Without aerospace (7.4) 8 - 12 x (3.7) 4-6.5 (cordiform cells -9) μm. The diameter of the colony is 50-90 μm. Sometimes, it is 95 μm wide. It is irregularly shaped, spherical, or elliptical. Cells surrounded by a colorless, thick mucilage sheath contain mucilage stems. They are found lined up and united. Cells are obovoid or rarely club-shaped. Cells that are usually light blue-green and rarely yellowish or dark olive green in color contain granules. When dividing, they take the form of a cordiform. It is 4-7.5 μm wide and 8-15 μm long. Colonies are microscopic (Fig. 7) [21, 31, 61, 78, 85, 127, 129, 336 - 362]. There are fuel recovery studies on this species [339, 353].

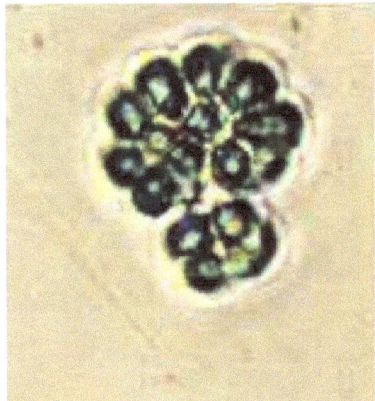

Fig. (7). *Gomphosphaeria aponina* (scaled with 10 μm) [363].

Merismopedia Tenuissima

It is 1.3-2 μm in size, colorless or blue-green in color, and surrounded by a special sheath. Colonies are rectangular, free-floating, usually 16-100 cells, cells tightly together. The colony, consisting of 32 cells, has 4 corners, it bends when the colony reaches a certain size. It is hemispherical during cleavage. Cell contents are homogeneously distributed. Cells can range from spherical shape to oval (Fig. **8**) [2, 32, 105, 106, 113, 119, 133, 135, 151, 152, 168, 246, 323, 330, 364 - 424]. There are fuel recovery studies on this species [396, 397].

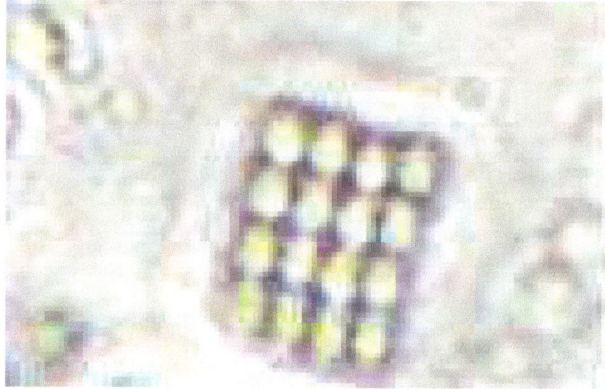

Fig. (8). *Merismopedia tenuissima* (scaled with 10 μm) [425].

Merismopedia minima

Cells are 0.4-0.8 (1.2) μm wide, spherical, and greenish or a light blue-green in color. Colonies are small, quadrat, regular shape with 4-32 cells (Fig. **9**) [24, 59, 91, 137, 172, 378, 422, 426, 427].

Fig. (9). *Merismopedia minima* (scaled with 10 μm) [434].

Merismopedia punctata

The mucilage is 2.5-3.6 μm wide without sheath and is light blue-green in color. Colonies are flat-shaped, with 64 or more (albeit less frequently) cells. Cells can be spherical, oval, or hemispherical. The mucilage sheath is distinct and colorless [428, 433] (Fig. **10**).

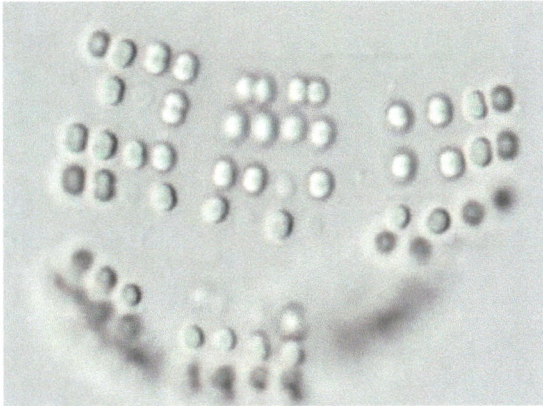

Fig. (10). *Merismopedia minima* (scaled with 10 μm) [435].

Cell contents are homogeneous, pale blue green in color. 3-6 μm wide, ellipsoid, spherical or hemispherical. Cells are usually arranged in a dense and orderly fashion. Colonies are mostly small, the number of cells can be up to 64 (Fig. **11**) [1, 12, 17, 20, 21, 50, 76, 90, 106, 117, 127, 129, 143, 150, 151, 160, 166, 177, 284, 323, 349, 369, 379, 386, 391, 395, 396, 400, 407, 416, 436 - 468]. There are fuel recovery studies on this species [396, 451, 467].

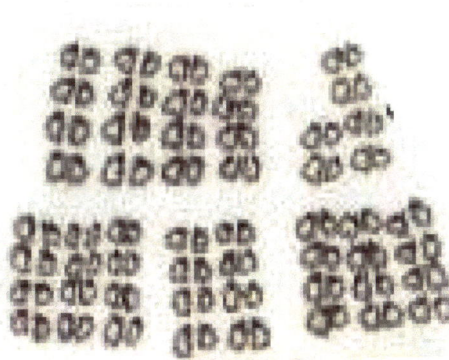

Fig. (11). *Merismopedia glauca* (scaled with 10 μm) [434].

Aphanocapsa incerta

The mucilage sheath is colorless or slightly olive green in color. Colonies are microscopically observed and are spherical, sometimes partially flattened, and rarely irregular. Colonies consist of a large number of irregular and densely clustered cells. Cells found together in clumps are olive green in color (Fig. **12**) [100, 101, 113, 115, 323, 372, 380, 398, 401, 427 - 431, 469 - 483].

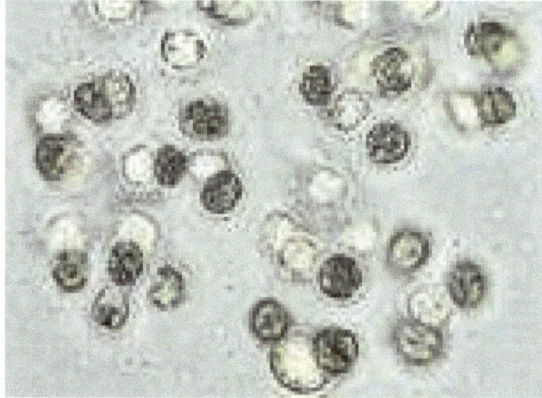

Fig. (12). *Aphanocapsa incerta* (scaled with 10 μm) [484].

Pseudanabaena limnetica

Cells are cylindrical. Single *trichomes* are straight or wavy and 1-2.5 μm wide (Fig. **13**) [2, 6, 101, 113, 115, 122, 133, 137, 177, 310, 321, 385, 391, 398, 401, 407, 420, 432, 465, 470, 483, 485 - 513]. There are fuel recovery studies on this species [506, 514, 515].

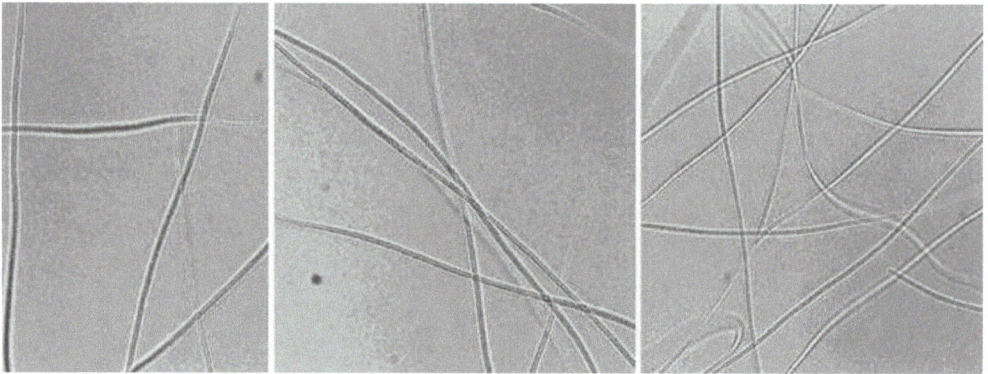

Fig. (13). *Pseudanabaena limnetica* (scaled with 10 μm) [516].

Oscillatoria limosa Ag

The trailing cell is round. The *trichome* is more or less flat. Its color varies from blackish blue-green to brown. The length of the cells is 1/3-1/6 more than their width (Fig. **14**) [30, 31, 61, 162, 517, 518]. There are fuel recovery studies on this species [292, 396, 519 - 522].

Fig. (14). *Oscillatoria limosa* (scaled with 10 μm) [523].

Oscillatoria rubescens

They float freely. Cells are 6-8 μm wide and granular, gas vacuole flat. Knitwear is straight, the end cell is head-shaped and convex. Trichome is dark reddish, sometimes violet in color (Fig. **15**) [82, 422, 478, 518, 524 - 590]. There are fuel recovery studies on this species [541, 542, 589 - 591].

Fig. (15). *Oscillatoria rubescens* algal bloom (scaled with 10 μm) [592].

Oscillatoria tenuis

Cells (4) -5-8- (10) μm wide and 2.5-3.2- (5) μm long. They are found straight or vaguely bent, especially in the anterior part. They do not taper towards the tip. Sometimes they can be found scattered among other algae. The outer membrane sometimes thickens and ends. The sheath is often found. The cell in the apical is not convex, smooth, and capitate. Trichomes are clustered and appear blue-green (Fig. **16**) [4 - 6, 21, 115, 149 - 152, 166, 217, 292, 324, 379, 395, 474, 496, 527, 548, 549, 593 - 647]. There are fuel recovery studies on this species [292, 626, 637, 648 - 652].

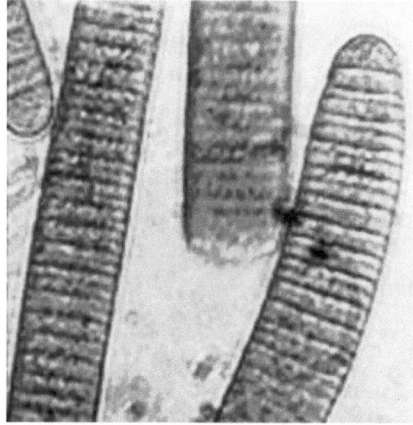

Fig. (16). *Oscillatoria tenuis* (scaled with 10 μm) [653].

Oscillatoria brevis

The granule content cannot be noticed. It does not narrow towards the apex. It is straight or oblique. The cells are 1.5-1.8 μm wide and are about 3 times the width. The apical cell is round and does not contain calyptra. Trichomes are solitary and planktonic or coexist with other algae in littoral regions (Fig. **17**) [2, 40, 75, 95, 121, 148, 152, 161, 168, 355, 496, 508, 637, 654 - 712]. There are fuel recovery studies on this species [655, 674, 705, 713 - 715].

Oscillatoria brevis

They contain granules throughout the septum. The cell body is 1.5-3 μm long. It is 4-6.5 μm wide. The knitwear is flat, and tapered on both sides and there is no narrowing in the cell joints (Fig. **18**) [30, 50, 95, 324, 408, 496, 521, 607, 618, 717 - 775]. There are fuel recovery studies on this species [521, 750].

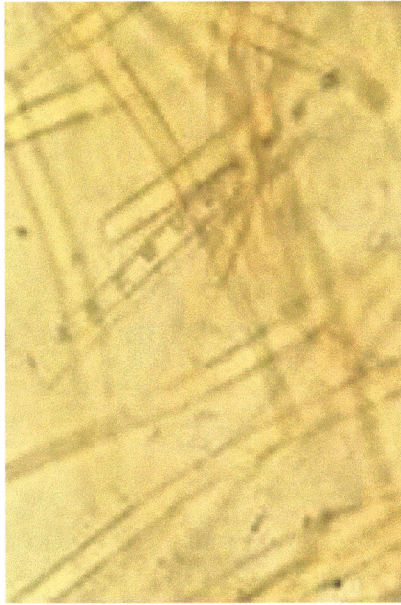

Fig. (17). *Oscillatoria limnetica* (scaled with 10 μm) [716].

Fig. (18). *Oscillatoria brevis* (scaled with 10 μm) [776].

Oscillatoria amphibia

The lengths of the cells are 2-4 times their width (4-8 μm). They form the density of blue-green plants. There is no thinning towards the apex. Apical cells are flat and broad oval in structure and have a convex outer membrane. The width of the cells that appear as a series of double granules along the length of the trichome and contain single granules on both sides, which do not narrow in the opposite cell walls is (1.5) -2-2.8- (4) μm. Its opposite walls are not bordered. Trichomes are straight or curled (Fig. **19**) [31, 61, 723, 777 - 794]. There are fuel recovery studies on this species [779, 785].

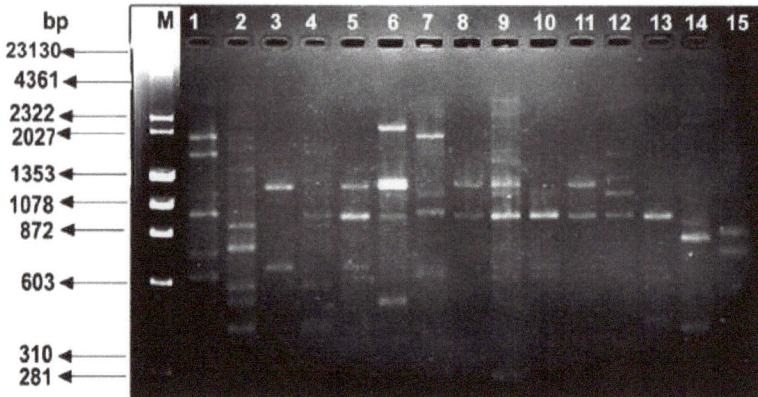

Fig. (19). Number four is the PCR fingerprint of *Oscillatoria amphibia* [795].

Oscillatoria prolifica

Cells are 2.5-5 μm wide and 4-6 μm long. The cell content is dense, sometimes granular and not narrowed at the opposite walls. *Apical* cell head is broadly flattened and calyptrained. The trichomes, clustered as a floating purple-black mass, extend straight to the top, but flexibly [128, 796 - 799].

Oscillatoria princeps

The compression at the end of the thread sometimes has a slightly thinned membrane, sometimes this membrane is absent in the last cell. They form mostly thalli and are flat, there is no narrowing in the cell walls, their width is 16-60 μm, mostly 25-50 μm, blue-green turned to dirty green, slightly tapering towards the tip and curved ends. Cells are 3.5–7 μm tall, the end cell is distinctly round. Trichomes are blue-green, more or less brownish, violet, or reddish (Fig. **20**) [73, 583, 800 - 808]. There are fuel recovery studies on this species [809, 810].

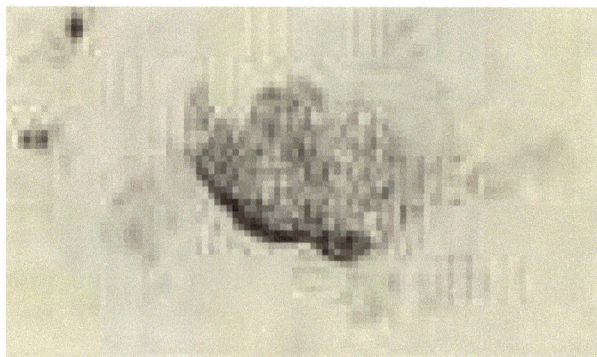

Fig. (20). *Oscillatoria princeps* (scaled with 10 μm) [811].

Lyngbya limnetica

Filaments are 2-2.2 μm wide. Cells are 6-12 μm long. Cells contain granules. Trichomes are 1-2- (2.5) μm wide. Apex parts are not thinned. It has a thin and colorless cover. Filaments are single and planktonic [82, 127, 796, 798, 812 - 814]. There are fuel recovery studies on this species [815, 816].

Lyngbya contorta

They form very large spirals and make them easy to distinguish. They are found in large groups or rarely mingle with other algae. They are wrapped in almost complete circular rings. There is no narrowing in the cell walls with a single granule or no granule on it, the last cell without thinning is round. Filaments are single, free-floating, brittle, and wound in regular spirals. The threads are 1–1.5 μm wide, the narrow sheath is colorless, the cells are 1–2 μm in diameter and 3–5 μm in length [127, 789, 790, 814, 817 - 819]. There is a fuel recovery study on this species [820].

Oscillatoria majör

Usually lamellae. Filaments are single and found amongst other algae. Sometimes they can be found in groups, but they do not form large communities. Filaments are 11-17 μm wide. Trichome is vaguely capitate in older ones. Cells (2) -3.5-4 μm in length and the width of the cells is 1 / 5-1 / 4 of their length. They contain homogeneous granules. It is thick (3-3.7 μm) and has a tight cover. The filaments are straight and have no trichomes or are slightly tapered towards the apex. Filaments are 22-26 μm wide [623, 821].

Lyngbya wollei

Usually lamellae, these can sometimes be found in groups, but they do not form large communities. Filaments are single and found amongst other algae. Trichome is vaguely capitate in older ones. Filaments are 11-17 μm wide. The filaments are straight and have no trichomes or are slightly tapered towards the apex. Cells (2) - 3.5-4 μm in length and the width of the cells is 1/5-1/4 of their length. They contain homogeneous granules. It is thick (3-3.7 μm) and has a tight cover. Filaments are 22-26 μm wide (Fig. **21**) [822 - 826]. There is a fuel recovery study on this species [827].

Fig. (21). Micro (a) and macro (b-f) photos of *Lyngbya wollei* [828].

Oscillatoria formosa

The shape of the cells is almost quadrilateral, the cells are 2.5-5 µm long, the cell length is 1/2 times the width, the last cell is wider than the others, the intermediate chamber is sometimes slightly granular, there is no calyptra or cap. Thallus blue-green, knitwear straight, there is some narrowing of the cell walls. Width 4–6 µm, bright blue-green, tapering and curving towards the tip (Fig. **22**) [829 - 833]. There are fuel recovery studies on this species [714, 834].

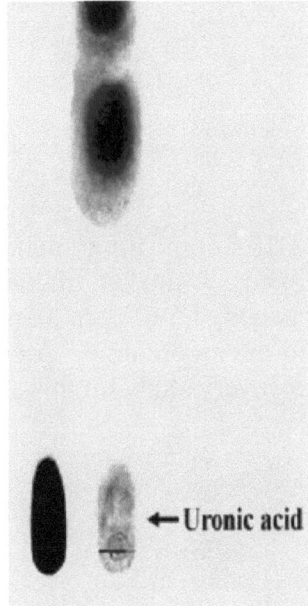

Fig. (22). A thin layer chromotogram of *Oscillatoria formosa* [835].

Phormidium ambiguum

Trichomes are curved and often entangled, with or without a very weakened ending, 5-7.5 µm wide, 1.5-3.4 µm long. The diagonal walls are somewhat narrow or not, the granules may be adjacent or the cells are shorter than their width, sometimes not nearly isodiametric. The sheaths are colorless, stable, thin, or gelatinous, sometimes thick and more or less lamellar. Sheaths Apical cell rounded, without callipse, not capitate, sometimes outer cell wall thickened *trichomes* are ranging from olive-green to blue-green, with granular cellular content (Fig. **23**) [408, 639, 640, 836 - 838]. There are fuel recovery studies on this species [839, 840].

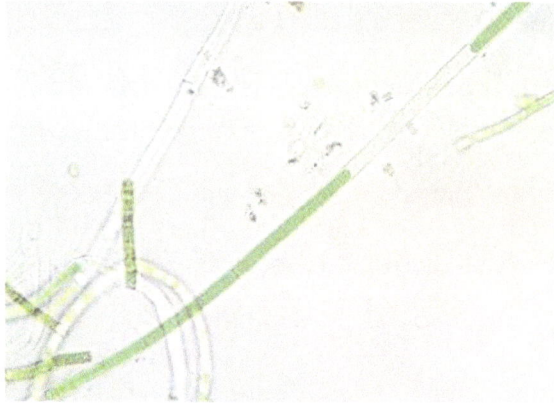

Fig. (23). *Phormidium ambiguum* (scaled with 10 μm) [841].

Anabaenopsis elenkinii

Akinets, somewhat ovoid, 8.3-10.5 μm in diameter, 9.3-12 μm in length; sometimes almost spherical, 8.3-10.7 μm in diameter. Cells 4.6-5.7 μm in diameter. Heterocyst spherical shaped, 4.6-6.7 μm diameter. Trichomes consist of an ellipsoid or elongating-ovoid cells containing pseudovakuol (Fig. **24**) [404, 418, 842 - 844]. There is a fuel recovery study on this species [844].

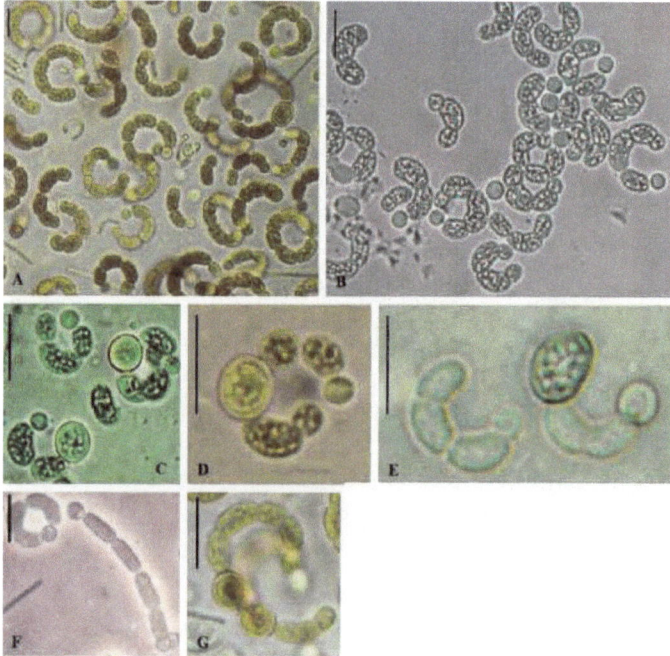

Fig. (24). *Anabaenopsis elenkinii* (a-g) (scaled with 10 μm) [845].

Anabaena affinis

Cells are of spherical shapes. They are plactonic or found alone and in free form in coastal flora with other algae. Trichomes, which can be straight or curved, are covered with a thin and often vague, broad mucilage sheath. Single and dispersed ones are 9.5-12 µm wide and 17-24- (26 µm) long. It is homogeneous or contains pseudovakuol (if present alone). The width of the cells is 5-6-7 µm. Heterocyte spheroid and slightly larger than vegetative cells. Gonads are mostly short-cylindrical, sometimes ovoid, and swollen at the poles. The wide ones are 7.5-10 µm wide. Trichomes are straight or curved (Fig. **25**) [846 - 850]. There are fuel recovery studies on this species [851, 852].

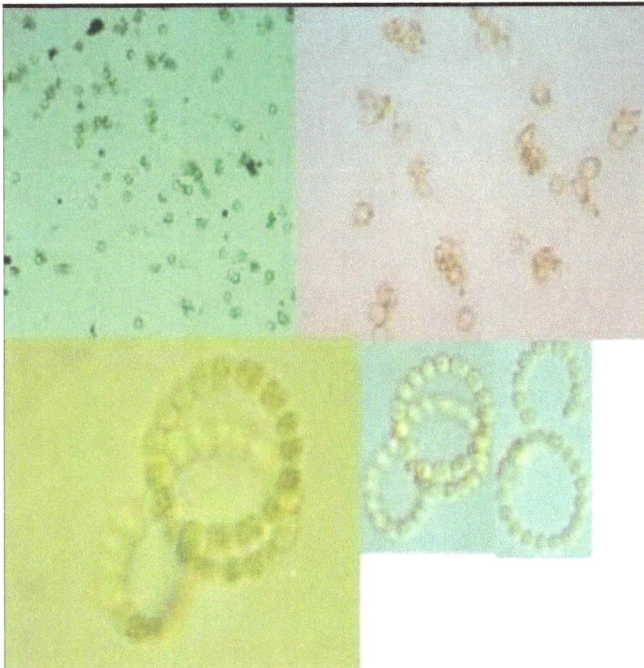

Fig. (25). *Anabaena affinis* (scaled with 10 µm) [853].

Anabaena flos-aquae

It is shaped from spherical to light cylinder. Heterocytes of spherical shape or somehow narrowed at their poles are 7-9 µm wide and 6-10 µm long. The granule-containing cells with distinct pseudovacuoles are (4) -5-6- (8 µm) in diameter and 6-8- (12 µm) in length. Pachtonic trichomes, which are very curved and twisted, sometimes irregularly spiral-shaped, form a single or together cluster. Gonads are generally spherical, cylindrical or sausage-shaped and (6) -8-12- (13 µm) wide and (20) -24-30 (50 µm) long, next to the filaments that are neighboring heterocytes, in single or sometimes multiple entanglements (Fig. **26**) [315, 668, 854 - 856]. There are fuel recovery studies on this species [521, 722, 750, 857 - 862].

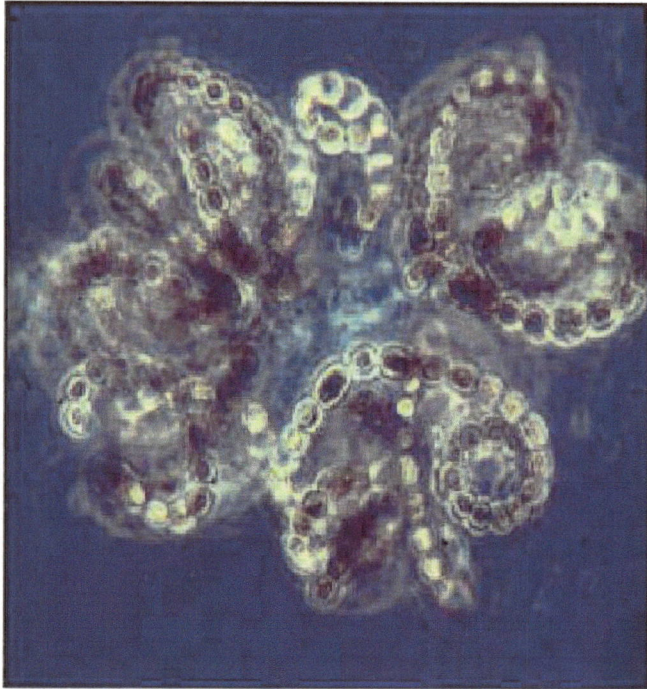

Fig. (26). *Anabaena flos-aquae* (scaled with 10 μm) [863].

Anabaena catenula

The colorless ones are sometimes cylindrical with slightly concave edges in the middle. It is 9-10 μm wide and 16-28 μm long. Cylinder (barrel) shaped cells are 6.5-8.8 μm wide. Cells in the terminal are oval. Spherical or ellipsoidal heterocytes are 9-12.5 μm wide. Akinetes are typically solitary and free from heterostics but are soft in the structure that rarely merges with heterocytes or forms 4 rows, and their color turns brown. Trichomes are curved, irregularly encircled mucilage sheaths and are matte blue-green in color (Fig. **27**) [87, 631, 799, 864 - 867]. There are fuel recovery studies on this species [631, 868].

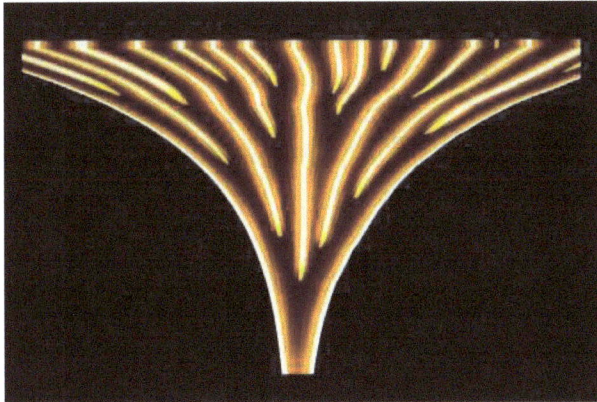

Fig. (27). *Anabaena catenula* (scaled with 10 μm) [869].

Spirulina Laxissima

The apex is distinctly rounded. The spirals are very loosely twisted. It is 4.5-5.3 μm wide and 17-22 μm long between two spirals. Trichomes are very thin and 0.7-0.8 μm in diameter (Fig. **28**) [812, 870 - 873]. There is a fuel recovery study on this species [840].

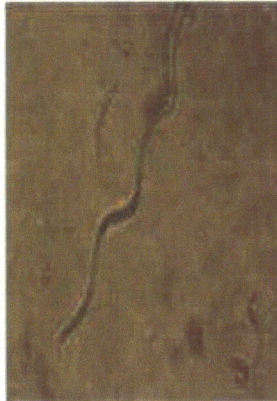

Fig. (28). *Spirulina laxissima* (scaled with 10 μm) [874].

Arthrospira majör

The trichome is 2.7-5 μm wide, 1.2-1.7 (-2) μm wide, blue-green spirals 2.5-4 μm wide, coiled in regular spirals [875 - 881].

Spirulina nordstedtii

Cells in knitwear contain a pale, pale, or bright blue-green color. It is in the form of a very close and regular spiral. The diameter of the trichomes is 2 μm, the distance between the spirals is 5 μm, and the spiral is 5 μm wide [882 - 891].

Plectonema notatum

Filaments are 1.7-2 µm long. The cells are cylindrical, small, and there are small but noticeable granules within the tight cross walls. Branching is double or single. It consists of free-living, single or thin mucous threads. Cells are 1.2-2 µm wide, 2-3 µm long, and the sheath is thin but distinct [9, 87, 892 - 896].

Cystoseira barbata

Tallus lengths can be 50-60 cm and sometimes even more, such as 150 cm. The branches are covered with main branchlets about 1 mm thick. In addition, forms with large thalluses can be seen in some regions. It is generally a simple algae species and has a woody, leafy structure. Round, simple, or prominently branched, 0.5-1 cm thick; the lower parts of the branches are formed by the parts leftover from the breakage of the branches. They can show branching in all directions (Fig. **29**) [897 - 906]. There is a fuel recovery study on this species [907].

Fig. (29). *Cystoseira barbata* [908].

Ulva rigida

It is a salinity-resistant species and can be found in both salty and brackish waters. Ulva rigida is a cosmopolitan species that naturally spread in parts of our coasts where nutrient elements such as nitrogen and phosphorus are abundant, especially in shallow and rocky areas. It is highly tolerant of stressful conditions. The upper part of the algae, which clings to the ground with a short stem-like foot, is quite wide and consists of two cell layers. Ulva species in the green algae group can adapt to different environmental conditions. It is commonly found where the amount of nutrients are abundant, and the wave intensity is low (Fig. **30**) [909 - 913]. There are fuel recovery studies on this species [909 - 911, 914 - 950].

Fig. (30). *Ulva rigida* [951].

Gracilaria verrucose

Gracilaria species are the most suitable algae for agar production, and agar yield and quality may vary depending on the growth rate, seasonal differences, and environmental factors. Algae has commercial importance because it contains agar-agar which is a polysaccharide in its cell walls. Gracilaria verrucosa species, which is included in the phylum of red algae (Rhodophyta), has a natural distribution in various parts of our coasts, especially in Izmir and Izmit bays. So much so that up to 70% of agar can be obtained from the raw material of algae (Fig. **31**) [952 - 957]. There are fuel recovery studies on this species [952 - 954].

Fig. (31). *Gracilaria verrucose* [958].

Chroococcus nägeli

Occasionally, gas vacuoles (in some planktonic species) and sometimes in species with large cells, chromotoplasm can be observed (tylakoids tend to accumulate near the cell wall). Colonies are generally microscopic, those with only a few cells are more or less spherical. The gelatinous colonial layer is delicate, fluid, homogeneous, colorless, or lamellar, limited and rarely colored. In rare cases, many cells can come together to form large macroscopic colonies. Cells are surrounded by their own sheaths. Cells are rarely spherical (newly reproduced cells), usually going from hemispherical to amorphous. They can be bright or light blue-green, yellowish, pinkish, or violet. They are single cell colonial species. These are generally the same shaped cells; they are homogeneous or lamellar structure. The contents of the cell are homogeneous or granular, and sometimes a few striking granules of the cell can be easily observed with light microscopy (Fig. **32**) [85, 134, 354, 959 - 963].

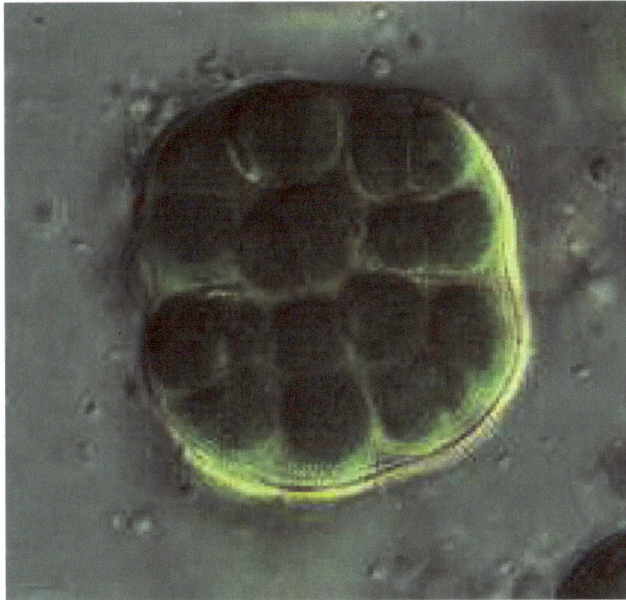

Fig. (32). *Chroococcus Nägeli* (scaled with 10 μm) [964].

Chroococcus dimidiatus

Cells can be easily distinguished by the large dark granules they contain. Their cells are single or usually 2 to 4 cells, rarely 8 cells come together and exist in groups. Their gelatinous or mucous matrix sheath is colorless. They live planktonically. It is found within the body of water. They swim freely. Their cells exist as a single cell in a round or ellipsoid shape. Colors are blue-green, yellowish, or olive green. The unsheathed length is 8–32 μm, and the sheathed is 13-25 μm [17, 85, 354, 408, 473, 668, 965, 966].

Pleurococcus minör

They are especially common in calm lake waters. They are easily distinguished by the smooth green color of the cells. The matrix in the cell sheath is quite abundant. Tallus slimy gelatinous structure, spherical-shaped cells 3-4 μm in diameter, dirty blue-green or olive green in color, rarely between 4 and 8, they are solitary or many cells exist together, their colorless sheath is quite thin, but very carefully can be seen visually when looked [877, 967 - 976].

Gloeocapsa kützing

Colonies are made up of groups of cells that many cells come together to form. The mucilage cover is delicate, but it stratifies over time and has clear borders. Colonies are microscopic structures, small or larger structures formed by irregularly gathered cells. Sometimes these large groups can even be macroscopic. They sometimes cover moist terrestrial substrates in layers with large groups they form. A few species are intense reddish, bluish, orange, or yellowish in color. These groups are connected by a mucilage sheath. They have a very wide and lamellar cover. Older colonies have many cells. Cells are found irregularly in colonies, there is more or less distance between cells in the colony. They are unicellular colonial forms [10, 85, 354, 960, 963, 976 - 979].

Gloeocapsa quaternata

The thallus is blackish or bright green, yellowish or brown in color, enlarged or shaped into lumps. They spread in the upper parts of moist soils. It forms a slimy layer on the upper surface of still and hot waters. The cells can be in single groups or they mostly form groups of 2 to 4 cells in a row, the number of cells in the colony sometimes increases up to 8, and the size of the colonies varies between 11-22 μm. Unsheathed cells are 3-4.5 μm in width, rarely can expand up to 5 μm in width, cells with sheaths are 7-11 μm wide in blue-green color, their sheath is often lamellar structure, colorless to red [17, 408, 823, 980 - 988].

Spirulina turpin Ex gomont

Its filaments are not branched and have no permanent sheaths. They rarely live individually, usually in piles or macroscopically, visibly covering the substrate in a thin layer. They are of filament structure. In trichomes: the width of the spirals is (2-) 2.5 (-4) μm. Trichomes are isopolars and are 0.5–3 μm wide, consist of cylindrical cells, no narrowing of the intermediate lamellae, which cannot be observed under light microscopy. The entire thread is regularly twisted in the form of a screw along the length of the trichome, the width of these screw-like crimps is the same in the same thread or even in threads of the same type (very rarely there are no spirals), the screw-like folds are very tight. There is no narrowing towards the ends of the yarn. Trichomes have a lot of movement ability (rotational motion). Their cells lack gas vacuoles and have conspicuous granules. The diagonal diameters of the cell are more or less equal. Cell size is greater than cell width. Usually, they have a homogeneous content. Their colors are bright blue-green, olive green, or pinkish. Thylakoids are lined up asymmetrically on the outer cell wall. Heterocytes and akinet are not found. The cells at the end of the thread are curled in a wide circle. There is no thickening or cap in the cell wall in the last cell. Reproduction occurs with mobile hormogones, which are formed by the fragmentation of the trichome, reproduction does not occur with necritic cells. Cell division occurs crosswise, in which division occurs vertically along the long axis of the trichome, and the daughter cells grow to the same size as the previous generation before dividing again. All cells of the thread can divide. The spirals touch each other or have very short intervals between them [789, 790, 977, 989]. There are fuel recovery studies on this species [714, 990 - 992].

Spirulina subsalsa oersted

Trichome 1-2 μm wide, color from blue-green to reddish-purple, the majority of which are irregularly coiled in dense spirals, rarely regularly curled, sometimes in loose spirals, thallus bright blue-green or yellow-green, single strands among other algae They are found with spirals 3–5 μm wide. Spirulina sp. species can move freely and make their way through other algae in the substrate. They can form long threads but are mostly found as single threads. It is not often observed that individuals of the same species come together. It has been determined that Spirulina species are always found next to Phormidium species in thermal waters. They live among dead leaves or other algae in still waters. The width of the trichome varies between 0.8 μm and 2 μm, while the spiral width varies between 1.7 μm and 3.5–4 μm. Its dimensions have also been reported in which the trichome structure is very different (Fig. **33**) [993-1000].

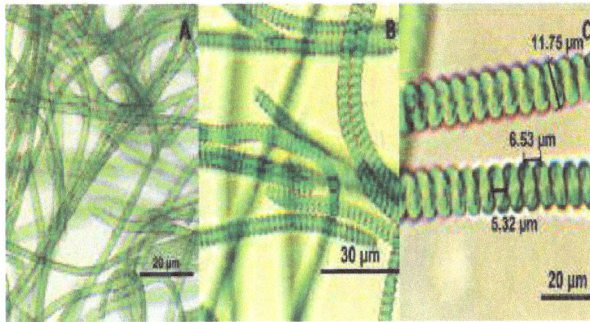

Fig. (33). *Spirulina subsalsa [1001].*

Spirulina subtilissima kützing

The trichome is 0.6-0.9 μm wide, coiled in regular spirals, bright bluish or yellowish, spirals 1.5-2.5 (-2.8) μm wide, spacing between spirals 1.25-2 μm. They are found together with other algae in still waters. Although they are among other algae species, individuals of the same species coexist in large groups. They live in waters that are still but the water temperature does not decrease [999, 1002, 1003].

Spirulina meneghiniana (zanardini) gomont

The trichome is 1.2–1.8 μm wide, flexible, coiled in irregular spirals, bright blue-green, thalli dark blue-green, spirals 3.2–5 μm wide. It is found planktonically mixed with other threads in calm flowing waters. It forms intricate knots with other algae species and threads of its kind. It is a planktonic species (Fig. **34**) [463, 870, 961, 1004-1006].

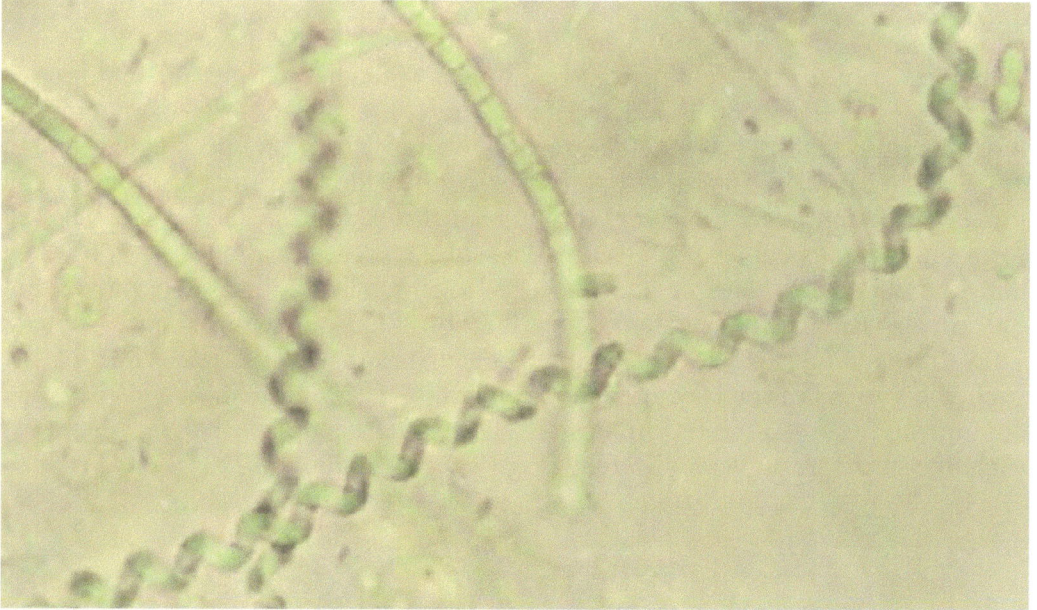

Fig. (34). *Spirulina meneghiniana* (scaled with 10 µm) [1007].

Spirulina Major kützing

The trichome is 1.2-1.7 (-2) µm wide, coiled in regular spirals, blue-green spirals 2.5-4 µm wide, and 2.7-5 µm wide. They can be found among other algae in salty waters, as well as in ponds, away from the springs where the water temperature drops in hot springs. They spread in moist soils, in still or very slow-flowing waters (Fig. **35**) [33, 34, 429, 1008-1010].

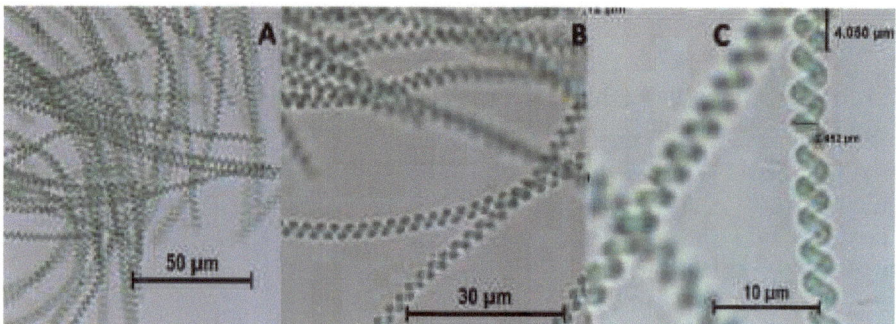

Fig. (35). *Spirulina majör* [1001].

Leptolyngbya

Isopolar yarns are twisted, wavy, or densely spiraled. In filamentous structure, the filaments are wound in long, single, or clusters and form thin layers (the layers they form are macroscopic and a few cm wide. The thin threads are 0.5-3.2 μm wide. Simple, thin but tight, usually colorless and has a facultative sheath with an open apical end Sheath is only one Very rarely two trichomes (in short pieces) can be contained within the same sheath Trichomes are thin, cylindrical in shape, usually narrowing towards the end. Rarely pseudo-branching (visible under certain conditions), usually giving only one side branch. The cell at the far end is round or conical. The intermediate lamellae of stationary trichomes may or may not be narrowed. Cylindrical-shaped cells have homogeneous content, no aerotopes. They have a small number of but distinct granules. The last cell has no thickened cell wall or caltype. They are pale blue-green, gray, olive green, yellowish, or reddish in color. Heterocyte and akinet are not found. The cell continues to develop until it reaches the same size as the previous generation before dividing again. Cell division occurs symmetrically by dividing into two. Typical necritic cells do not cause the trichome to break apart, but sometimes hormogon detachment from the thread after cell death in the thread. Reproduction occurs by mobile hormones and these hormones are released from the end of the trichome. Several species are known to be found in thermal and mineral springs or aetophytic on rocks and walls. Leptolyngbya species are widely found in soil, freshwater as periphyton and metaphyton, and in halophile (marine) biotopes. They thrive well in a cultural environment and are well suited to culture. Their characteristic species live endolytically inside the mucilage structures of the colonial layers formed by other algae species. In terms of the simplicity of its morphology, its taxonomy is not fully established. The species used in the biological description of this genus is "Plectonema boryanum" (= Leptolyngbya boryanum). Many species are identified by different surnames. These names are: 'Lyngbya', 'Phormidium' and 'Plectonema'. The genus, which is quite common, has very different definitions, but they all need a description of their distinctive characteristics and taxonomic review (Fig. **36**) [408, 1011-1014]. There are fuel recovery studies on this species [542, 628, 834, 1015-1050].

Leptolyngbya tenuis

A little narrowing is observed in the cell walls, the thread becomes thinner towards the tip, 1-2 μm wide and light blue-green, the sheath is thin, loose, they are dyed violet purple with a mixture of chlorine-zinc-iodine. Thin, non-dense thalli are light blue-green in color. The sheath is enlarged, the trichome is straight or slightly curled, the threads are densely intertwined. The last cell is fully cone-shaped, with no calyptra. The cell size is 2.5–5 μm, 3 times the width, the intermediate chamber is granule-free, and the cell walls are often not easily observed. They are especially common in thermal waters and are among the important types of thermal waters. They live on moist surfaces, among other algae, in fresh and saltwater (Fig. **37**) [17, 408, 633, 1014, 1052].

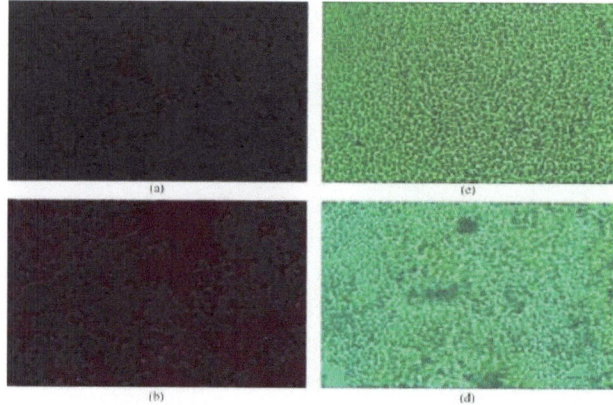

Fig. (36). *Leptolyngbya species* (a-d) (scaled with 10 μm) [1051].

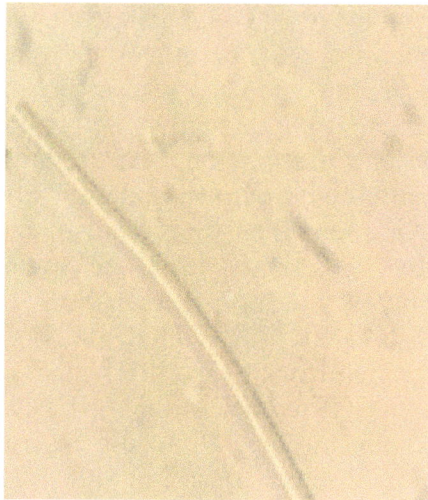

Fig. (37). *Leptolyngbya tenuis* (scaled with 10 μm) [178].

Planktolyngbya

They are planktonic species; they live in mesotrophic dams. The habitat of a few species is limited to tropical and temperate zones. False branching is seen rarely. Generally, no narrowing is observed towards the end. Trichomes are cylindrical, isopolar, straight, wavy, screw-shaped or spirally curved, narrow, about 3 μm wide. Cells are single rows. Rarely, cell lengths are shorter than their widths, usually, cell lengths are longer than widths. Indistinctly, there may or may not be narrowing of the intermediate lamellae. It is always motionless. Cells are cylindrical shaped. They either have no aerotopes or facultative polar aerotopes. Sometimes it has single granules. Thylakoids are lined along the cell wall to the edge of the cell. The last cell is circular or narrowed and rounded. There is no time frame. Cell contents are pale gray-blue, blue-green, yellowish, or olive green in color. Cell division occurs by crossover division, takes place along the long axis of the trichome. The division by splitting the trichome into pieces does not occur with the help of necritic cells and the resulting hormogons are immobile. Young cells grow to the same size as the previous generation before dividing for a second. All cells can divide. In filamentous structures, filaments are single, thin, plain, colorless, and have tight sheath (Fig. **38**) [514, 789, 790, 1024, 1053, 1054]. There are fuel recovery studies on this species [648, 1050, 1055].

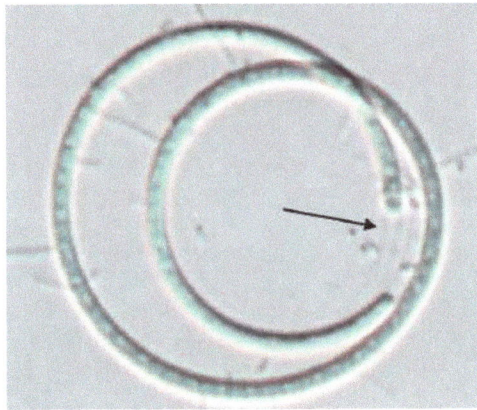

Fig. (38). *Planktolyngbya* (scaled with 10 µm) [435].

Planktolyngbya subtilis

It spreads in small pools, moist soils, lakes, and dams used for decorative purposes. The cells are 1-1.5 µm wide, the length ratio of quadrangular ones is 1/3, rarely this ratio is 1/8, the cell size varies between 1-3 µm, there is a pale blue-green granule in the cell wall that cannot be narrowed there is no thinning towards the tip in the last round cell. Filaments can be straight or slightly curved or twisted, free-floating in single filaments, 1-2 µm wide, colorless sheath thin or narrow, not dyed blue with chlorine-zinc-iodine (Fig. **39**) [790, 1024, 1054, 1056, 1057]. There are fuel recovery studies on this species [1058-1071].

Fig. (39). *Planktolyngbya subtilis* [1072]

Limnothrix

The most common species (L. redekei) is widespread in temperate zones. Its filaments are straight, slightly curved, or irregularly wound in the form of screws. The thread consisting of a large number of cells is isopolars. It has no sheath or has a fairly thin colorless sheath, and this sheath is a facultative state for many species. Trichomes are cylindrical, the cells that make up the trichome are isodiametric or longer than their width. The yarn either does not move at all or has very little, reduced motion. There is no or very slight narrowing in the intermediate lamellae. Cell walls are very thin. The width of the thread can be up to 16 μm. The thread does not have a calyptra at the tip but rarely has conical plasmic bumps. As you move towards the end of the thread, no thinning is seen, and it usually ends with a cylindrical but sometimes conical cell. Cells are pale blue-green, yellowish, or pinkish in color. They can create gas vesicles if they wish. The composition of phycobilin may vary (the ratio of phycoerythrin/phycocyanin may change). Aerotopes settle at the end of the cells or combine along the long axis in the cell to form a large central aerotop. Thylakoids sometimes take a slightly waveform and settle parallel along the cell wall. Heterocyte and akinet are not found. Reproduction by disintegrating the trichome does not occur with the help of necritic cells, the hormogons are immobile. All cells can divide (cell division occurs by dividing in two), daughter cells reach the same size as the previous generation without dividing again, and divide after growing. They are planktonic or thicoplanktonic species. They spread from eutrophic to mesotroph in water tanks, dams, eutrophic lakes, and ponds (Fig. **40**). They have a filament structure and their filaments are free-floating singly [326, 408, 514, 640, 1066].

Limnothrix amphigranulata

It is easy to distinguish with prominent granules on either side of the cell wall. The trichome is flat, the cell walls constricted clearly, the tip is not tapered, is 1.75-2 μm wide, slightly more or twice the length of their length, cells 2.5-5 μm tall, with two gas vacuoles in the intermediate chamber, pale blue-green, end cell round, no calyptra. They are found in planktonic form in rivers and lakes.

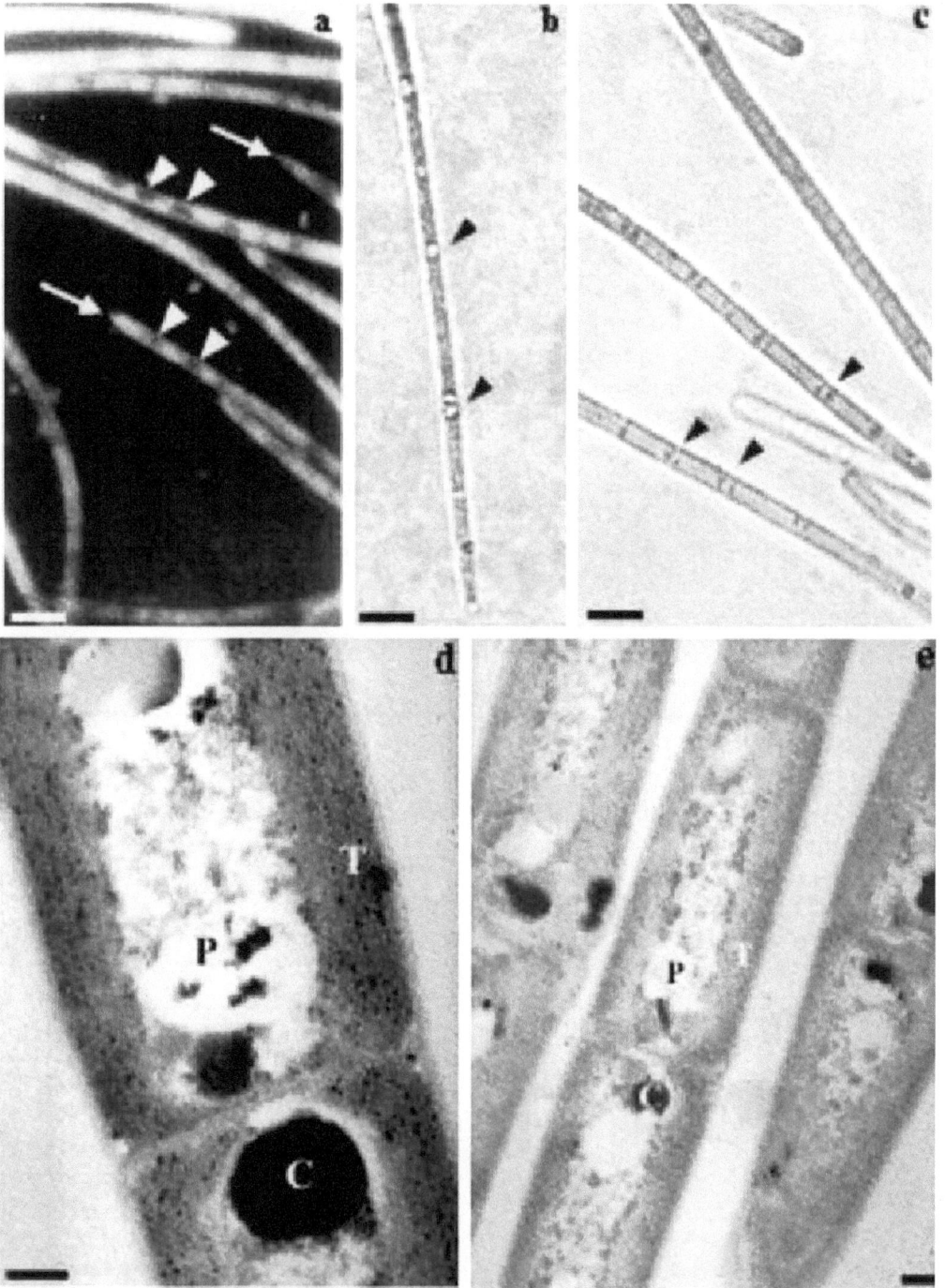

Fig. (40). *Limnothrix species* (a-e) (scaled with 10 μm) [1073].

Jaaginema anagnostidis

Trichomes are isopolars and often wavy or curled. Several species are known to live in mineral, thermal, or saltwater. Sometimes they form macroscopically visible layers. Trichomes are always unsheathed. The sheath is narrow, thin, about 0.53 μm wide, cells in a single row. They are always motionless. Its cells are cylindrical, elongated, longer than its length (most of the time), without aerotopes, sometimes with a single granule (rarely in the intermediate lamella). Generally, there is no narrowing towards the tip. There is no or only slight narrowing of the intermediate lamellae. Cell division occurs by diagonal division, divisions take place in the vertical plane along the long axis of the trichome, and the daughter cells grow until they reach the same size as the previous generation before dividing again. The last cell is rounded or tapered, pointed or conical in shape, and never has a calyptra. Cell contents are pale blue-green, gray, yellowish, or olive green in color. Some species have chromatic adaptation ability (the rate of phycobil may vary). Being mainly benthic organisms, they thrive in a variety of aquatic biotopes. The division, which occurs with the fragmentation of the trichome, does not occur with the help of neritic cells and the hormocytes formed are immobile. They are commonly found as metaphytons in ponds and lakes, in dams and reservoirs rich in aquatic plants. It is filamentous in nature; the filaments are usually freely clustered as single or small colonies (mixed or intertwined).

Jaaginema pseudogeminata

Thalli are pale or dirty blue-green in color. Trichomes are curled, light blue-green, the ends are not tapered. The width of the thread is 1.3-2.2 μm, the length of the cells is close to the width, or it may be slightly longer or shorter, about 2.6 μm in length, there is no narrowing of the cell walls, the cell walls are thin and without granules, the last cell is round, caliptra there is no anther. Among other algae, they are found on leaves and submerged rocks, and especially in moist soils.

Oscillatoria vaucher

Genome size 2.50–4.38 x 109 daltons and stromatolite fossils belong to the early precambian period. The last cells take the form of a broad circle. Sometimes the trichome has a bastion or a narrow head. They mostly have sensitive and fragile structures. They form well-visible layers. They are rarely found singly or in small groups. Often, they are replaced by sliding, friction movements. Apart from being unsheathed, they can be very thin under difficult conditions or cultural conditions. They are colorless, their tight sheath opens at the ends as the yarn grows. They can be of quite different sizes, from microscopic to macroscopic. Their trichomes are isopolar, straight, or slightly curved, usually, they can be from 8 μ to 60 mm in width. Intermediate lamellae can be narrowed or not. The trichome becomes thinner towards the end. The cells do not have gas vacuoles, but they have small granules or granules that are free inside the cell. The fragments detached from the trichome are shorter than the trichome, and mobile hormogons are formed by detachment from the trichome or due to necritic cells. The cells in a single row in the trichome are short cylindrical or disc-shaped (cell size less than cell width). Their colors are blue-green, brownish, or pinkish. Thylakoids are usually irregularly lined up along the cell wall. Pseudo-branching is seen. They do not have heterocysts and akinets. Cell division crossover occurs. When the cleavage occurs in rapid succession, zones (layers) growing irregularly sometimes form. Their species find habitats in very different substrates (sand, stone, mud, *etc.*), they spread in shallow water biotopes, littoral, and benthic parts of freshwater, brackish or marine habitats, and moist soils. Other studies on the genus Oscillatoria are recombination studies. Some species have a planktonic role, floating in macroscopic layers or in small groups of threads with cells that have developed gas vacuoles. This common blue-green algae genus has spread all over the world, but a few species are geographically restricted. They are fairly common species consisting of planktonic or non-planktonic branched filaments.

Oscillatoria princeps vaucher ex gomont

Trichomes are blue-green, more or less brownish, violet, or reddish. Cells are 3.5–7 μm tall, the end cell is distinctly round. They form mostly thalli and are flat, there is no narrowing in the cell walls, their width is 16-60 μm, mostly 25-50 μm, blue-green turned dirty green, slightly tapering towards the tip and curved ends. Dead (necrite) cells between the cells are seen in the thread, then ruptures were observed in the thread with thinning of the wall at these points. It spreads in freshwaters, on the surface of the water in the sea, and in very humid soils. The anther at the end of the thread sometimes has a slightly thinned membrane, sometimes this membrane is absent in the last cell. The threads can be seen with the naked eye if you look at the thallus carefully. It is a cosmopolitan species. The anther, which almost all of its threads have, is observed. Intracellular granule structure can be seen easily [4 1074, 1075].

Oscillatoria proteus skuja

Cell walls are highly granular, often granular, cell size sometimes 1 / 2–1 / 3–1 / 5 times its width, cell length 2- 3.7 μm, content pale olive green or blue-green, head cell conical roundish. Trichomes are among other planktonic algae. It is more or less straight, sometimes curved, tapering to the very tip, and slightly curved or curved, 6–7 μm wide. Trichomes of the species are found freely on the surface of the water.

Oscillatoria boryana bory

With the dark green thallus color and thread structure easily observed even with the naked eye, species identification is quite easy. By forming large groups, they are found as planktonic in parts where the hot spring water flows slowly, and the water is clear. They spread in uncontaminated wastewater channels, on rocks along rivers, and in moist soils. The tip of the trichome or the whole is screw-like curved, rarely flat, narrowed cell walls, 6-8 μm wide, sometimes very slightly granular cell wall, cells 4-6 μm tall, end cell rounded or more or less tapered, no cap and calyptra.

Oscillatoria calcuttensis Biswas

They float as layers in stagnant waters, especially in rice fields, on the water surface of migration, pools, and ponds. It is blue-green in color. The end cell can be conical and pointed. Cell length can be 2–5 times the width, cells 6–10 μm in length, it is a species characterized by the presence of 3 granules in the intermediate lamella. The trichome is leathery brown, the trichomes are arranged in a parallel straight line, there is no narrowing in the intermediate lamellae, it weakens slightly towards the end of the trichome, the end of the trichome is curved or curved. It does not have a title. They can form large slimy algae layers on slow-flowing waters.

Lyngbya C. agardh ex gomont

They are widely found around the world, mostly halophilic species. Several species live in planktonically soliterm trichomes. They have a tight sheath, the sheath is sometimes stratified (thickening) and brownish, the sheath opens at the ends. Filaments can rarely be unsheathed (usually only breeding hormogons); they do not a branch or pseudo-branches with slightly short protruding sporadical branches. The coiled filaments are rarely intermixed to form free clusters, usually, these clusters are macroscopic, they can be a few cm or dm wide. They consist of barrel-shaped discoid or short cylindrical cells, narrowing of the intermediate lamellae may or may not be visible, thinning towards the end of the thread may or may not be seen, they are immobile or can move very little. Heterocyte and akinet are not found. Reproduction occurs by short-motile hormogons, divided into necritic cells and trichomes. Trichomes are isopolars, straight or slightly wavy, and sometimes quite large (5.5–60 μm). Cell division occurs crosswise, division occurs vertically along the long axis of the trichome, often rapid successive division occurs. It is usually sparsely granular, sometimes most of these granules are intermediate lamellae. Cell contents are blue-green, olive green, yellowish, brownish, or pinkish in color. Its cells are very short and always less in cell size, selectively gas vacuoles are absent or seldom present. The cell is surrounded by thylakoids, there is a little extra content in the cell, and this content causes the outer cell wall to thicken or take the shape of an anther. On rare occasions, filaments live in complex clusters as metaphytons in some aquatic plants that live in not very contaminated waters. It is filamentous in structure, filaments are thick, rarely soliterm, filaments are usually tightly bound, they form stratification, sometimes they form layers like skin [789, 790, 1076, 1077].

Lyngbya contorta lemmermann

In the hot springs sampled, they are located in large clusters at the points where the water flow slows down, forming small ponds and among other algae. They are wrapped in almost complete circular rings. The threads are 1–1.5 µm wide, the narrow sheath is colorless, the cells are 1–2 µm wide, 3–5 µm tall. Filaments are single, free-floating, wound in fragile and regular spirals. They form very large spirals and make them easy to distinguish. They are found in large groups or rarely mingle with other algae. They live planktonically in lakes and stagnant waters. There is no narrowing in the cell walls with a single granule or no granule on it, the last cell without thinning is round [1078].

Phormidium kützing ex gomont

In Phormidium, young individuals are bright green, old sheaths are brownish, those that store minerals are golden green and dark red, those that dry in the sun are yellow-pink and white. Some species are known to be found in highly extreme areas (thermal springs, desert sands, *etc.*). They are layered in smooth and smooth leathery layers. They can be of very different sizes microscopically and sometimes macroscopically. It is divided into 3 subgenres according to the frequency of sheath formation: Gomontinema (Anagnostidis and Komarek), Pormidium, and Hansgirgia (Anagnostidis and Komarek). Usually, they cover the surface of the substrate like a cover. Sheath formation occurs selectively at different frequencies, only suboptimal conditions (*subg. Gomontinema*), depending on the change of environmental factors (subg. Phormidium) or regularly under all conditions (in separate, single living trichomes-subg. Hansgirgia); The sheaths are tubular (thin tube), tight, colorless, the sheath is attached to the trichome, no stratification is observed, the sheath is open at the ends of the trichome, each sheath contains only one trichome. Trichome is usually 2–12 (14) µm wide. They never branch out. The thread consists of cylindrical or barrel-like cells, the diagonal diameters of the cells may be more or less equal or slightly more or less than the cell length width. There may or may not be narrowing in the intermediate lamellae. Trichomes are isopolars, the yarn is more or less straight, they can be spirally curled or wavy. The thread does not become thinner towards the end and the end may be twisted aside or screwed towards the end. Whether it is with or without a sheath, it is always mobile (wave, friction, creep) In suboptimal conditions (but never characteristic of planktonic species) they sometimes have granular contents or have conspicuous granules lined up along the length of the cell, or the granules are arranged along intermediate lamellae. Cells do not have gas vacuoles or have special gas vacuoles of large sizes. Cell contents are usually blue-green, rarely brownish, pinkish, or purple in color. Towards the end, the thread became thinner, or the tip was sharp. Sometimes it also has a cap thread. Cell division occurs transversely, perpendicular to the long axis of the trichome. The last cells in the thread are widely curled and the end of the thread takes the form of a circle. Thylakoids line vertically along the cell wall. Young cells continue their development without dividing until they reach the same size as the previous generation. All cells of the thread, except the cells at the ends, are capable of dividing. Sometimes it occurs in poorly observed meristematic zones. Reproduction occurs by mobile hormogons of various lengths. Hormogones are formed from the end of the trichome with the help of necritic cells or by breaking up the whole trichome (mostly by necritic cells). It is filamentous (threadlike), filaments do not branch, rarely live soliterm, they are generally sensitive. Heterocyte and akinet are not found. Some species are found in the littoral zone of the sea, rarely as single filaments, usually in layers on different terrestrial or marine substrates (soil, wet rocks, mud, aquatic plants, rocks, and wood material in still waters and streams) [789, 790, 959, 989, 1079-1082].

Phormidium chlorinum (kützing)

It is characterized by a large granule located inside the cell. They inhabit planktonically in lakes, ponds or lagoons, moist soils, and filter tanks. Thalli rather thin, yellowish-green, trichome flat or curved, cell wall not narrowed or slightly constricted, 3.5-4 μm wide, sometimes up to 6 μm, no gas vacuoles, cells slightly longer or shorter in width, cells 3 7–8 μm tall, cell wall without granules, no calyptra.

Phormidium willei

They spread in humid soils, small ponds, pools, and ponds. In the last cell, the cell is curled in a circle, without sheath thickening. The trichome is pale blue-green, grayish blue-green, curved at the tip, no narrowing of the cell walls 2.4-3.6 μm in width, no thinning at the tip, no cap, cell lengths can be from 1.3 to twice the width, cell the wall is without granules [3, 437, 993, 1014, 1083-1085].

Phormidium terebriforme

Thalus matte blue, the ends of the trichomes are curved and slightly thin, the cell walls do not narrow, 4-6.5 μm wide, 2.5-6 μm tall, the last cell is round, no caliptra and cap. They are formed as green algae layers in still waters. They spread in lakes, rocky and canal sides, and used drainage waters and moist soils [136, 993].

Phormidium amphibium

They also thrive in salty lakes, moist soils, and underwater objects. The trichome is straight or curved and there is no thinning towards the tip. Talus is dark blue-green in color. There is no narrowing of the cell walls. The lengths of the cells can be 2–3 times their width and the cell lengths are 4–8.5 μm. Its cells are 2-3 (3.5) μm wide, pale blue-green in color. There is no caliptra. They are found as microphytoplankton throughout rivers. There are two granules in the middle part of the yarn. The last cell has no title, and the last cell has a round shape. They live planktonically in lakes and dam waters in water tanks [101].

Phormidium formosum

It is found on soils that have been wetted or submerged with water for long periods, on objects underwater, and on moist soil and rocks. Width 4–6 μm, bright blue-green, tapering, and curving towards the tip. Thalli is blue-green, trichome is flat, and there is some narrowing of cell walls. They spread in water tanks, artificial lakes, over obstacles in the riverbed, wastewater channels, and salty lakes. The shape of the cells is almost quadrangular, the cell length is 1/2 times its width, the cells are 2.5-5 μm long, the intermediate chamber is sometimes slightly granular, the last cell is wider than the others, there is no cap, no calyptra. They are found in rivers as microphytoplankton (Fig. **41**) [122, 292, 324, 384, 595, 831, 993, 1000, 1008, 1086-1094]. There is a fuel recovery study on this species [292].

Fig. (41). *Phormidium formosum* (scaled with 10 μm) [1095].

Phormidium splendens

They settle in the still waters of lakes, poolside, where water flows from waterfalls. The cell length can be 2–4 times the width, or the cells can be in the form of a quadrilateral, 3–9 μm in length. It occurs as planktonic in waters or develops on underwater objects. The partition is often granular, the end of the thread is more or less curled, the last cell is almost rectangular, often without calyptra. Thallus bright blue-green or olive green, trichome straight or curved, there is no narrowing of the cell walls, the thread gradually becomes thinner towards the tip, its cells are 2–3 μm wide.

Calothrix thermalis

The thallus is enlarged, musilated and soft, with a blue-green or olive-green color. The filaments are intertwined, and easily pliable, the filaments can be up to 3 mm in length, the threads form dense clumps. The filaments are 8–10 μm wide and the bottom of the thread is swollen. Its sheath is thick, hyaline, or yellowish in color at the bottom and is not lamellar. The cells of the thread are 5-8 μm wide, the cell length is usually 1/3 more than the width, blue-green color, has a heterocyst, and is located at the bottom, but rarely is located between the cells in the thread, the heterocyst is ellipsoidal or hemispherical. They live in the concrete walls of freshwater channels, and in the concrete walls of canals where water flows from thermal water sources [46, 152, 162, 168, 995, 1092, 1096, 1097].

Calothrix marchica lemmermann

Filaments can be straight or slightly curled. The filaments are singular, and the base width is around 5-6 μm. First, they have a tight and colorless coat. The sheath is not painted blue with the chlorine-zinc-iodine solution. Towards the tip, it becomes thinner and ends in hair. Narrowing of the intermediate lamellae is observed. The base width of the yarn is 4–5.5 μm. Cell size is approximately equal to its width or ½ - ¼ more than its width. The last cell has a conical shape, the tip may be slightly tapered. It has single heterosis. Basal is almost cylindrical or semi-cylindrical in shape. The thread is 4–5.5 μm wide. They spread in moist soils and moist walls, water tanks, pools, and settlements [28, 159, 1098].

Mastigocladus cf. laminosus

Other species need revision. The type species M. laminosus is a typical obligate thermophilic species and lives in all hot springs around the world (including Antarctica). Akinetes are very rare, they are found only in old trichomes. Thallus structure, soft form, spongy layers, sometimes with small lime crystals, smooth or gelatinous surface. The filaments formed by the cells lined up in single rows form inverted 'V' shaped short branching, although not as much as the true side branching pseudo-branching, the sheath is thin and tight or fluid, is located between the cells in the heterocyst thread, hormogon formation is unknown. They are dirty blue-green or olive green in color. Perhaps this depends on specific environmental conditions (45-60 °C, 7.5 <Ph, aerobic, low salinity), and all other environmental conditions, especially other species living in cold waters, are evident. Thallus consists of more or less densely mixed filaments. Trichomes are irregularly curled, have a thin, distinctive and colorless coat. There is true branching, and they are 'T' or 'Y' shaped, often unilateral. Cell division is crosswise or longitudinal before branching. At some stages, branching is very rare or almost absent. Pseudo-branching is a selective phenomenon but not common the structures it creates are quite frequent and even stratified from time to time. Branching generally becomes a continuous thinning towards the tip. Heterocytes are intercalated, solitary or (rarely) in pairs. Reproduction occurs by hormogones or separations in the trichome It varies a lot, M. laminosus is forgiven, one of the reasons for this is that it is genotypically confused with other prokaryotes.

CONCLUSION

The main characteristics and requirements of algae species in biofuel production are summarized below.

1. Algae (single-celled microalgae) are microscopic photosynthetic creatures that live on land and in the sea. They are considered by scientists as the creatures that enable life to begin and migrate to land, and it is stated that they have existed for about 5 billion years. It is known that approximately 30,000 species of algae, which are estimated to have between 200,000 and 2-3 million species in nature, have been recorded so far, and only a few of them can be commercialized.

2. Seaweeds have extremely rich carbohydrate, protein, and especially fatty acid content. Algae with high nutritional value are the most important source of nutrients, vitamins, and trace elements for aquatic organisms. They also provide essential pigments for the development of coloration in fish and other aquatic creatures.

CONSENT FOR PUBLICATION

Not applicable.

CONFLICT OF INTEREST

The authosr declare no conflict of interest, financial or otherwise.

ACKNOWLEDGEMENT

Declared none.

REFERENCES

[1]　Harris JM, Vinobaba P, Kularatne RKA, Ellawala K. *C. Spatial* and temporal distribution of cyanobacteria in Batticaloa Lagoon. J Environ Sci (China) 2016; 47: 211-8.
[http://dx.doi.org/10.1016/j.jes.2016.01.020] [PMID: 27593288]

[2]　Mohamed ZA, Al Shehri AM. Microcystin production in epiphytic cyanobacteria on submerged macrophytes. Toxicon 2010; 55(7): 1346-52.
[http://dx.doi.org/10.1016/j.toxicon.2010.02.007] [PMID: 20167231]

[3]　Morris TE, Visscher PT, O'Leary MJ, Fearns PRCS, Collins LB. The biogeomorphology of Shark Bay's microbialite coasts. Earth Sci Rev 2020; 205: 102921.
[http://dx.doi.org/10.1016/j.earscirev.2019.102921]

[4]　Narchonai G, Arutselvan C, LewisOscar F, Thajuddin N. Deciphering the microalgal diversity and water quality assessment of two urban temple ponds in Pondicherry, India. Biocatal Agric Biotechnol 2019; 22: 101427.
[http://dx.doi.org/10.1016/j.bcab.2019.101427]

[5]　Narchonai G, Arutselvan C, LewisOscar F, Thajuddin N. Enhancing starch accumulation/production in *Chlorococcum humicola* through sulfur limitation and 2,4-D treatment for butanol production. Biotechnol Rep (Amst) 2020; 28: e00528.
[http://dx.doi.org/10.1016/j.btre.2020.e00528] [PMID: 32995316]

[6]　Nassar MZA, Gharib SM. Spatial and temporal patterns of phytoplankton composition in *Burullus Lagoon*, Southern Mediterranean Coast, Egypt. Egypt J Aquat Res 2014; 40(2): 133-42.
[http://dx.doi.org/10.1016/j.ejar.2014.06.004]

[7]　Pérez-Lloréns JL. Microalgae: From staple foodstuff to avant-garde cuisine. Int J Gastron Food Sci 2020; 21: 100221.
[http://dx.doi.org/10.1016/j.ijgfs.2020.100221]

[8]　Santhakumari S, Jayakumar R, Logalakshmi R, *et al. In vitro* and *in vivo* effect of 2,6-Di-tert-but-l-4-methylphenol as an antibiofilm agent against quorum sensing mediated biofilm formation of *Vibrio* spp. Int J Food Microbiol 2018; 281: 60-71.
[http://dx.doi.org/10.1016/j.ijfoodmicro.2018.05.024] [PMID: 29864697]

[9]　Sharma NK, Rai AK. Algal Particles in the Atmosphere II. Reference Module in Earth Systems and Environmental Sciences. Elsevier 2017.
[http://dx.doi.org/10.1016/B978-0-12-409548-9.10508-1]

[10]　Silva SMF, Pienaar RN. Marine cyanophytes from the Western Cape, South Africa: Chroococcales. S Afr J Bot 1999; 65(1): 32-49.
[http://dx.doi.org/10.1016/S0254-6299(15)30938-8]

[11]　Bhatnagar A, Makandar MB, Garg MK, Bhatnagar M. Community structure and diversity of cyanobacteria and green algae in the soils of Thar Desert (India). J Arid Environ 2008; 72(2): 73-83.
[http://dx.doi.org/10.1016/j.jaridenv.2007.05.007]

[12]　Çelekli A, Kayhan S, Çetin T. First assessment of lakes' water quality in Aras River catchment (Turkey); Application of phytoplankton metrics and multivariate approach. Ecol Indic 2020; 117: 106706.
[http://dx.doi.org/10.1016/j.ecolind.2020.106706]

[13]　Dasauni K, Nailwal TK. Biodiversity of microbial life: Indian Himalayan region. In: De Mandal S, Bhatt P, Eds. Recent Advancements in Microbial Diversity. Academic Press 2020; pp. 1-17.
[http://dx.doi.org/10.1016/B978-0-12-821265-3.00001-3]

[14] Esposito G, Teta R, Marrone R, *et al.* A fast detection strategy for cyanobacterial blooms and associated cyanotoxins (FDSCC) reveals the occurrence of lyngbyatoxin A in campania (South Italy). Chemosphere 2019; 225: 342-51.
[http://dx.doi.org/10.1016/j.chemosphere.2019.02.201] [PMID: 30884295]

[15] Foekema EM, Kaag NHBM, Kramer KJM, Long K. Mesocosm validation of the marine no effect concentration of dissolved copper derived from a species sensitivity distribution. Sci Total Environ 2015; 521-522: 173-82.
[http://dx.doi.org/10.1016/j.scitotenv.2015.03.086] [PMID: 25829294]

[16] He D, Simoneit BRT, Jara B, Jaffé R. Occurrence and distribution of monomethylalkanes in the freshwater wetland ecosystem of the Florida Everglades. Chemosphere 2015; 119: 258-66.
[http://dx.doi.org/10.1016/j.chemosphere.2014.06.054] [PMID: 25033241]

[17] Kaštovský J, Fučíková K, Veselá J, Brewer Carías C, Vegas-Vilarrúbia T. Algae. In: Rull V, Vegas-Vilarrúbia T, Huber O, Señaris C, Eds. Biodiversity of Pantepui. Academic Press 2019; pp. 95-120.
[http://dx.doi.org/10.1016/B978-0-12-815591-2.00005-7]

[18] Raja R, Coelho A, Hemaiswarya S, Kumar P, Carvalho IS, Alagarsamy A. Applications of microalgal paste and powder as food and feed: An update using text mining tool. Beni Suef Univ J Basic Appl Sci 2018; 7(4): 740-7.
[http://dx.doi.org/10.1016/j.bjbas.2018.10.004]

[19] Sivasubramanian V. Phycoremediation and Business Prospects. In: Prasad MNV, Ed. Bioremediation and Bioeconomy. Elsevier 2016; pp. 421-56.
[http://dx.doi.org/10.1016/B978-0-12-802830-8.00017-4]

[20] Wijeyaratne WMDN, Nanayakkara DBM. Monitoring of water quality variation trends in a tropical urban wetland system located within a Ramsar wetland city: A GIS and phytoplankton based assessment. Environ Nanotechnol Monit Manag 2020; 14: 100323.
[http://dx.doi.org/10.1016/j.enmm.2020.100323]

[21] Alkhalaf IA, Hübener T, Porembski S. Prey spectra of aquatic Utricularia species (Lentibulariaceae) in northeastern Germany: The role of planktonic algae. Flora (Jena) 2009; 204(9): 700-8.
[http://dx.doi.org/10.1016/j.flora.2008.09.008]

[22] Çelekli A, Öztürk B, Kapı M. Relationship between phytoplankton composition and environmental variables in an artificial pond. Algal Res 2014; 5: 37-41.
[http://dx.doi.org/10.1016/j.algal.2014.05.002]

[23] Chen J, Lu Y, Ye X, Emam M, Zhang H, Wang H. Current advances in *Vibrio harveyi quorum* sensing as drug discovery targets. Eur J Med Chem 2020; 207: 112741.
[http://dx.doi.org/10.1016/j.ejmech.2020.112741] [PMID: 32871343]

[24] Haroon AM, Hussian AEM. Ecological assessment of the macrophytes and phytoplankton in El-Rayah Al-Behery, River Nile, Egypt. Egypt J Aquat Res 2017; 43(3): 195-203.
[http://dx.doi.org/10.1016/j.ejar.2017.08.002]

[25] Kidron GJ, Barinova S, Vonshak A. The effects of heavy winter rains and rare summer rains on biological soil crusts in the Negev Desert. Catena 2012; 95: 6-11.
[http://dx.doi.org/10.1016/j.catena.2012.02.021]

[26] Krings M, Harper CJ. Deciphering interfungal relationships in the 410-million-yr-old Rhynie chert: Brijax amictus gen. et sp. nov. (Chytridiomycota) colonizing the walls of glomeromycotan acaulospores. Rev Palaeobot Palynol 2020; 281: 104287.
[http://dx.doi.org/10.1016/j.revpalbo.2020.104287]

[27] Lu C, Yang Z, Liu J, *et al.* Chlorpyrifos inhibits nitrogen fixation in rice-vegetated soil containing Pseudomonas stutzeri A1501. Chemosphere 2020; 256: 127098.
[http://dx.doi.org/10.1016/j.chemosphere.2020.127098] [PMID: 32470732]

[28] Mandal S, Rath J. Algal colonization and its ecophysiology on the fine sculptures of terracotta

monuments of Bishnupur, West Bengal, India. Int Biodeterior Biodegradation 2013; 84: 291-9.
[http://dx.doi.org/10.1016/j.ibiod.2012.05.034]

[29]   Papendorf O, König GM, Wright AD. Hierridin B and 2,4-dimethoxy-6-heptadecyl-phenol, secondary metabolites from the cyanobacterium *Phormidium ectocarpi* with antiplasmodial activity. Phytochemistry 1998; 49(8): 2383-6.
[http://dx.doi.org/10.1016/S0031-9422(98)00440-3] [PMID: 9887530]

[30]   Barinova SS, Yehuda G, Nevo E. Comparative analysis of algal communities in the rivers of northern and southern Israel as bearing on ecological consequences of climate change. J Arid Environ 2010; 74(7): 765-76.
[http://dx.doi.org/10.1016/j.jaridenv.2009.03.001]

[31]   Coutinho R, Seeliger U. The horizontal distribution of the benthic algal flora in the Patos Lagoon estuary, Brazil, in relation to salinity, substratum and wave exposure. J Exp Mar Biol Ecol 1984; 80(3): 247-57.
[http://dx.doi.org/10.1016/0022-0981(84)90153-9]

[32]   Cutrim MVJ, Ferreira FS, Duarte dos Santos AK, *et al.* Trophic state of an urban coastal lagoon (northern Brazil), seasonal variation of the phytoplankton community and environmental variables. Estuar Coast Shelf Sci 2019; 216: 98-109.
[http://dx.doi.org/10.1016/j.ecss.2018.08.013]

[33]   El-Serehy HA, Abdallah HS, Al-Misned FA, Al-Farraj SA, Al-Rasheid KA. Assessing water quality and classifying trophic status for scientifically based managing the water resources of the Lake Timsah, the lake with salinity stratification along the Suez Canal. Saudi J Biol Sci 2018; 25(7): 1247-56.
[http://dx.doi.org/10.1016/j.sjbs.2018.05.022] [PMID: 30505166]

[34]   El-Serehy HA, Abdallah HS, Al-Misned FA, Irshad R, Al-Farraj SA, Almalki ES. Aquatic ecosystem health and trophic status classification of the Bitter Lakes along the main connecting link between the Red Sea and the Mediterranean. Saudi J Biol Sci 2018; 25(2): 204-12.
[http://dx.doi.org/10.1016/j.sjbs.2017.12.004] [PMID: 29472766]

[35]   Zaady E, Arbel S, Barkai D, Sarig S. Long-term impact of agricultural practices on biological soil crusts and their hydrological processes in a semiarid landscape. J Arid Environ 2013; 90: 5-11.
[http://dx.doi.org/10.1016/j.jaridenv.2012.10.021]

[36]   Zhang Y, Hoffmann L. Blue-green algal mats of the salinas in San-ya, Hai-nan Island (China): Structure, taxonomic composition, and implications for the interpretation of Precambrian stromatolites. Precambrian Res 1992; 56(3-4): 275-90.
[http://dx.doi.org/10.1016/0301-9268(92)90105-W]

[37]   Zhang YM, Chen J, Wang L, Wang XQ, Gu ZH. The spatial distribution patterns of biological soil crusts in the Gurbantunggut Desert, Northern Xinjiang, China. J Arid Environ 2007; 68(4): 599-610.
[http://dx.doi.org/10.1016/j.jaridenv.2006.06.012]

[38]   Zhang YM, Wang HL, Wang XQ, Yang WK, Zhang DY. The microstructure of microbiotic crust and its influence on wind erosion for a sandy soil surface in the Gurbantunggut Desert of Northwestern China. Geoderma 2006; 132(3-4): 441-9.
[http://dx.doi.org/10.1016/j.geoderma.2005.06.008]

[39]   Zhao Y, Xu M, Belnap J. Potential nitrogen fixation activity of different aged biological soil crusts from rehabilitated grasslands of the hilly Loess Plateau, China. J Arid Environ 2010; 74(10): 1186-91.
[http://dx.doi.org/10.1016/j.jaridenv.2010.04.006]

[40]   Álvarez S, Díaz P, López-Archilla AI, Guerrero MC. Phytoplankton composition and dynamics in three shallow temporary salt lakes (Monegros, Spain). J Arid Environ 2006; 65(4): 553-71.
[http://dx.doi.org/10.1016/j.jaridenv.2005.09.001]

[41]   Bhowmick S, Mazumdar A, Moulick A, Adam V. Algal metabolites: An inevitable substitute for antibiotics. Biotechnol Adv 2020; 43: 107571.

[http://dx.doi.org/10.1016/j.biotechadv.2020.107571] [PMID: 32505655]

[42] Bird CW, Lynch JM, Pirt SJ, Reid WW. The identification of hop-22(29)-ene in prokaryotic organisms. Tetrahedron Lett 1971; 12(34): 3189-90.
[http://dx.doi.org/10.1016/S0040-4039(01)97127-8]

[43] Chellappa NT, Chellappa T, Câmara FRA, Rocha O, Chellappa S. Impact of stress and disturbance factors on the phytoplankton communities in Northeastern Brazil reservoir. Limnologica 2009; 39(4): 273-82.
[http://dx.doi.org/10.1016/j.limno.2009.06.006]

[44] Friedl T, Lorenz M. Meeting report: Second European Phycological Congress, Montecatini Terme, Italy, September 20-26, 1999. Protist 1999; 150(4): 357-62.
[http://dx.doi.org/10.1016/S1434-4610(99)70036-2] [PMID: 10714769]

[45] López-Sandoval Ó, Montejano G, Carmona J, Cantoral E, Becerra-Absalón I. Diversidad algal de un ambiente extremo: El manantial geotermal Los Hervideros, México. Rev Mex Biodivers 2016; 87(1): 1-9.
[http://dx.doi.org/10.1016/j.rmb.2016.01.004]

[46] Mohamed ZA. Toxic cyanobacteria and cyanotoxins in public hot springs in Saudi Arabia. Toxicon 2008; 51(1): 17-27.
[http://dx.doi.org/10.1016/j.toxicon.2007.07.007] [PMID: 17825867]

[47] Montánchez I, Kaberdin VR. Vibrio harveyi: A brief survey of general characteristics and recent epidemiological traits associated with climate change. Mar Environ Res 2020; 154: 104850.
[http://dx.doi.org/10.1016/j.marenvres.2019.104850] [PMID: 32056705]

[48] Sandoval Pérez AL, Camargo-Ricalde SL, Montaño NM, *et al.* Biocrusts, inside and outside resource islands of Mimosa luisana (Leguminosae), improve soil carbon and nitrogen dynamics in a tropical semiarid ecosystem. Eur J Soil Biol 2016; 74: 93-103.
[http://dx.doi.org/10.1016/j.ejsobi.2016.03.006]

[49] Choudhary P, Srivastava RK. Techno-economic case study: Bio-fixation of industrial emissions at an Indian oil and gas plant. J Clean Prod 2020; 266: 121820.
[http://dx.doi.org/10.1016/j.jclepro.2020.121820]

[50] de Jonge VN. Algal vegetations on salt-marshes along the western Dutch Wadden Sea. Neth J Sea Res 1976; 10(2): 262-83.
[http://dx.doi.org/10.1016/0077-7579(76)90018-1]

[51] Fall R, Lewin RA, Fall LR. Intracellular coagulation inhibits the extraction of proteins from Prochloron. Phytochemistry 1983; 22(11): 2365-8.
[http://dx.doi.org/10.1016/0031-9422(83)80121-6]

[52] Goh BHH, Ong HC, Cheah MY, Chen WH, Yu KL, Mahlia TMI. Sustainability of direct biodiesel synthesis from microalgae biomass: A critical review. Renew Sustain Energy Rev 2019; 107: 59-74.
[http://dx.doi.org/10.1016/j.rser.2019.02.012]

[53] Gülz PG, Faxel P, Feige GB, Schmitz B, Egge H. Die kohlenwasserstoffe der blaualgen-flechte Peltigera canina (L.) Willd. Z Pflanzenphysiol 1978; 89(2): 159-67.
[http://dx.doi.org/10.1016/S0044-328X(78)80056-7]

[54] Hamawand I, Yusaf T, Hamawand S. Growing algae using water from coal seam gas industry and harvesting using an innovative technique: A review and a potential. Fuel 2014; 117: 422-30.
[http://dx.doi.org/10.1016/j.fuel.2013.09.040]

[55] McCormick PV, O'Dell MB, Shuford RBE III, Backus JG, Kennedy WC. Periphyton responses to experimental phosphorus enrichment in a subtropical wetland. Aquat Bot 2001; 71(2): 119-39.
[http://dx.doi.org/10.1016/S0304-3770(01)00175-9]

[56] Schneider H, Gelpi E, Bennett EO, Oró J. Fatty acids of geochemical significance in microscopic algae. Phytochemistry 1970; 9(3): 613-7.

[http://dx.doi.org/10.1016/S0031-9422(00)85701-5]

[57]   Soule T, Garcia-Pichel F. Cyanobacteria. In: Schmidt TM, Ed. Encyclopedia of Microbiology. 4th ed. Oxford: Academic Press 2019; pp. 799-817.

[58]   Tiwari B, Kharwar S, Tiwari DN. Pesticides and Rice Agriculture. In: Mishra AK, Tiwari DN, Rai AN, Eds. Cyanobacteria. Academic Press 2019; pp. 303-25.
        [http://dx.doi.org/10.1016/B978-0-12-814667-5.00015-5]

[59]   Coute A, Tell G, Therezien Y. Cyanophyceae (Cyanobacteria) aérophiles de Nouvelle-Calédonie. Cryptogam, Algol 1999; 20(4): 301-44.
        [http://dx.doi.org/10.1016/S0181-1568(00)88147-7]

[60]   Coutinho ML, Miller AZ, Macedo MF. Biological colonization and biodeterioration of architectural ceramic materials: An overview. J Cult Herit 2015; 16(5): 759-77.
        [http://dx.doi.org/10.1016/j.culher.2015.01.006]

[61]   Coutinho R, Seeliger U. Seasonal occurrence and growth of benthic algae in the Patos Lagoon estuary, Brazil. Estuar Coast Shelf Sci 1986; 23(6): 889-900.
        [http://dx.doi.org/10.1016/0272-7714(86)90079-X]

[62]   Garcia-Pichel F. Cyanobacteria. In: Schaechter M, Ed. Encyclopedia of Microbiology. 3rd ed. Oxford: Academic Press 2009; pp. 107-24.
        [http://dx.doi.org/10.1016/B978-012373944-5.00250-9]

[63]   Hayatsu R, McBeth RL, Scott RG, Botto RE, Winans RE. Artificial coalification study: Preparation and characterization of synthetic mecerals. Org Geochem 1984; 6: 463-71.
        [http://dx.doi.org/10.1016/0146-6380(84)90069-X]

[64]   Juntono . Blue-green algae in rice soils of Jogjakarta, central Java. Soil Biol Biochem 1973; 5(1): 91-5.
        [http://dx.doi.org/10.1016/0038-0717(73)90095-3]

[65]   Saliot A. Natural Hydrocarbons in Sea Water Alain Saliot. In: Duursma EK, Dawson R, Eds. Elsevier Oceanography Series. Elsevier 1981; Vol. 31: pp. 327-74.

[66]   Shiea J, Brassell SC, Ward DM. Mid-chain branched mono- and dimethyl alkanes in hot spring cyanobacterial mats: A direct biogenic source for branched alkanes in ancient sediments? Org Geochem 1990; 15(3): 223-31.
        [http://dx.doi.org/10.1016/0146-6380(90)90001-G]

[67]   Tornabene TG. Microbial Formation Of Hydrocarbons. In: Schlegel HG, Barnea J, Eds. Microbial Energy Conversion. Pergamon 1977; pp. 281-99.
        [http://dx.doi.org/10.1016/B978-0-08-021791-8.50029-9]

[68]   Yen TF. Structural Aspects of Organic Components in Oil Shales. In: Yen TF, Chilingarian GV, Eds. Developments in Petroleum Science. Elsevier 1976; Vol. 5: pp. 129-48.

[69]   Adey WH, Loveland K. Chapter 22 - Estuaries: Ecosystem Modeling and Restoration. In: Adey WH, Loveland K, Eds. Dynamic Aquaria. Third Edition. London: Academic Press 2007; pp. 405-41.

[70]   Alcocer J, Hammer UT. Saline lake ecosystems of Mexico. Aquat Ecosyst Health Manage 1998; 1(3-4): 291-315.
        [http://dx.doi.org/10.1080/14634989808656925]

[71]   Collins L, Alvarez D, Chauhan A. 15 - Phycoremediation Coupled with Generation of Value-Added Products. In: Das S, Ed. Microbial Biodegradation and Bioremediation. Oxford: Elsevier 2014; pp. 341-87.
        [http://dx.doi.org/10.1016/B978-0-12-800021-2.00015-7]

[72]   Kimble BJ, Maxwell JR, Philp RP, *et al.* Tri- and tetraterpenoid hydrocarbons in the Messel oil shale. Geochim Cosmochim Acta 1974; 38(7): 1165-81.
        [http://dx.doi.org/10.1016/0016-7037(74)90011-8]

[73]   Moul ET, Buell HF. 24 - Algae of the Pine Barrens. In: Forman RTT, Ed. Pine Barrens. Academic

Press 1979; pp. 425-40.

[74]   Reed WE. Biogeochemistry of Mono Lake, California. Geochim Cosmochim Acta 1977; 41(9): 1231-45.
       [http://dx.doi.org/10.1016/0016-7037(77)90069-2]

[75]   Stal L, Moezelaar R. Fermentation in cyanobacteria. FEMS Microbiol Rev 1997; 21(2): 179-211.
       [http://dx.doi.org/10.1016/S0168-6445(97)00056-9]

[76]   Stal LJ, Gemerden H, Krumbein WE. Structure and development of a benthic marine microbial mat. FEMS Microbiol Lett 1985; 31(2): 111-25.
       [http://dx.doi.org/10.1111/j.1574-6968.1985.tb01138.x]

[77]   Wetzel RG. 8 - Structure And Productivity Of Aquatic Ecosystems. In: Wetzel RG, Ed. Limnology. 3rd ed. San Diego: Academic Press 2001; pp. 129-50.
       [http://dx.doi.org/10.1016/B978-0-08-057439-4.50012-5]

[78]   Zismann L, Berdugo V, Kimor B. The food and feeding habits of early stages of grey mullets in the Haifa Bay region. Aquaculture 1975; 6(1): 59-75.
       [http://dx.doi.org/10.1016/0044-8486(75)90089-7]

[79]   Index To Plant Names. Inorganic Nutrition of Plants. Steward FC, Ed. Academic Press 1963; pp. 786-96.
       [http://dx.doi.org/10.1016/B978-0-12-395600-2.50023-1]

[80]   Adey WH, Finn M, Kangas P, Lange L, Luckett C, Spoon DM. A Florida Everglades Mesocosm — model veracity after four years of self-organization. Ecol Eng 1996; 6(1-3): 171-224.
       [http://dx.doi.org/10.1016/0925-8574(95)00057-7]

[81]   Erwin J. Comparative Biochemistry of Fatty Acids in Eukaryotic Microorganisms. In: Erwin JA, Ed. Lipids and Biomembranes of Eukaryotic Microorganisms. Academic Press 1973; pp. 41-143.
       [http://dx.doi.org/10.1016/B978-0-12-242050-4.50008-2]

[82]   Fogg GE, Stewart WDP, Fay P, Walsby AE. Freshwater ecology. In: Fogg GE, Stewart WDP, Fay P, Walsby AE, Eds. The Blue-green Algae. Academic Press 1973; pp. 255-80.
       [http://dx.doi.org/10.1016/B978-0-12-261650-1.50016-2]

[83]   Knoll AH, Golubic S. Anatomy and taphonomy of a precambrian algal stromatolite. Precambrian Res 1979; 10(1-2): 115-51.
       [http://dx.doi.org/10.1016/0301-9268(79)90022-6]

[84]   Kolattukudy PE, Walton TJ. The biochemistry of plant cuticular lipids. Prog Chem Fats Other Lipids 1972; 13(3): 119-75.
       [PMID: 4602868]

[85]   Komárek J. 3 - Coccoid And Colonial Cyanobacteria. In: Wehr JD, Sheath RG, Eds. Freshwater Algae of North America. Burlington: Academic Press 2003; pp. 59-116.
       [http://dx.doi.org/10.1016/B978-012741550-5/50004-0]

[86]   Moore RE. Constituents of Blue-Green Algae. In: Scheuer PJ, Ed. Marine Natural Products. Academic Press 1981; pp. 1-52.
       [http://dx.doi.org/10.1016/B978-0-12-624004-7.50008-1]

[87]   Nicholas DJD. Chapter 3 - Inorganic Nutrient Nutrition of Microorganisms. In: Steward FC, Ed. Inorganic Nutrition of Plants. Academic Press 1963; pp. 363-447.
       [http://dx.doi.org/10.1016/B978-0-12-395600-2.50015-2]

[88]   Mathimani T, Beena Nair B, Ranjith kumar R. Evaluation of microalga for biodiesel using lipid and fatty acid as a marker – A central composite design approach. Journal of the Energy Institute 2016; 89(3): 436-46.
       [http://dx.doi.org/10.1016/j.joei.2015.02.010]

[89]   Sharma J, Kumar SS, Bishnoi NR, Pugazhendhi A. Enhancement of lipid production from algal

biomass through various growth parameters. J Mol Liq 2018; 269: 712-20.
[http://dx.doi.org/10.1016/j.molliq.2018.08.103]

[90]   Belykh OI, Tikhonova IV, Kuzmin AV, *et al.* First detection of benthic cyanobacteria in Lake Baikal producing paralytic shellfish toxins. Toxicon 2016; 121: 36-40.
[http://dx.doi.org/10.1016/j.toxicon.2016.08.015] [PMID: 27569199]

[91]   Cocquyt C, Vyverman W. Phytoplankton in Lake Tanganyika: A Comparison of Community Composition and Biomass off Kigoma with Previous Studies 27 Years Ago. J Great Lakes Res 2005; 31(4): 535-46.
[http://dx.doi.org/10.1016/S0380-1330(05)70282-3]

[92]   Heynig VH. Interessante phytoplankter aus Gewässern des Bezirks halle (DDR). V. Arch Protistenkd 1987; 134(2-3): 179-90.
[http://dx.doi.org/10.1016/S0003-9365(87)80072-6]

[93]   Riaño B, Hernández D, García-González MC. Microalgal-based systems for wastewater treatment: Effect of applied organic and nutrient loading rate on biomass composition. Ecol Eng 2012; 49: 112-7.
[http://dx.doi.org/10.1016/j.ecoleng.2012.08.021]

[94]   Schauerte W, Lay JP, Klein W, Korte F. Influence of 2,4,6-trichlorophenol and pentachlorophenol on the biota of aquatic systems. Chemosphere 1982; 11(1): 71-9.
[http://dx.doi.org/10.1016/0045-6535(82)90095-9]

[95]   Scholz B, Liebezeit G. Screening for biological activities and toxicological effects of 63 phytoplankton species isolated from freshwater, marine and brackish water habitats. Harmful Algae 2012; 20: 58-70.
[http://dx.doi.org/10.1016/j.hal.2012.07.007]

[96]   Singh SP, Kashyap AK. Manganese toxicity and mutagenesis in two blue-green algae. Environ Exp Bot 1978; 18(1): 47-53.
[http://dx.doi.org/10.1016/0098-8472(78)90020-5]

[97]   Coelho S, Pérez-Ruzafa A, Gamito S. Phytoplankton community dynamics in an intermittently open hypereutrophic coastal lagoon in southern Portugal. Estuar Coast Shelf Sci 2015; 167: 102-12.
[http://dx.doi.org/10.1016/j.ecss.2015.07.022]

[98]   El-Karim MSA. Epipelic algal distribution in Ismailia Canal and the possible use of diatoms as bioindicators and a biomonitoring tool. Egypt J Aquat Res 2014; 40(4): 385-93.
[http://dx.doi.org/10.1016/j.ejar.2014.12.005]

[99]   El-Karim MSA. Survey to compare phytoplankton functional approaches: How can these approaches assess River Nile water quality in Egypt? Egypt J Aquat Res 2015; 41(3): 247-55.
[http://dx.doi.org/10.1016/j.ejar.2015.07.002]

[100] Kormas KA, Gkelis S, Vardaka E, Moustaka-Gouni M. Morphological and molecular analysis of bloom-forming Cyanobacteria in two eutrophic, shallow Mediterranean lakes. Limnologica 2011; 41(3): 167-73.
[http://dx.doi.org/10.1016/j.limno.2010.10.003]

[101] Kozak A, Celewicz-Gołdyn S, Kuczyńska-Kippen N. Cyanobacteria in small water bodies: The effect of habitat and catchment area conditions. Sci Total Environ 2019; 646: 1578-87.
[http://dx.doi.org/10.1016/j.scitotenv.2018.07.330] [PMID: 30235642]

[102] Krienitz L, Dadheech PK, Fastner J, Kotut K. The rise of potentially toxin producing cyanobacteria in Lake Naivasha, Great African Rift Valley, Kenya. Harmful Algae 2013; 27: 42-51.
[http://dx.doi.org/10.1016/j.hal.2013.04.005]

[103] Laplace-Treyture C, Feret T. Performance of the Phytoplankton Index for Lakes (IPLAC): A multimetric phytoplankton index to assess the ecological status of water bodies in France. Ecol Indic 2016; 69: 686-98.
[http://dx.doi.org/10.1016/j.ecolind.2016.05.025]

[104] Pandey LK. In situ assessment of metal toxicity in riverine periphytic algae as a tool for biomonitoring of fluvial ecosystems. Environmental Technology & Innovation 2020; 18: 100675.
[http://dx.doi.org/10.1016/j.eti.2020.100675]

[105] Pełechata A, Pełechaty M, Pukacz A. Winter temperature and shifts in phytoplankton assemblages in a small Chara-lake. Aquat Bot 2015; 124: 10-8.
[http://dx.doi.org/10.1016/j.aquabot.2015.03.001]

[106] Sathicq MB, Bauer DE, Gómez N. Influence of El Niño Southern Oscillation phenomenon on coastal phytoplankton in a mixohaline ecosystem on the southeastern of South America: Río de la Plata estuary. Mar Pollut Bull 2015; 98(1-2): 26-33.
[http://dx.doi.org/10.1016/j.marpolbul.2015.07.017] [PMID: 26183307]

[107] Barbiero RP, Tuchman ML. Results from the U.S. EPA's Biological Open Water Surveillance Program of the Laurentian Great Lakes: II. Deep Chlorophyll Maxima. J Great Lakes Res 2001; 27(2): 155-66.
[http://dx.doi.org/10.1016/S0380-1330(01)70629-6]

[108] Bowling LC, Blais S, Sinotte M. Heterogeneous spatial and temporal cyanobacterial distributions in Missisquoi Bay, Lake Champlain: An analysis of a 9 year data set. J Great Lakes Res 2015; 41(1): 164-79.
[http://dx.doi.org/10.1016/j.jglr.2014.12.012]

[109] Carrera-Martinez D, Mateos-Sanz A, Lopez-Rodas V, Costas E. Adaptation of microalgae to a gradient of continuous petroleum contamination. Aquat Toxicol 2011; 101(2): 342-50.
[http://dx.doi.org/10.1016/j.aquatox.2010.11.009] [PMID: 21216344]

[110] Gillett ND, Steinman AD. An analysis of long-term phytoplankton dynamics in Muskegon Lake, a Great Lakes Area of Concern. J Great Lakes Res 2011; 37(2): 335-42.
[http://dx.doi.org/10.1016/j.jglr.2011.01.009]

[111] Malam Issa O, Trichet J, Défarge C, Couté A, Valentin C. Morphology and microstructure of microbiotic soil crusts on a tiger bush sequence (Niger, Sahel). Catena 1999; 37(1-2): 175-96.
[http://dx.doi.org/10.1016/S0341-8162(99)00052-1]

[112] Malve O, Laine M, Haario H, Kirkkala T, Sarvala J. Bayesian modelling of algal mass occurrences—using adaptive MCMC methods with a lake water quality model. Environ Model Softw 2007; 22(7): 966-77.
[http://dx.doi.org/10.1016/j.envsoft.2006.06.016]

[113] Pełechata A, Pełechaty M, Pukacz A. Factors influencing cyanobacteria community structure in Chara-lakes. Ecol Indic 2016; 71: 477-90.
[http://dx.doi.org/10.1016/j.ecolind.2016.07.022]

[114] Rolland A, Bertrand F, Maumy M, Jacquet S. Assessing phytoplankton structure and spatio-temporal dynamics in a freshwater ecosystem using a powerful multiway statistical analysis. Water Res 2009; 43(13): 3155-68.
[http://dx.doi.org/10.1016/j.watres.2009.03.049] [PMID: 19520411]

[115] Tokodi N, Drobac D, Meriluoto J, *et al.* Cyanobacterial effects in Lake Ludoš, Serbia - Is preservation of a degraded aquatic ecosystem justified? Sci Total Environ 2018; 635: 1047-62.
[http://dx.doi.org/10.1016/j.scitotenv.2018.04.177] [PMID: 29710560]

[116] Watson SB, Brownlee B, Satchwill T, Hargesheimer EE. Quantitative analysis of trace levels of geosmin and MIB in source and drinking water using headspace SPME. Water Res 2000; 34(10): 2818-28.
[http://dx.doi.org/10.1016/S0043-1354(00)00027-0]

[117] Brown CL, Manny BA. Nearshore Phytoplankton of Hammond Bay, Lake Huron. J Great Lakes Res 1983; 9(4): 523-9.
[http://dx.doi.org/10.1016/S0380-1330(83)71925-8]

[118] Friedrich G, Pohlmann M. Long-term plankton studies at the lower Rhine/Germany. Limnologica 2009; 39(1): 14-39.
[http://dx.doi.org/10.1016/j.limno.2008.03.006]

[119] Mallin MA, Stone KL, Pamperl MA. Phytoplankton community assessments of seven southeast U.S. cooling reservoirs. Water Res 1994; 28(3): 665-73.
[http://dx.doi.org/10.1016/0043-1354(94)90146-5]

[120] Meichtry de Zaburlín N, Eugenio Vogler R, Martín Llano V, Mabel Martens IS. Fitoplancton del embalse Yacyretá (Argentina-Paraguay) a una década de su llenado. Rev Mex Biodivers 2013; 84(1): 225-39.
[http://dx.doi.org/10.7550/rmb.26831]

[121] Reavie ED, Barbiero RP, Allinger LE, Warren GJ. Phytoplankton trends in the Great Lakes, 2001–2011. J Great Lakes Res 2014; 40(3): 618-39.
[http://dx.doi.org/10.1016/j.jglr.2014.04.013]

[122] Reavie ED, Jicha TM, Angradi TR, Bolgrien DW, Hill BH. Algal assemblages for large river monitoring: Comparison among biovolume, absolute and relative abundance metrics. Ecol Indic 2010; 10(2): 167-77.
[http://dx.doi.org/10.1016/j.ecolind.2009.04.009]

[123] Shimoda Y, Rao YR, Watson S, Arhonditsis GB. Optimizing the complexity of phytoplankton functional group modeling: An allometric approach. Ecol Inform 2016; 31: 1-17.
[http://dx.doi.org/10.1016/j.ecoinf.2015.11.001]

[124] Woszczyk M, Bechtel A, Gratzer R, *et al.* Composition and origin of organic matter in surface sediments of Lake Sarbsko: A highly eutrophic and shallow coastal lake (northern Poland). Org Geochem 2011; 42(9): 1025-38.
[http://dx.doi.org/10.1016/j.orggeochem.2011.07.002]

[125] A1 - Algae. Handbook of Environmental Data and Ecological Parameters. JØRgensen SE, Ed. Pergamon 1979; pp. 2-87.

[126] Barbiero RP, Tuchman ML. Results from the U.S. EPA's Biological Open Water Surveillance Program of the Laurentian Great Lakes: I. Introduction and Phytoplankton Results. J Great Lakes Res 2001; 27(2): 134-54.
[http://dx.doi.org/10.1016/S0380-1330(01)70628-4]

[127] Eaton SW, Kardos LP. The Limnology of Canandaigua Lake. In: Bloomfield JA, Ed. Lakes of New York State. Academic Press 1978; pp. 225-311.
[http://dx.doi.org/10.1016/B978-0-12-107301-5.50010-7]

[128] Harr TE, Fuhs GW, Green DM, Hetling LJ, Smith SB, Allen SP. Limnology of Canadarago Lake**This chapter is dedicated to the memory of Thomas E. Harr who began the compilation of this material from numerous reports and wholly contributed the material for the two large introductory sections and for part of the third. In preparing the final text we have sought to preserve his style of writing because we feel that in this manner we can preserve and convey to others a valid image of his pleasant and balanced, yet powerful, personality. In: Bloomfield JA, Ed. Ecology of the Lakes of East-Central New York. Academic Press 1980; pp. 129-264.
[http://dx.doi.org/10.1016/B978-0-12-107303-9.50009-7]

[129] Hopkins GJ, Lea C. A ten year study of phytoplankton biomass and composition in the nanticoke region of Long Point Bay, Lake Erie. J Great Lakes Res 1982; 8(3): 428-38.
[http://dx.doi.org/10.1016/S0380-1330(82)71984-7]

[130] Mills EL, Forney JL, Clady MD, Schaffner WR. Oneida Lake. In: Bloomfield JA, Ed. Lakes of New York State. Academic Press 1978; pp. 367-451.
[http://dx.doi.org/10.1016/B978-0-12-107302-2.50011-X]

[131] Wang P, Wang C. 4.8 - Water Quality in Taihu Lake and the Effects of the Water Transfer from the

Yangtze River to Taihu Lake Project. In: Ahuja S, Ed. Comprehensive Water Quality and Purification. Waltham: Elsevier 2014; pp. 136-61.
[http://dx.doi.org/10.1016/B978-0-12-382182-9.00071-2]

[132] Zohary T, Yacobi Y, Alster A, Fishbein T, Lippman S, Tibor G. Phytoplankton 2014; 6: 161-90.

[133] Chomérat N, Garnier R, Bertrand C, Cazaubon A. Seasonal succession of cyanoprokaryotes in a hypereutrophic oligo-mesohaline lagoon from the South of France. Estuar Coast Shelf Sci 2007; 72(4): 591-602.
[http://dx.doi.org/10.1016/j.ecss.2006.11.008]

[134] Delfino DO, Wanderley MD, Silva e Silva LH, Feder F, Lopes FAS. Sedimentology and temporal distribution of microbial mats from Brejo do Espinho, Rio de Janeiro, Brazil. Sediment Geol 2012; 263-264: 85-95.
[http://dx.doi.org/10.1016/j.sedgeo.2011.08.009]

[135] Devercelli M, O'Farrell I. Factors affecting the structure and maintenance of phytoplankton functional groups in a nutrient rich lowland river. Limnologica 2013; 43(2): 67-78.
[http://dx.doi.org/10.1016/j.limno.2012.05.001]

[136] Nowicka-Krawczyk P, Żelazna-Wieczorek J. Dynamics in cyanobacterial communities from a relatively stable environment in an urbanised area (ambient springs in Central Poland). Sci Total Environ 2017; 579: 420-9.
[http://dx.doi.org/10.1016/j.scitotenv.2016.11.080] [PMID: 27876389]

[137] Pazouki P, Prévost M, McQuaid N, et al. Breakthrough of cyanobacteria in bank filtration. Water Res 2016; 102: 170-9.
[http://dx.doi.org/10.1016/j.watres.2016.06.037] [PMID: 27343842]

[138] Rivera-Aguilar V, Montejano G, Rodríguez-Zaragoza S, Durán-Díaz A. Distribution and composition of cyanobacteria, mosses and lichens of the biological soil crusts of the Tehuacán Valley, Puebla, México. J Arid Environ 2006; 67(2): 208-25.
[http://dx.doi.org/10.1016/j.jaridenv.2006.02.013]

[139] Banerjee S, Pal R. Morphotaxonomic study of blue green algae from pristine areas of West Bengal with special reference to sem studies of different morphotypes and four new reports. Phytomorphology 2017; 67(3&4): 67-83.

[140] Bao VWW, Leung KMY, Qiu JW, Lam MHW. Acute toxicities of five commonly used antifouling booster biocides to selected subtropical and cosmopolitan marine species. Mar Pollut Bull 2011; 62(5): 1147-51.
[http://dx.doi.org/10.1016/j.marpolbul.2011.02.041] [PMID: 21420693]

[141] Brindhadevi K, Mathimani T, Rene ER, Shanmugam S, Chi NTL, Pugazhendhi A. Impact of cultivation conditions on the biomass and lipid in microalgae with an emphasis on biodiesel. Fuel 2021; 284: 119058.
[http://dx.doi.org/10.1016/j.fuel.2020.119058]

[142] Deepthi AS, Ray JG. Algal associates and the evidence of cyanobacterial nitrogen fixation in the velamen roots of epiphytic orchids. Glob Ecol Conserv 2020; 22: e00946.
[http://dx.doi.org/10.1016/j.gecco.2020.e00946]

[143] Goyal VC, Singh O, Singh R, et al. Ecological health and water quality of village ponds in the subtropics limiting their use for water supply and groundwater recharge. J Environ Manage 2021; 277: 111450.
[http://dx.doi.org/10.1016/j.jenvman.2020.111450] [PMID: 33031997]

[144] Koziróg A, Otlewska A, Piotrowska M, et al. Colonising organisms as a biodegradation factor affecting historical wood materials at the former concentration camp of Auschwitz II – Birkenau. Int Biodeterior Biodegradation 2014; 86: 171-8.
[http://dx.doi.org/10.1016/j.ibiod.2013.08.004]

[145] Mohamed ZA, Al Shehri AM. Cyanobacteria and their toxins in treated-water storage reservoirs in Abha city, Saudi Arabia. Toxicon 2007; 50(1): 75-84.
[http://dx.doi.org/10.1016/j.toxicon.2007.02.021] [PMID: 17451766]

[146] Zhang AQ, Leung KMY, Kwok KWH, Bao VWW, Lam MHW. Toxicities of antifouling biocide Irgarol 1051 and its major degraded product to marine primary producers. Mar Pollut Bull 2008; 57(6-12): 575-86.
[http://dx.doi.org/10.1016/j.marpolbul.2008.01.021] [PMID: 18314144]

[147] Zhang AQ, Zhou GJ, Lam MHW, Leung KMY. Toxicities of the degraded mixture of Irgarol 1051 to marine organisms. Chemosphere 2019; 225: 565-73.
[http://dx.doi.org/10.1016/j.chemosphere.2019.03.038] [PMID: 30901651]

[148] Al-Hussieny AA, Alraheem EA, Lafta HY, *et al.* Algae personification toxicity by GC–MASS and treatment by using material potassium permanganate in exposed basin. Egyptian Journal of Petroleum 2017; 26(3): 835-42.
[http://dx.doi.org/10.1016/j.ejpe.2016.10.020]

[149] He L, Lin Z, Wang Y, *et al.* Facilitating harmful algae removal in fresh water *via* joint effects of multi-species algicidal bacteria. J Hazard Mater 2021; 403: 123662.
[http://dx.doi.org/10.1016/j.jhazmat.2020.123662] [PMID: 32846260]

[150] Lv J, Wu H, Chen M. Effects of nitrogen and phosphorus on phytoplankton composition and biomass in 15 subtropical, urban shallow lakes in Wuhan, China. Limnologica 2011; 41(1): 48-56.
[http://dx.doi.org/10.1016/j.limno.2010.03.003]

[151] Miao X, Wang S, Liu M, *et al.* Changes in the phytoplankton community structure of the Backshore Wetland of Expo Garden, Shanghai from 2009 to 2010. Aquac Fish 2019; 4(5): 198-204.
[http://dx.doi.org/10.1016/j.aaf.2019.02.004]

[152] Mohamed ZA. Cyanotoxins in Egypt and Saudi Arabia. In: Nriagu J, Ed. Encyclopedia of Environmental Health. 2nd ed. Oxford: Elsevier 2019; pp. 796-804.
[http://dx.doi.org/10.1016/B978-0-12-409548-9.11579-9]

[153] Sheikh MA, Juma FS, Staehr P, *et al.* Occurrence and distribution of antifouling biocide Irgarol-1051 in coral reef ecosystems, Zanzibar. Mar Pollut Bull 2016; 109(1): 586-90.
[http://dx.doi.org/10.1016/j.marpolbul.2016.05.035] [PMID: 27234364]

[154] Zerrifi SEA, Kasrati A, Redouane EM, *et al.* Essential oils from Moroccan plants as promising ecofriendly tools to control toxic cyanobacteria blooms. Ind Crops Prod 2020; 143: 111922.
[http://dx.doi.org/10.1016/j.indcrop.2019.111922]

[155] Deng X, Gao K, Sun J. Physiological and biochemical responses of Synechococcus sp. PCC7942 to Irgarol 1051 and diuron. Aquat Toxicol 2012; 122-123: 113-9.
[http://dx.doi.org/10.1016/j.aquatox.2012.06.004] [PMID: 22789406]

[156] Douterelo I, Perona E, Mateo P. Use of cyanobacteria to assess water quality in running waters. Environ Pollut 2004; 127(3): 377-84.
[http://dx.doi.org/10.1016/j.envpol.2003.08.016] [PMID: 14638298]

[157] Falkiewicz-Dulik M, Janda K, Wypych G. 9 - Environmental Fate Of Biostabilizers. In Handbook of Material Biodegradation, Biodeterioration, and Biostablization. Second Edition. ChemTec Publishing 2015; pp. 405-16.

[158] Howe PL, Reichelt-Brushett AJ, Clark MW, Seery CR. Toxicity estimates for diuron and atrazine for the tropical marine cnidarian Exaiptasia pallida and in- hospite Symbiodinium spp. using PAM chlorophyll- a fluorometry. J Photochem Photobiol B 2017; 171: 125-32.
[http://dx.doi.org/10.1016/j.jphotobiol.2017.05.006] [PMID: 28501690]

[159] Lombardozzi V, Castrignanò T, D'Antonio M, Casanova Municchia A, Caneva G. An interactive database for an ecological analysis of stone biopitting. Int Biodeterior Biodegradation 2012; 73: 8-15.
[http://dx.doi.org/10.1016/j.ibiod.2012.04.016]

[160] Marynowski L, Rahmonov O, Smolarek-Lach J, Rybicki M, Simoneit BRT. Origin and significance of saccharides during initial pedogenesis in a temperate climate region. Geoderma 2020; 361: 114064.
[http://dx.doi.org/10.1016/j.geoderma.2019.114064]

[161] Mohamed ZA, Al Shehri AM. Microcystins in groundwater wells and their accumulation in vegetable plants irrigated with contaminated waters in Saudi Arabia. J Hazard Mater 2009; 172(1): 310-5.
[http://dx.doi.org/10.1016/j.jhazmat.2009.07.010] [PMID: 19640645]

[162] Sompong U, Hawkins PR, Besley C, Peerapornpisal Y. The distribution of cyanobacteria across physical and chemical gradients in hot springs in northern Thailand. FEMS Microbiol Ecol 2005; 52(3): 365-76.
[http://dx.doi.org/10.1016/j.femsec.2004.12.007] [PMID: 16329921]

[163] Sousa AP, Nunes B. Standard and biochemical toxicological effects of zinc pyrithione in Daphnia magna and Daphnia longispina. Environ Toxicol Pharmacol 2020; 80: 103402.
[http://dx.doi.org/10.1016/j.etap.2020.103402] [PMID: 32693026]

[164] Ariño X, Saiz-Jimenez C. Deterioration of the Elephant Tomb (Necropolis of Carmona, Seville, Spain). Int Biodeterior Biodegradation 1997; 40(2-4): 233-9.
[http://dx.doi.org/10.1016/S0964-8305(97)00034-6]

[165] Gallego-Cartagena E, Morillas H, Maguregui M, *et al.* A comprehensive study of biofilms growing on the built heritage of a Caribbean industrial city in correlation with construction materials. Int Biodeterior Biodegradation 2020; 147: 104874.
[http://dx.doi.org/10.1016/j.ibiod.2019.104874]

[166] Guo Q, Ma K, Yang L, Cai Q, He K. A comparative study of the impact of species composition on a freshwater phytoplankton community using two contrasting biotic indices. Ecol Indic 2010; 10(2): 296-302.
[http://dx.doi.org/10.1016/j.ecolind.2009.06.002]

[167] Marcon AE, Navoni JA, de Oliveira Galvão MF, *et al.* Mutagenic potential assessment associated with human exposure to natural radioactivity. Chemosphere 2017; 167: 36-43.
[http://dx.doi.org/10.1016/j.chemosphere.2016.09.136] [PMID: 27705811]

[168] Mohamed ZA. Cyanotoxins in Egypt and Saudi Arabia. In: Nriagu JO, Ed. Encyclopedia of Environmental Health. Burlington: Elsevier 2011; pp. 872-80.
[http://dx.doi.org/10.1016/B978-0-444-52272-6.00625-5]

[169] Nowicka-Krawczyk P, Żelazna-Wieczorek J, Otlewska A, *et al.* Diversity of an aerial phototrophic coating of historic buildings in the former Auschwitz II-Birkenau concentration camp. Sci Total Environ 2014; 493: 116-23.
[http://dx.doi.org/10.1016/j.scitotenv.2014.05.113] [PMID: 24937497]

[170] Sharma LK, Naik R, Pandey PC. 12 - Efficacy of hyperspectral data for monitoring and assessment of wetland ecosystem. In: Pandey PC, Srivastava PK, Balzter H, Bhattacharya B, Petropoulos GP, Eds. Hyperspectral Remote Sensing. Elsevier 2020; pp. 221-46.
[http://dx.doi.org/10.1016/B978-0-08-102894-0.00007-3]

[171] Tomaselli L, Lamenti G, Bosco M, Tiano P. Biodiversity of photosynthetic micro-organisms dwelling on stone monuments. Int Biodeterior Biodegradation 2000; 46(3): 251-8.
[http://dx.doi.org/10.1016/S0964-8305(00)00078-0]

[172] Yu Q, Chen Y, Liu Z, Zhu D, Wang H. Longitudinal variations of phytoplankton compositions in lake-to-river systems. Limnologica 2017; 62: 173-80.
[http://dx.doi.org/10.1016/j.limno.2016.02.007]

[173] Carteau D, Vallée-Réhel K, Linossier I, *et al.* Development of environmentally friendly antifouling paints using biodegradable polymer and lower toxic substances. Prog Org Coat 2014; 77(2): 485-93.
[http://dx.doi.org/10.1016/j.porgcoat.2013.11.012]

[174] Hamala JA, Kollig HP. The effects of atrazine on periphyton communities in controlled laboratory

ecosystems. Chemosphere 1985; 14(9): 1391-408.
[http://dx.doi.org/10.1016/0045-6535(85)90159-6]

[175] Janda K. 6.13 - Petroleum Products. In: Falkiewicz-Dulik M, Janda K, Wypych G, Eds. Handbook of Material Biodegradation, Biodeterioration, and Biostablization. Second Edition. ChemTec Publishing 2015; pp. 257-375.

[176] Ledger ME, Brown LE, Edwards FK, Hudson LN, Milner AM, Woodward G. Chapter Six - Extreme Climatic Events Alter Aquatic Food Webs: A Synthesis of Evidence from a Mesocosm Drought Experiment. In: Woodward G, O'Gorman EJ, Eds. Advances in Ecological Research. Academic Press 2013; 48: pp. 343-95.

[177] Tian C, Pei H, Hu W, Xie J. Variation of cyanobacteria with different environmental conditions in Nansi Lake, China. J Environ Sci (China) 2012; 24(8): 1394-402.
[http://dx.doi.org/10.1016/S1001-0742(11)60964-9] [PMID: 23513680]

[178] Arguelles E. Descriptive study of some epiphytic algae growing on Hydrilla verticillata (L.f) Royle (Hydrocharitaceae) found in the shallow freshwater lake Laguna de Bay (Philippines). Taxonomy and Diversity of Epiphytic Algae from Submerged Aquatic Macrophytes found in Laguna de Bay and its Tributary Rivers 2019; 23: 15-28.
[http://dx.doi.org/10.21608/EJABF.2019.29300]

[179] Brindhadevi K, Anto S, Rene ER, *et al.* Effect of reaction temperature on the conversion of algal biomass to bio-oil and biochar through pyrolysis and hydrothermal liquefaction. Fuel 2021; 285: 119106.
[http://dx.doi.org/10.1016/j.fuel.2020.119106]

[180] Brink J, Marx S. Harvesting of Hartbeespoort Dam micro-algal biomass through sand filtration and solar drying. Fuel 2013; 106: 67-71.
[http://dx.doi.org/10.1016/j.fuel.2012.10.034]

[181] Duman F, Sahin U, Atabani AE. Harvesting of blooming microalgae using green synthetized magnetic maghemite (γ-Fe2O3) nanoparticles for biofuel production. Fuel 2019; 256: 115935.
[http://dx.doi.org/10.1016/j.fuel.2019.115935]

[182] Guo Y, Zhao Y, Wang S, Jiang C, Zhang J. Relationship between the zeta potential and the chemical agglomeration efficiency of fine particles in flue gas during coal combustion. Fuel 2018; 215: 756-65.
[http://dx.doi.org/10.1016/j.fuel.2017.11.005]

[183] Imanparast S, Faramarzi MA, Hamedi J. Production of a cyanobacterium-based biodiesel by the heterogeneous biocatalyst of SBA-15@oleate@lipase. Fuel 2020; 279: 118580.
[http://dx.doi.org/10.1016/j.fuel.2020.118580]

[184] Javed F, Aslam M, Rashid N, *et al.* Microalgae-based biofuels, resource recovery and wastewater treatment: A pathway towards sustainable biorefinery. Fuel 2019; 255: 115826.
[http://dx.doi.org/10.1016/j.fuel.2019.115826]

[185] Kanda H, Li P, Ikehara T, Yasumoto-Hirose M. Lipids extracted from several species of natural blue–green microalgae by dimethyl ether: Extraction yield and properties. Fuel 2012; 95: 88-92.
[http://dx.doi.org/10.1016/j.fuel.2011.11.064]

[186] Sekar M, Mathimani T, Alagumalai A, *et al.* A review on the pyrolysis of algal biomass for biochar and bio-oil – Bottlenecks and scope. Fuel 2021; 283: 119190.
[http://dx.doi.org/10.1016/j.fuel.2020.119190]

[187] Vargas-Estrada L, Torres-Arellano S, Longoria A, Arias DM, Okoye PU, Sebastian PJ. Role of nanoparticles on microalgal cultivation: A review. Fuel 2020; 280: 118598.
[http://dx.doi.org/10.1016/j.fuel.2020.118598]

[188] Fan G, Du B, Zhou J, Yan Z, You Y, Luo J. Porous self-floating 3D Ag2O/g-C3N4 hydrogel and photocatalytic inactivation of *Microcystis aeruginosa* under visible light. Chem Eng J 2021; 404: 126509.

[http://dx.doi.org/10.1016/j.cej.2020.126509]

[189] Gao Y, Lu J, Orr PT, Chuang A, Franklin HM, Burford MA. Enhanced resistance of co-existing toxigenic and non-toxigenic *Microcystis aeruginosa* to pyrogallol compared with monostrains. Toxicon 2020; 176: 47-54.
[http://dx.doi.org/10.1016/j.toxicon.2020.01.013] [PMID: 32103795]

[190] Huang Y, Pan H, Liu H, Xi Y, Ren D. Characteristics of growth and microcystin production of *Microcystis aeruginosa* exposed to low concentrations of naphthalene and phenanthrene under different pH values. Toxicon 2019; 169: 103-8.
[http://dx.doi.org/10.1016/j.toxicon.2019.09.004] [PMID: 31494204]

[191] Shahmohamadloo RS, Simmons DBD, Sibley PK. Shotgun proteomics analysis reveals sub-lethal effects in Daphnia magna exposed to cell-bound microcystins produced by *Microcystis aeruginosa*. Comp Biochem Physiol Part D Genomics Proteomics 2020; 33: 100656.
[http://dx.doi.org/10.1016/j.cbd.2020.100656] [PMID: 32035333]

[192] Siedlewicz G, Żak A, Sharma L, Kosakowska A, Pazdro K. Effects of oxytetracycline on growth and chlorophyll a fluorescence in green algae (Chlorella vulgaris), diatom (Phaeodactylum tricornutum) and cyanobacteria (*Microcystis aeruginosa* and Nodularia spumigena). Oceanologia 2020; 62(2): 214-25.
[http://dx.doi.org/10.1016/j.oceano.2019.12.002]

[193] Symes E, van Ogtrop FF. The effect of pre-industrial and predicted atmospheric CO_2 concentrations on the development of diazotrophic and non-diazotrophic cyanobacterium: Dolichospermum circinale and *Microcystis aeruginosa*. Harmful Algae 2019; 88: 101536.
[http://dx.doi.org/10.1016/j.hal.2018.10.005] [PMID: 31582161]

[194] Wu X, Liu J, Zhu JJ. Sono-Fenton hybrid process on the inactivation of *Microcystis aeruginosa*: Extracellular and intracellular oxidation. Ultrason Sonochem 2019; 53: 68-76.
[http://dx.doi.org/10.1016/j.ultsonch.2018.12.034] [PMID: 30600211]

[195] Wu Y, Wan L, Zhang W, Ding H, Yang W. Resistance of cyanobacteria *Microcystis aeruginosa* to erythromycin with multiple exposure. Chemosphere 2020; 249: 126147.
[http://dx.doi.org/10.1016/j.chemosphere.2020.126147] [PMID: 32062559]

[196] Zhang D, Ye Q, Zhang F, *et al.* Flocculating properties and potential of Halobacillus sp. strain H9 for the mitigation of *Microcystis aeruginosa* blooms. Chemosphere 2019; 218: 138-46.
[http://dx.doi.org/10.1016/j.chemosphere.2018.11.082] [PMID: 30471494]

[197] Zhang S, Chen Y, Zang X, Zhang X. Harvesting of *Microcystis aeruginosa* using membrane filtration: Influence of pore structure on fouling kinetics, algogenic organic matter retention and cake formation. Algal Res 2020; 52: 102112.
[http://dx.doi.org/10.1016/j.algal.2020.102112]

[198] Czarny K, Krawczyk B, Szczukocki D. Toxic effects of bisphenol A and its analogues on cyanobacteria Anabaena variabilis and *Microcystis aeruginosa*. Chemosphere 2021; 263: 128299.
[http://dx.doi.org/10.1016/j.chemosphere.2020.128299] [PMID: 33297241]

[199] Deng J, Fu D, Hu W, Lu X, Wu Y, Bryan H. Physiological responses and accumulation ability of *Microcystis aeruginosa* to zinc and cadmium: Implications for bioremediation of heavy metal pollution. Bioresour Technol 2020; 303: 122963.
[http://dx.doi.org/10.1016/j.biortech.2020.122963] [PMID: 32050124]

[200] Jiang Y, Liu Y, Zhang J. Mechanisms for the stimulatory effects of a five-component mixture of antibiotics in *Microcystis aeruginosa* at transcriptomic and proteomic levels. J Hazard Mater 2021; 406: 124722.
[http://dx.doi.org/10.1016/j.jhazmat.2020.124722] [PMID: 33296757]

[201] Kuang X, Peng L, Chen A, Zeng Q, Luo S, Shao J. Enhancement mechanisms of copper(II) adsorption onto kaolinite by extracellular polymeric substances of *Microcystis aeruginosa* (cyanobacterium). Int Biodeterior Biodegradation 2019; 138: 8-14.

[http://dx.doi.org/10.1016/j.ibiod.2018.12.009]

[202] Liu Y, Cui M, Zhang J, Gao B. Impacts of antibiotic contaminants on *Microcystis aeruginosa* during potassium permanganate treatment. Harmful Algae 2020; 92: 101741.
[http://dx.doi.org/10.1016/j.hal.2020.101741] [PMID: 32113608]

[203] Oehrle S, Rodriguez-Matos M, Cartamil M, Zavala C, Rein KS. Toxin composition of the 2016 *Microcystis aeruginosa* bloom in the St. Lucie Estuary, Florida. Toxicon 2017; 138: 169-72.
[http://dx.doi.org/10.1016/j.toxicon.2017.09.005] [PMID: 28899665]

[204] Vasudevan S, Arulmoorthy MP, Suresh R. Isolation, purification and structural elucidation of secondary metabolites from *Microcystis aeruginosa* bloom from Muttukadu estuary and its *in vitro* antibacterial, antioxidant and anticancer potency. S Afr J Bot 2020; 132: 59-67.
[http://dx.doi.org/10.1016/j.sajb.2020.03.030]

[205] Vieira Lemos R, Tsujimura S, Ledezma P, Tokunou Y, Okamoto A, Freguia S. Extracellular electron transfer by *Microcystis aeruginosa* is solely driven by high pH. Bioelectrochemistry 2021; 137: 107637.
[http://dx.doi.org/10.1016/j.bioelechem.2020.107637] [PMID: 32898791]

[206] Wang Y, Zhao M, Liu T. Extraction of allelochemicals from poplar alkaline peroxide mechanical pulping effluents and their allelopathic effects on *Microcystis aeruginosa*. Journal of Bioresources and Bioproducts 2020; 5(4): 276-82.
[http://dx.doi.org/10.1016/j.jobab.2020.10.006]

[207] Wu D, Wang T, Wang J, Jiang L, Yin Y, Guo H. Size-dependent toxic effects of polystyrene microplastic exposure on *Microcystis aeruginosa* growth and microcystin production. Sci Total Environ 2021; 761: 143265.
[http://dx.doi.org/10.1016/j.scitotenv.2020.143265] [PMID: 33257060]

[208] Cruces E, Barrios AC, Cahue YP, Januszewski B, Gilbertson LM, Perreault F. Similar toxicity mechanisms between graphene oxide and oxidized multi-walled carbon nanotubes in *Microcystis aeruginosa*. Chemosphere 2021; 265: 129137.
[http://dx.doi.org/10.1016/j.chemosphere.2020.129137] [PMID: 33288276]

[209] El-Sheekh MM, Alwaleed EA, Kassem WMA, Saber H. Antialgal and antiproliferative activities of the algal silver nanoparticles against the toxic cyanobacterium *Microcystis aeruginosa* and human tumor colon cell line. Environ Nanotechnol Monit Manag 2020; 14: 100352.
[http://dx.doi.org/10.1016/j.enmm.2020.100352]

[210] Lian H, Xiang P, Xue Y, Jiang Y, Li M, Mo J. Efficiency and mechanisms of simultaneous removal of *Microcystis aeruginosa* and microcystins by electrochemical technology using activated carbon fiber/nickel foam as cathode material. Chemosphere 2020; 252: 126431.
[http://dx.doi.org/10.1016/j.chemosphere.2020.126431] [PMID: 32208197]

[211] Park KY, Choi SY, Ahn S, Kweon JH. Disinfection by-product formation potential of algogenic organic matter from *Microcystis aeruginosa*: Effects of growth phases and powdered activated carbon adsorption. J Hazard Mater 2021; 408: 124864.
[http://dx.doi.org/10.1016/j.jhazmat.2020.124864] [PMID: 33387716]

[212] Xie E, Su Y, Deng S, Kontopyrgou M, Zhang D. Significant influence of phosphorus resources on the growth and alkaline phosphatase activities of *Microcystis aeruginosa*. Environ Pollut 2021; 268(Pt B): 115807.
[http://dx.doi.org/10.1016/j.envpol.2020.115807] [PMID: 33096390]

[213] Yu RP, Wang LP, Zhao CK, Wu SF, Song QJ. Determination of volatile metabolites in *Microcystis aeruginosa* using headspace-solid phase microextraction arrow combined with gas chromatography-mass spectrometry. Chin J Anal Chem 2020; 48(6): 750-6.
[http://dx.doi.org/10.1016/S1872-2040(20)60026-3]

[214] Zhang J, Jiang L, Wu D, Yin Y, Guo H. Effects of environmental factors on the growth and microcystin production of *Microcystis aeruginosa* under TiO_2 nanoparticles stress. Sci Total Environ

2020; 734: 139443.
[http://dx.doi.org/10.1016/j.scitotenv.2020.139443] [PMID: 32454338]

[215] Zhao W, Zhao P, Tian Y, Shen C, Li Z, Jin C. Transport and retention of *Microcystis aeruginosa* in porous media: Impacts of ionic strength, flow rate, media size and pre-oxidization. Water Res 2019; 162: 277-87.
[http://dx.doi.org/10.1016/j.watres.2019.07.001] [PMID: 31284157]

[216] Zhu Y, Cheng S, Wang P, *et al.* A possible environmental-friendly removal of *Microcystis aeruginosa* by using pyroligneous acid. Ecotoxicol Environ Saf 2020; 205: 111159.
[http://dx.doi.org/10.1016/j.ecoenv.2020.111159] [PMID: 32829212]

[217] Bernat-Quesada F, Álvaro M, García H, Navalón S. Impact of chlorination and pre-ozonation on disinfection by-products formation from aqueous suspensions of cyanobacteria: *Microcystis aeruginosa*, Anabaena aequalis and Oscillatoria tenuis. Water Res 2020; 183: 116070.
[http://dx.doi.org/10.1016/j.watres.2020.116070] [PMID: 32622236]

[218] Fan G, Zhou J, Zheng X, Luo J, Hong L, Qu F. Fast photocatalytic inactivation of *Microcystis aeruginosa* by metal-organic frameworks under visible light. Chemosphere 2020; 239: 124721.
[http://dx.doi.org/10.1016/j.chemosphere.2019.124721] [PMID: 31493752]

[219] Li S, Dao GH, Tao Y, *et al.* The growth suppression effects of UV-C irradiation on *Microcystis aeruginosa* and Chlorella vulgaris under solo-culture and co-culture conditions in reclaimed water. Sci Total Environ 2020; 713: 136374.
[http://dx.doi.org/10.1016/j.scitotenv.2019.136374] [PMID: 31955073]

[220] Li S, Tao Y, Dao GH, Hu HY. Synergetic suppression effects upon the combination of UV-C irradiation and berberine on *Microcystis aeruginosa* and Scenedesmus obliquus in reclaimed water: Effectiveness and mechanisms. Sci Total Environ 2020; 744: 140937.
[http://dx.doi.org/10.1016/j.scitotenv.2020.140937] [PMID: 32711324]

[221] Liu W, Yang J, Tian Y, *et al.* An in situ extractive fermentation strategy for enhancing prodigiosin production from Serratia marcescens BWL1001 and its application to inhibiting the growth of *Microcystis aeruginosa*. Biochem Eng J 2020; 107836.

[222] Menezes I, Maxwell-McQueeney D, Capelo-Neto J, Pestana CJ, Edwards C, Lawton LA. Oxidative stress in the cyanobacterium *Microcystis aeruginosa* PCC 7813: Comparison of different analytical cell stress detection assays. Chemosphere 2021; 269: 128766.
[http://dx.doi.org/10.1016/j.chemosphere.2020.128766] [PMID: 33143884]

[223] Tian C, Zhao Y-X. Dosage and pH dependence of coagulation with polytitanium salts for the treatment of *Microcystis aeruginosa*-laden and Microcystis wesenbergii-laden surface water: The influence of basicity. J Water Process Eng 2020; 101726.

[224] Wang D, Ao Y, Wang P. Effective inactivation of *Microcystis aeruginosa* by a novel Z-scheme composite photocatalyst under visible light irradiation. Sci Total Environ 2020; 746: 141149.
[http://dx.doi.org/10.1016/j.scitotenv.2020.141149] [PMID: 32763606]

[225] Wang S, Li Q, Huang S, Zhao W, Zheng Z. Single and combined effects of microplastics and lead on the freshwater algae *Microcystis aeruginosa*. Ecotoxicol Environ Saf 2021; 208: 111664.
[http://dx.doi.org/10.1016/j.ecoenv.2020.111664] [PMID: 33396174]

[226] Duman F, Atelge MR, Kaya M, *et al.* A novel *Microcystis aeruginosa* supported manganese catalyst for hydrogen generation through methanolysis of sodium borohydride. Int J Hydrogen Energy 2020; 45(23): 12755-65.
[http://dx.doi.org/10.1016/j.ijhydene.2020.01.068]

[227] Franco ES, Pádua VL, Rodriguez MDVR, *et al.* A simple liquid-liquid extraction-gas chromatography-mass spectrometry method for the determination of haloacetic acids in environmental samples: Application in water with *Microcystis aeruginosa* cells. Microchem J 2019; 150: 104088.
[http://dx.doi.org/10.1016/j.microc.2019.104088]

[228] Hernando M, De Troch M, de la Rosa F, Giannuzzi L. Fatty acid response of the invasive bivalve Limnoperna fortunei fed with *Microcystis aeruginosa* exposed to high temperature. Comp Biochem Physiol C Toxicol Pharmacol 2021; 240: 108925.
[http://dx.doi.org/10.1016/j.cbpc.2020.108925] [PMID: 33130072]

[229] Jia N, Yang Y, Yu G, *et al.* Interspecific competition reveals *Raphidiopsis raciborskii* as a more successful invader than *Microcystis aeruginosa*. Harmful Algae 2020; 97: 101858.
[http://dx.doi.org/10.1016/j.hal.2020.101858] [PMID: 32732052]

[230] Pan D, Pavagadhi S, Umashankar S, *et al.* Resource partitioning strategies during toxin production in *Microcystis aeruginosa* revealed by integrative omics analysis. Algal Res 2019; 42: 101582.
[http://dx.doi.org/10.1016/j.algal.2019.101582]

[231] Wan L, Wu Y, Zhang B, Yang W, Ding H, Zhang W. Effects of moxifloxacin and gatifloxacin stress on growth, photosynthesis, antioxidant responses, and microcystin release in *Microcystis aeruginosa*. J Hazard Mater 2021; 409: 124518.
[http://dx.doi.org/10.1016/j.jhazmat.2020.124518] [PMID: 33191018]

[232] Wang B, Zhang Y, Qin Y, Li H. Removal of *Microcystis aeruginosa* and control of algal organic matter by Fe(II)/peroxymonosulfate pre-oxidation enhanced coagulation. Chem Eng J 2021; 403: 126381.
[http://dx.doi.org/10.1016/j.cej.2020.126381]

[233] Xu S, Liu Y, Zhang J, Gao B. Proteomic mechanisms for the combined stimulatory effects of glyphosate and antibiotic contaminants on *Microcystis aeruginosa*. Chemosphere 2021; 267: 129244.
[http://dx.doi.org/10.1016/j.chemosphere.2020.129244] [PMID: 33321278]

[234] Yuan R, Li Y, Li J, Ji S, Wang S, Kong F. The allelopathic effects of aqueous extracts from Spartina alterniflora on controlling the *Microcystis aeruginosa* blooms. Sci Total Environ 2020; 712: 136332.
[http://dx.doi.org/10.1016/j.scitotenv.2019.136332] [PMID: 31935546]

[235] Zhang M, Steinman AD, Xue Q, Zhao Y, Xu Y, Xie L. Effects of erythromycin and sulfamethoxazole on *Microcystis aeruginosa*: Cytotoxic endpoints, production and release of microcystin-LR. J Hazard Mater 2020; 399: 123021.
[http://dx.doi.org/10.1016/j.jhazmat.2020.123021] [PMID: 32937707]

[236] Álvarez SD, Kruk C, Martínez de la Escalera G, Montes MA, Segura AM, Piccini C. Morphology captures toxicity in *Microcystis aeruginosa* complex: Evidence from a wide environmental gradient☐. Harmful Algae 2020; 97: 101854.
[http://dx.doi.org/10.1016/j.hal.2020.101854] [PMID: 32732048]

[237] Chen H, Tsai KP, Liu Y, *et al.* Characterization of Dissolved Organic Matter from Wildfire-induced *Microcystis aeruginosa* Blooms controlled by Copper Sulfate as Disinfection Byproduct Precursors Using APPI(-) and ESI(-) FT-ICR MS. Water Res 2021; 189: 116640.
[http://dx.doi.org/10.1016/j.watres.2020.116640] [PMID: 33260105]

[238] Duan Y, Xiong D, Li Y, Dong H, Wang W, Zhang J. Changes in the gastrointestinal microbial community of Lateolabrax maculatus in a naturally occurring *Microcystis aeruginosa* bloom environment. Aquaculture 2020; 528: 735444.
[http://dx.doi.org/10.1016/j.aquaculture.2020.735444]

[239] Gong W, Fan Y, Xie B, *et al.* Immobilizing *Microcystis aeruginosa* and powdered activated carbon for the anaerobic digestate effluent treatment. Chemosphere 2020; 244: 125420.
[http://dx.doi.org/10.1016/j.chemosphere.2019.125420] [PMID: 31790994]

[240] Gonsior M, Powers LC, Williams E, *et al.* The chemodiversity of algal dissolved organic matter from lysed *Microcystis aeruginosa* cells and its ability to form disinfection by-products during chlorination. Water Res 2019; 155: 300-9.
[http://dx.doi.org/10.1016/j.watres.2019.02.030] [PMID: 30852317]

[241] Peng Y, Zhang Z, Kong Y, *et al.* Effects of ultrasound on *Microcystis aeruginosa* cell destruction and

release of intracellular organic matter. Ultrason Sonochem 2020; 63: 104909.
[http://dx.doi.org/10.1016/j.ultsonch.2019.104909] [PMID: 31945559]

[242] Qi J, Song Y, Liang J, *et al.* Growth inhibition of *Microcystis aeruginosa* by sand-filter prevalent manganese-oxidizing bacterium. Separ Purif Tech 2021; 256: 117808.
[http://dx.doi.org/10.1016/j.seppur.2020.117808]

[243] Sampognaro L, Eirín K, Martínez de la Escalera G, Piccini C, Segura A, Kruk C. Experimental evidence on the effects of temperature and salinity in morphological traits of the *Microcystis aeruginosa* complex. J Microbiol Methods 2020; 175: 105971.
[http://dx.doi.org/10.1016/j.mimet.2020.105971] [PMID: 32544485]

[244] Zhang K, Pan R, Luo Z, Zhang T, Fan J. Interspecific competition between *Microcystis aeruginosa* and Pseudanadaena and their production of T&O compounds. Chemosphere 2020; 252: 126509.
[http://dx.doi.org/10.1016/j.chemosphere.2020.126509] [PMID: 32224357]

[245] Zhang X, Lu X, Li H. Isolation and identification of a novel allelochemical from Ruppia maritima extract against the cyanobacteria *Microcystis aeruginosa*. Environmental Technology & Innovation 2021; 21: 101301.
[http://dx.doi.org/10.1016/j.eti.2020.101301]

[246] Barešová M, Načeradská J, Novotná K, Čermáková L, Pivokonský M. The impact of preozonation on the coagulation of cellular organic matter produced by *Microcystis aeruginosa* and its toxin degradation. J Environ Sci (China) 2020; 98: 124-33.
[http://dx.doi.org/10.1016/j.jes.2020.05.031] [PMID: 33097143]

[247] Caracciolo AB, Dejana L, Fajardo C, *et al.* A new fluorescent oligonucleotide probe for *in situ* identification of *Microcystis aeruginosa* in freshwater. Microchem J 2019; 148: 503-13.
[http://dx.doi.org/10.1016/j.microc.2019.05.017]

[248] Fan G, Zhan J, Luo J, *et al.* Fabrication of heterostructured Ag/AgCl@g-C$_3$N$_4$@UIO-66(NH$_2$) nanocomposite for efficient photocatalytic inactivation of *Microcystis aeruginosa* under visible light. J Hazard Mater 2021; 404(Pt B): 124062.
[http://dx.doi.org/10.1016/j.jhazmat.2020.124062] [PMID: 33068992]

[249] He Y, Ma J, Joseph V, *et al.* Potassium regulates the growth and toxin biosynthesis of *Microcystis aeruginosa*. Environ Pollut 2020; 267: 115576.
[http://dx.doi.org/10.1016/j.envpol.2020.115576] [PMID: 32898730]

[250] Schmidt KC, Jackrel SL, Smith DJ, Dick GJ, Denef VJ. Genotype and host microbiome alter competitive interactions between *Microcystis aeruginosa* and Chlorella sorokiniana. Harmful Algae 2020; 99: 101939.
[http://dx.doi.org/10.1016/j.hal.2020.101939] [PMID: 33218432]

[251] Wu D, Yang S, Du W, Yin Y, Zhang J, Guo H. Effects of titanium dioxide nanoparticles on *Microcystis aeruginosa* and microcystins production and release. J Hazard Mater 2019; 377: 1-7.
[http://dx.doi.org/10.1016/j.jhazmat.2019.05.013] [PMID: 31129339]

[252] Wu X, Xu G, Zhu JJ. Sonochemical synthesis of Fe$_3$O$_4$/carbon nanotubes using low frequency ultrasonic devices and their performance for heterogeneous sono-persulfate process on inactivation of *Microcystis aeruginosa*. Ultrason Sonochem 2019; 58: 104634.
[http://dx.doi.org/10.1016/j.ultsonch.2019.104634] [PMID: 31450346]

[253] Xu H, Liu Y, Tang Z, Li H, Li G, He Q. Methane production in harmful algal blooms collapsed water: The contribution of non-toxic *Microcystis aeruginosa* outweighs that of the toxic variety. J Clean Prod 2020; 276: 124280.
[http://dx.doi.org/10.1016/j.jclepro.2020.124280]

[254] Zhao P, Liu S, Huang W, *et al.* Influence of eugenol on algal growth, cell physiology of cyanobacteria *Microcystis aeruginosa* and its interaction with signaling molecules. Chemosphere 2020; 255: 126935.
[http://dx.doi.org/10.1016/j.chemosphere.2020.126935] [PMID: 32387731]

[255] Zhou L, Cai M, Zhang X, Cui N, Chen G, Zou G. In-situ nitrogen-doped black TiO2 with enhanced visible-light-driven photocatalytic inactivation of *Microcystis aeruginosa* cells: Synthesization, performance and mechanism. Appl Catal B 2020; 272: 119019.
[http://dx.doi.org/10.1016/j.apcatb.2020.119019]

[256] Brêda-Alves F, Militão FP, de Alvarenga BF, *et al.* Clethodim (herbicide) alters the growth and toxins content of *Microcystis aeruginosa* and *Raphidiopsis raciborskii*. Chemosphere 2020; 243: 125318.
[http://dx.doi.org/10.1016/j.chemosphere.2019.125318] [PMID: 31995862]

[257] de la Rosa F, De Troch M, Malanga G, Hernando M. Differential sensitivity of fatty acids and lipid damage in *Microcystis aeruginosa* (cyanobacteria) exposed to increased temperature. Comp Biochem Physiol C Toxicol Pharmacol 2020; 235: 108773.
[http://dx.doi.org/10.1016/j.cbpc.2020.108773] [PMID: 32360213]

[258] Liu Y, Yang M, Zheng L, *et al.* Antioxidant responses of triangle sail mussel Hyriopsis cumingii exposed to toxic *Microcystis aeruginosa* and thermal stress. Sci Total Environ 2020; 743: 140754.
[http://dx.doi.org/10.1016/j.scitotenv.2020.140754] [PMID: 32758840]

[259] Liu Y, Yang Q, Zhu M, *et al.* Endocytosis in *Microcystis aeruginosa* accelerates the synthesis of microcystins in the presence of lanthanum(III). Harmful Algae 2020; 93: 101791.
[http://dx.doi.org/10.1016/j.hal.2020.101791] [PMID: 32307072]

[260] Zhang X, Lian B. Carbonation of heat-activated serpentine driven by *Microcystis aeruginosa* PCC7806. Algal Res 2020; 50: 101995.
[http://dx.doi.org/10.1016/j.algal.2020.101995]

[261] Zhao M, Qu D, Shen W, Li M. Effects of dissolved organic matter from different sources on *Microcystis aeruginosa* growth and physiological characteristics. Ecotoxicol Environ Saf 2019; 176: 125-31.
[http://dx.doi.org/10.1016/j.ecoenv.2019.03.085] [PMID: 30925328]

[262] Zhu Y, Ling J, Li L, Guan X. The effectiveness of bisulfite-activated permanganate technology to enhance the coagulation efficiency of *Microcystis aeruginosa*. Chin Chem Lett 2020; 31(6): 1545-9.
[http://dx.doi.org/10.1016/j.cclet.2019.12.036]

[263] Kim H, Jo BY, Kim HS. Effect of different concentrations and ratios of ammonium, nitrate, and phosphate on growth of the blue-green alga (cyanobacterium) *Microcystis aeruginosa* isolated from the Nakdong River, Korea. Algae 2017; 32(4): 275-84.
[http://dx.doi.org/10.4490/algae.2017.32.10.23]

[264] Abdelhaleem A, Chu W, Farzana S. Diphenamid photodegradation using Fe(III) impregnated N-doped TiO₂/sulfite/visible LED process: Influence of wastewater matrix, kinetic modeling, and toxicity evaluation. Chemosphere 2020; 256: 127094.
[http://dx.doi.org/10.1016/j.chemosphere.2020.127094] [PMID: 32559888]

[265] Chalifour A, Chin WY, Leung PY, Cheung SG, Tam NFY. Effect of light on the transformation of BDE-47 by living and autoclaved cultures of Microcystis flos-aquae and Chlorella vulgaris. Chemosphere 2019; 233: 140-8.
[http://dx.doi.org/10.1016/j.chemosphere.2019.05.189] [PMID: 31170584]

[266] Hofbauer B, Jüttner F. Occurrence of isopropylthio compounds in the aquatic ecosystem (Lake Neusiedl, Austria) as a chemical marker for Microcystis flos-aquae. FEMS Microbiol Lett 1988; 53(2): 113-21.
[http://dx.doi.org/10.1111/j.1574-6968.1988.tb02654.x]

[267] Ono F, Nishihira N, Hada Y, *et al.* Strong tolerance of blue-green alga Microcystis flos-aquae to very high pressure. J Phys Chem Solids 2015; 84: 57-62.
[http://dx.doi.org/10.1016/j.jpcs.2014.08.008]

[268] Wan J, Guo P, Peng X, Wen K. Effect of erythromycin exposure on the growth, antioxidant system and photosynthesis of Microcystis flos-aquae. J Hazard Mater 2015; 283: 778-86.

[http://dx.doi.org/10.1016/j.jhazmat.2014.10.026] [PMID: 25464321]

[269] Wang M, Zhang Y, Guo P. Effect of florfenicol and thiamphenicol exposure on the photosynthesis and antioxidant system of Microcystis flos-aquae. Aquat Toxicol 2017; 186: 67-76.
[http://dx.doi.org/10.1016/j.aquatox.2017.02.022] [PMID: 28257901]

[270] Wu X, Wu H, Wang S, *et al.* Effect of propionamide on the growth of Microcystis flos-aquae colonies and the underlying physiological mechanisms. Sci Total Environ 2018; 630: 526-35.
[http://dx.doi.org/10.1016/j.scitotenv.2018.02.217] [PMID: 29486445]

[271] Xu Q, Yang L, Yang W, *et al.* Volatile organic compounds released from Microcystis flos-aquae under nitrogen sources and their toxic effects on Chlorella vulgaris. Ecotoxicol Environ Saf 2017; 135: 191-200.
[http://dx.doi.org/10.1016/j.ecoenv.2016.09.027] [PMID: 27741460]

[272] Zang X, Wang L, Xiao J, Zhang H, Li L, Zhang X. Harvesting colonial Microcystis flos-aquae using two-stage filtration: Influence of pre-filtration on harvesting performance. Separ Purif Tech 2020; 245: 116736.
[http://dx.doi.org/10.1016/j.seppur.2020.116736]

[273] Zang X, Zhang H, Liu Q, Li L, Li L, Zhang X. Harvesting of Microcystis flos-aquae using chitosan coagulation: Influence of proton-active functional groups originating from extracellular and intracellular organic matter. Water Res 2020; 185: 116272.
[http://dx.doi.org/10.1016/j.watres.2020.116272] [PMID: 32810743]

[274] Akbar S, Du J, Lin H, Kong X, Sun S, Tian X. Understanding interactive inducible defenses of Daphnia and its phytoplankton prey. Harmful Algae 2017; 66: 47-56.
[http://dx.doi.org/10.1016/j.hal.2017.05.003] [PMID: 28602253]

[275] Chen W, Peng L, Wan N, Song L. Mechanism study on the frequent variations of cell-bound microcystins in cyanobacterial blooms in Lake Taihu: Implications for water quality monitoring and assessments. Chemosphere 2009; 77(11): 1585-93.
[http://dx.doi.org/10.1016/j.chemosphere.2009.09.037] [PMID: 19853885]

[276] Feng W, Liu S, Li C, *et al.* Algal uptake of hydrophilic and hydrophobic dissolved organic nitrogen in the eutrophic lakes. Chemosphere 2019; 214: 295-302.
[http://dx.doi.org/10.1016/j.chemosphere.2018.09.070] [PMID: 30265937]

[277] Jia Y, Han G, Wang C, *et al.* The efficacy and mechanisms of fungal suppression of freshwater harmful algal bloom species. J Hazard Mater 2010; 183(1-3): 176-81.
[http://dx.doi.org/10.1016/j.jhazmat.2010.07.009] [PMID: 20675050]

[278] Jia Y, Wang Q, Chen Z, Jiang W, Zhang P, Tian X. Inhibition of phytoplankton species by co-culture with a fungus. Ecol Eng 2010; 36(10): 1389-91.
[http://dx.doi.org/10.1016/j.ecoleng.2010.06.017]

[279] Liu Y, Xu Y, Wang Z, *et al.* Dominance and succession of Microcystis genotypes and morphotypes in Lake Taihu, a large and shallow freshwater lake in China. Environ Pollut 2016; 219: 399-408.
[http://dx.doi.org/10.1016/j.envpol.2016.05.021] [PMID: 27209340]

[280] Luo L, Duan N, Wang XC, Guo W, Ngo HH. New thermodynamic entropy calculation based approach towards quantifying the impact of eutrophication on water environment. Sci Total Environ 2017; 603-604: 86-93.
[http://dx.doi.org/10.1016/j.scitotenv.2017.06.069] [PMID: 28623794]

[281] Ma H, Cui F, Fan Z, Liu Z, Zhao Z. Efficient control of Microcystis blooms by promoting biological filter-feeding in raw water. Ecol Eng 2012; 47: 71-5.
[http://dx.doi.org/10.1016/j.ecoleng.2012.06.026]

[282] Zhou Q, Chen W, Zhang H, *et al.* A flow cytometer based protocol for quantitative analysis of bloom-forming cyanobacteria (Microcystis) in lake sediments. J Environ Sci (China) 2012; 24(9): 1709-16.
[http://dx.doi.org/10.1016/S1001-0742(11)60993-5] [PMID: 23520881]

[283] Zhu W, Dai X, Li M. Relationship between extracellular polysaccharide (EPS) content and colony size of Microcystis is colonial morphology dependent. Biochem Syst Ecol 2014; 55: 346-50.
[http://dx.doi.org/10.1016/j.bse.2014.04.009]

[284] Abd El-Hady HH, Fathey SA, Ali GH, Gabr YG. Biochemical profile of phytoplankton and its nutritional aspects in some khors of Lake Nasser, Egypt. Egyptian Journal of Basic and Applied Sciences 2016; 3(2): 187-93.
[http://dx.doi.org/10.1016/j.ejbas.2016.03.002]

[285] Cai Y, Kong F, Shi L, Yu Y. Spatial heterogeneity of cyanobacterial communities and genetic variation of Microcystis populations within large, shallow eutrophic lakes (Lake Taihu and Lake Chaohu, China). J Environ Sci (China) 2012; 24(10): 1832-42.
[http://dx.doi.org/10.1016/S1001-0742(11)61007-3] [PMID: 23520854]

[286] Chalifour A, Tam NFY. Tolerance of cyanobacteria to the toxicity of BDE-47 and their removal ability. Chemosphere 2016; 164: 451-61.
[http://dx.doi.org/10.1016/j.chemosphere.2016.08.109] [PMID: 27604061]

[287] Guo Z, Guo A, Guo Q, *et al.* Decomposition of dexamethasone by gamma irradiation: Kinetics, degradation mechanisms and impact on algae growth. Chem Eng J 2017; 307: 722-8.
[http://dx.doi.org/10.1016/j.cej.2016.08.138]

[288] Jiang YJ, He W, Liu WX, *et al.* The seasonal and spatial variations of phytoplankton community and their correlation with environmental factors in a large eutrophic Chinese lake (Lake Chaohu). Ecol Indic 2014; 40: 58-67.
[http://dx.doi.org/10.1016/j.ecolind.2014.01.006]

[289] Ma J. Differential sensitivity of three cyanobacterial and five green algal species to organotins and pyrethroids pesticides. Sci Total Environ 2005; 341(1-3): 109-17.
[http://dx.doi.org/10.1016/j.scitotenv.2004.09.028] [PMID: 15833245]

[290] Ma J, Lu N, Qin W, Xu R, Wang Y, Chen X. Differential responses of eight cyanobacterial and green algal species, to carbamate insecticides. Ecotoxicol Environ Saf 2006; 63(2): 268-74.
[http://dx.doi.org/10.1016/j.ecoenv.2004.12.002] [PMID: 16677910]

[291] Sidelev S, Zubishina A, Chernova E. Distribution of microcystin-producing genes in Microcystis colonies from some Russian freshwaters: Is there any correlation with morphospecies and colony size? Toxicon 2020; 184: 136-42.
[http://dx.doi.org/10.1016/j.toxicon.2020.06.005] [PMID: 32540220]

[292] Vieira KR, Pinheiro PN, Zepka LQ. Volatile organic compounds from microalgae. In: Jacob-Lopes E, Maroneze MM, Queiroz MI, Zepka LQ, Eds. Handbook of Microalgae-Based Processes and Products. Academic Press 2020; pp. 659-86.
[http://dx.doi.org/10.1016/B978-0-12-818536-0.00024-5]

[293] Eguzozie KU, Mavumengwana V, Kayitesi E. Incidence of microcystins (Hepatotoxin) in floating scums in the Swartspruit River, South Africa. Toxicol Lett 2016; 258: S184.
[http://dx.doi.org/10.1016/j.toxlet.2016.06.1679]

[294] Gaget V, Humpage AR, Huang Q, Monis P, Brookes JD. Benthic cyanobacteria: A source of cylindrospermopsin and microcystin in Australian drinking water reservoirs. Water Res 2017; 124: 454-64.
[http://dx.doi.org/10.1016/j.watres.2017.07.073] [PMID: 28787682]

[295] Guidetti Vendruscolo R, Bittencourt Fagundes M, Jacob-Lopes E, Wagner R. Analytical strategies for using gas chromatography to control and optimize microalgae bioprocessing. Curr Opin Food Sci 2019; 25: 73-81.
[http://dx.doi.org/10.1016/j.cofs.2019.02.008]

[296] Lezcano MÁ, Agha R, Cirés S, Quesada A. Spatial-temporal survey of Microcystis oligopeptide chemotypes in reservoirs with dissimilar waterbody features and their relation to genetic variation.

Harmful Algae 2019; 81: 77-85.
[http://dx.doi.org/10.1016/j.hal.2018.11.009] [PMID: 30638501]

[297]   Liu J, Yu Q, Ye B, *et al.* Programmed cell death of Chlamydomonas reinhardtii induced by three cyanobacterial volatiles β-ionone, limonene and longifolene. Sci Total Environ 2021; 762: 144539.
[http://dx.doi.org/10.1016/j.scitotenv.2020.144539] [PMID: 33360449]

[298]   Mowe MAD, Porojan C, Abbas F, *et al.* Rising temperatures may increase growth rates and microcystin production in tropical Microcystis species. Harmful Algae 2015; 50: 88-98.
[http://dx.doi.org/10.1016/j.hal.2015.10.011] [PMID: 28366396]

[299]   Nandini S, Araiza-Vázquez DA, Sarma SSS. Moina macrocopa demographic response to harmful cyanobacteria. Ecohydrol Hydrobiol 2020.

[300]   Violi JP, Mitrovic SM, Colville A, Main BJ, Rodgers KJ. Prevalence of β-methylamino-L-alanine (BMAA) and its isomers in freshwater cyanobacteria isolated from eastern Australia. Ecotoxicol Environ Saf 2019; 172: 72-81.
[http://dx.doi.org/10.1016/j.ecoenv.2019.01.046] [PMID: 30682636]

[301]   Williamson N, Kobayashi T, Outhet D, Bowling LC. Survival of cyanobacteria in rivers following their release in water from large headwater reservoirs. Harmful Algae 2018; 75: 1-15.
[http://dx.doi.org/10.1016/j.hal.2018.04.004] [PMID: 29778219]

[302]   Zhou Z, Zhang Z, Feng L, *et al.* Adverse effects of levofloxacin and oxytetracycline on aquatic microbial communities. Sci Total Environ 2020; 734: 139499.
[http://dx.doi.org/10.1016/j.scitotenv.2020.139499] [PMID: 32464375]

[303]   Barón-Sola Á, Sanz-Alférez S, del Campo FF. First evidence of accumulation in cyanobacteria of guanidinoacetate, a precursor of the toxin cylindrospermopsin. Chemosphere 2015; 119: 1099-104.
[http://dx.doi.org/10.1016/j.chemosphere.2014.08.046] [PMID: 25460748]

[304]   Gkelis S, Papadimitriou T, Zaoutsos N, Leonardos I. Anthropogenic and climate-induced change favors toxic cyanobacteria blooms: Evidence from monitoring a highly eutrophic, urban Mediterranean lake. Harmful Algae 2014; 39: 322-33.
[http://dx.doi.org/10.1016/j.hal.2014.09.002]

[305]   Ma J, Wang P, Ren L, Wang X, Paerl HW. Using alkaline phosphatase activity as a supplemental index to optimize predicting algal blooms in phosphorus-deficient lakes: A case study of Lake Taihu, China. Ecol Indic 2019; 103: 698-712.
[http://dx.doi.org/10.1016/j.ecolind.2019.04.043]

[306]   Mowe MAD, Porojan C, Abbas F, *et al.* Corrigendum to "Rising temperatures may increase growth rates and microcystin production in tropical Microcystis species" [Harmful Algae 50 88-98]. Harmful Algae 2017; 63: 205-6.
[http://dx.doi.org/10.1016/j.hal.2016.07.007] [PMID: 28366396]

[307]   Su M, Gaget V, Giglio S, Burch M, An W, Yang M. Establishment of quantitative PCR methods for the quantification of geosmin-producing potential and Anabaena sp. in freshwater systems. Water Res 2013; 47(10): 3444-54.
[http://dx.doi.org/10.1016/j.watres.2013.03.043] [PMID: 23622984]

[308]   Wei L, Li X, Yi J, Yang Z, Wang Q, Ma W. A simple approach for the efficient production of hydrogen from Taihu Lake Microcystis spp. blooms. Bioresour Technol 2013; 139: 136-40.
[http://dx.doi.org/10.1016/j.biortech.2013.04.026] [PMID: 23648763]

[309]   Wu Y, Guo P, Zhang X, Zhang Y, Xie S, Deng J. Effect of microplastics exposure on the photosynthesis system of freshwater algae. J Hazard Mater 2019; 374: 219-27.
[http://dx.doi.org/10.1016/j.jhazmat.2019.04.039] [PMID: 31005054]

[310]   Zervou SK, Gkelis S, Kaloudis T, Hiskia A, Mazur-Marzec H. New microginins from cyanobacteria of Greek freshwaters. Chemosphere 2020; 248: 125961.
[http://dx.doi.org/10.1016/j.chemosphere.2020.125961] [PMID: 32059332]

[311] Zheng H, Sun C, Hou X, Wu M, Yao Y, Li F. Pyrolysis of Arundo donax L. to produce pyrolytic vinegar and its effect on the growth of dinoflagellate Karenia brevis. Bioresour Technol 2018; 247: 273-81.
[http://dx.doi.org/10.1016/j.biortech.2017.09.049] [PMID: 28950136]

[312] Zhu P, Zhang BF, Wu JH, *et al.* Sensitive and rapid detection of microcystin synthetase E Gene (mcyE) by loop-mediated isothermal amplification: A new assay for detecting the potential microcystin-producing Microcystis in the aquatic ecosystem. Harmful Algae 2014; 37: 8-16.
[http://dx.doi.org/10.1016/j.hal.2014.04.018]

[313] Chen S, Li J, Feng W, *et al.* Biochemical responses of the freshwater microalga Dictyosphaerium sp. upon exposure to three sulfonamides. J Environ Sci (China) 2020; 97: 141-8.
[http://dx.doi.org/10.1016/j.jes.2020.05.018] [PMID: 32933729]

[314] El-Sheekh MM, Haroon AM, Sabae S. Seasonal and spatial variation of aquatic macrophytes and phytoplankton community at El-Quanater El-Khayria River Nile, Egypt. Beni Suef Univ J Basic Appl Sci 2018; 7(3): 344-52.
[http://dx.doi.org/10.1016/j.bjbas.2018.03.002]

[315] González-Pleiter M, Cirés S, Hurtado-Gallego J, Leganés F, Fernández-Piñas F, Velázquez D. Ecotoxicological Assessment of Antibiotics in Freshwater Using Cyanobacteria. In: Mishra AK, Tiwari DN, Rai AN, Eds. Cyanobacteria. Academic Press 2019; pp. 399-417.
[http://dx.doi.org/10.1016/B978-0-12-814667-5.00020-9]

[316] Kim YJ, Park HK, Kim IS. Invasion and toxin production by exotic nostocalean cyanobacteria (Cuspidothrix, Cylindrospermopsis, and Sphaerospermopsis) in the Nakdong River, Korea. Harmful Algae 2020; 100: 101954.
[http://dx.doi.org/10.1016/j.hal.2020.101954] [PMID: 33298363]

[317] Ma J, Brookes JD, Qin B, *et al.* Environmental factors controlling colony formation in blooms of the cyanobacteria Microcystis spp. in Lake Taihu, China. Harmful Algae 2014; 31: 136-42.
[http://dx.doi.org/10.1016/j.hal.2013.10.016] [PMID: 28040102]

[318] Pierattini EC, Francini A, Raffaelli A, Sebastiani L. Morpho-physiological response of *Populus alba* to erythromycin: A timeline of the health status of the plant. Sci Total Environ 2016; 569-570: 540-7.
[http://dx.doi.org/10.1016/j.scitotenv.2016.06.152] [PMID: 27366984]

[319] Schrader KK, Dennis ME. Cyanobacteria and earthy/musty compounds found in commercial catfish (Ictalurus punctatus) ponds in the Mississippi Delta and Mississippi–Alabama Blackland Prairie. Water Res 2005; 39(13): 2807-14.
[http://dx.doi.org/10.1016/j.watres.2005.04.044] [PMID: 15993924]

[320] Wang C, Yu J, Guo Q, *et al.* Occurrence of swampy/septic odor and possible odorants in source and finished drinking water of major cities across China. Environ Pollut 2019; 249: 305-10.
[http://dx.doi.org/10.1016/j.envpol.2019.03.041] [PMID: 30901644]

[321] Yang G, Tang X, Wilhelm SW, *et al.* Intermittent disturbance benefits colony size, biomass and dominance of Microcystis in Lake Taihu under field simulation condition. Harmful Algae 2020; 99: 101909.
[http://dx.doi.org/10.1016/j.hal.2020.101909] [PMID: 33218435]

[322] Brill JL, Belanger SE, Barron MG, *et al.* Derivation of algal acute to chronic ratios for use in chemical toxicity extrapolations. Chemosphere 2021; 263: 127804.
[http://dx.doi.org/10.1016/j.chemosphere.2020.127804] [PMID: 33297001]

[323] Kruk M, Jaworska B, Jabłońska-Barna I, Rychter A. How do differences in the nutritional and hydrological background influence phytoplankton in the Vistula Lagoon during a hot summer day? Oceanologia 2016; 58(4): 341-52.
[http://dx.doi.org/10.1016/j.oceano.2016.05.004]

[324] Lee J, Rai PK, Jeon YJ, Kim KH, Kwon EE. The role of algae and cyanobacteria in the production and

release of odorants in water. Environ Pollut 2017; 227: 252-62.
[http://dx.doi.org/10.1016/j.envpol.2017.04.058] [PMID: 28475978]

[325]   Liu L, Wu W, Zhang J, Lv P, Xu L, Yan Y. Progress of research on the toxicology of antibiotic pollution in aquatic organisms. Acta Ecol Sin 2018; 38(1): 36-41.
[http://dx.doi.org/10.1016/j.chnaes.2018.01.006]

[326]   Nishanth S, Bharti A, Gupta H, Gupta K, Gulia U, Prasanna R. Cyanobacterial extracellular polymeric substances (EPS): Biosynthesis and their potential applications. In: Das S, Dash HR, Eds. Microbial and Natural Macromolecules. Academic Press 2021; pp. 349-69.
[http://dx.doi.org/10.1016/B978-0-12-820084-1.00015-6]

[327]   Wang Z, Huang K, Zhou P, Guo H. A hybrid neural network model for cyanobacteria bloom in Dianchi Lake. Procedia Environ Sci 2010; 2: 67-75.
[http://dx.doi.org/10.1016/j.proenv.2010.10.010]

[328]   Zhao Y, Tang R. Improvement of organisms by biomimetic mineralization: A material incorporation strategy for biological modification. Acta Biomater 2020.
[PMID: 32629191]

[329]   Zheng X, Zhang W, Yuan Y, *et al.* Growth inhibition, toxin production and oxidative stress caused by three microplastics in *Microcystis aeruginosa*. Ecotoxicol Environ Saf 2021; 208: 111575.
[http://dx.doi.org/10.1016/j.ecoenv.2020.111575] [PMID: 33396101]

[330]   Bowling LC, Zamyadi A, Henderson RK. Assessment of *in situ* fluorometry to measure cyanobacterial presence in water bodies with diverse cyanobacterial populations. Water Res 2016; 105: 22-33.
[http://dx.doi.org/10.1016/j.watres.2016.08.051] [PMID: 27592302]

[331]   Lána J, Hofman J, Bláha L. Can cyanobacterial biomass applied to soil affect survival and reproduction of springtail Folsomia candida? Ecotoxicol Environ Saf 2011; 74(4): 840-3.
[http://dx.doi.org/10.1016/j.ecoenv.2010.11.003] [PMID: 21176962]

[332]   Lv Y, Li Y, Liu X, Xu K. The tolerance mechanism and accumulation characteristics of Phragmites australis to sulfamethoxazole and ofloxacin. Chemosphere 2020; 253: 126695.
[http://dx.doi.org/10.1016/j.chemosphere.2020.126695] [PMID: 32278902]

[333]   Zhang Y, Zhang X, Guo R, *et al.* Effects of florfenicol on growth, photosynthesis and antioxidant system of the non-target organism Isochrysis galbana. Comp Biochem Physiol C Toxicol Pharmacol 2020; 233: 108764.
[http://dx.doi.org/10.1016/j.cbpc.2020.108764] [PMID: 32294556]

[334]   Oncel SS. Biohydrogen from Microalgae, Uniting Energy, Life, and Green Future. In: Kim S-K, Ed. Handbook of Marine Microalgae. Boston: Academic Press 2015; pp. 159-96.
[http://dx.doi.org/10.1016/B978-0-12-800776-1.00011-X]

[335]   Olrik K, Olrik K, Cronberg G, Annadotter H. Lake Phytoplankton Responss to Global Climate Changes. In: Goldman CR, Kumagai M, Robarts RD, Eds. Climatic Change and Global Warming of Inland Waters Impacts and Mitigtion for Ecosystems and Societies - Wiley-Blackwell 2013. 2013; pp. 173-99.

[336]   Díaz-Pardo E, Vazquez G, López-López E. The phytoplankton community as a bioindicator of health conditions of Atezca Lake, Mexico. Aquat Ecosyst Health Manage 1998; 1(3-4): 257-66.
[http://dx.doi.org/10.1080/14634989808656922]

[337]   Eng-Wilmot DL, Martin DF. Short-term effects on Artemia salina of aponin and Gomphosphaeria aponina in unialgal cultures and in mixed cultures with gymnodinium breve. J Pharm Sci 1979; 68(8): 963-6.
[http://dx.doi.org/10.1002/jps.2600680812] [PMID: 113525]

[338]   Haande S, Rohrlack T, Semyalo RP, *et al.* Phytoplankton dynamics and cyanobacterial dominance in Murchison Bay of Lake Victoria (Uganda) in relation to environmental conditions. Limnologica 2011; 41(1): 20-9.

[http://dx.doi.org/10.1016/j.limno.2010.04.001]

[339]	Lem NW, Glick BR. Biotechnological uses of cyanobacteria. Biotechnol Adv 1985; 3(2): 195-208.
	[http://dx.doi.org/10.1016/0734-9750(85)90291-5] [PMID: 14544049]

[340]	Marshall HG. The composition of phytoplankton within the Chesapeake Bay plume and adjacent waters off the Virginia coast, U.S.A. Estuar Coast Shelf Sci 1982; 15(1): 29-43.
	[http://dx.doi.org/10.1016/0272-7714(82)90034-8]

[341]	Martin D, Gonzalez MH. Effects of salinity on synthesis of DNA, acidic polysaccharide, and growth in the blue-green alga, Gomphosphaeria aponina. Water Res 1978; 12(11): 951-5.
	[http://dx.doi.org/10.1016/0043-1354(78)90078-7]

[342]	Matthews RS. Artemia salina as a test organism for measuring superoxide-mediated toxicity. Free Radic Biol Med 1995; 18(5): 919-22.
	[http://dx.doi.org/10.1016/0891-5849(94)00205-X] [PMID: 7797101]

[343]	McCoy LF Jr, Martin DF. The influence of Gomphosphaeria aponina on the growth of Gymnodinium breve and the effect of aponin on the ichthyotoxicity of Gymnodinium breve. Chem Biol Interact 1977; 17(1): 17-24.
	[http://dx.doi.org/10.1016/0009-2797(77)90068-0] [PMID: 406055]

[344]	Vázquez G, Favila ME. Status of the health conditions of subtropical Atezca Lake. Aquat Ecosyst Health Manage 1998; 1(3-4): 245-55.
	[http://dx.doi.org/10.1080/14634989808656921]

[345]	Fabregas J, Herrero C, Abalde J, Liaño R, Cabezas B. Biomass production and biochemical variability of the marine microalga Dunaliella tertiolecta (Butcher) with high nutrient concentrations. Aquaculture 1986; 53(3-4): 187-99.
	[http://dx.doi.org/10.1016/0044-8486(86)90350-9]

[346]	Fabregas J, Herrero C, Cabezas B, Abalde J. Mass culture and biochemical variability of the marine microalga Tetraselmis suecica Kylin (Butch) with high nutrient concentrations. Aquaculture 1985; 49(3-4): 231-44.
	[http://dx.doi.org/10.1016/0044-8486(85)90082-1]

[347]	Fabregas J, Herrero C, Cabezas B, Abalde J. Biomass production and biochemical composition in mass cultures of the marine microalga Isochrysis galbana Parke at varying nutrient concentrations. Aquaculture 1986; 53(2): 101-13.
	[http://dx.doi.org/10.1016/0044-8486(86)90280-2]

[348]	Granéli E, Weberg M, Salomon PS. Harmful algal blooms of allelopathic microalgal species: The role of eutrophication. Harmful Algae 2008; 8(1): 94-102.
	[http://dx.doi.org/10.1016/j.hal.2008.08.011]

[349]	Markert B, Pedrozo F, Geller W, *et al.* A contribution to the study of the heavy-metal and nutritional element status of some lakes in the southern Andes of Patagonia (Argentina). Sci Total Environ 1997; 206(1): 1-15.
	[http://dx.doi.org/10.1016/S0048-9697(97)00218-0]

[350]	Pappas JL, Stoermer EF. Effects of inorganic nitrogen enrichment on Lake Huron Phytoplankton: An experimental study. J Great Lakes Res 1995; 21(2): 178-91.
	[http://dx.doi.org/10.1016/S0380-1330(95)71030-9]

[351]	Singh SK, Kaur R, Bansal A, Kapur S, Sundaram S. Biotechnological exploitation of cyanobacteria and microalgae for bioactive compounds. In: Verma ML, Chandel AK, Eds. Biotechnological Production of Bioactive Compounds. Elsevier 2020; pp. 221-59.
	[http://dx.doi.org/10.1016/B978-0-444-64323-0.00008-4]

[352]	Steidinger KA. Historical perspective on Karenia brevis red tide research in the Gulf of Mexico. Harmful Algae 2009; 8(4): 549-61.
	[http://dx.doi.org/10.1016/j.hal.2008.11.009]

[353]   de la Noue J, de Pauw N. The potential of microalgal biotechnology: A review of production and uses of microalgae. Biotechnol Adv 1988; 6(4): 725-70.
[http://dx.doi.org/10.1016/0734-9750(88)91921-0] [PMID: 14550037]

[354]   Komárek J, Johansen JR. Chapter 3 - Coccoid Cyanobacteria. In: Wehr JD, Sheath RG, Kociolek JP, Eds. Freshwater Algae of North America. Second Edition. Boston: Academic Press 2015; pp. 75-133.

[355]   Makarewicz JC. Phytoplankton biomass and species composition in Lake Erie, 1970 to 1987. J Great Lakes Res 1993; 19(2): 258-74.
[http://dx.doi.org/10.1016/S0380-1330(93)71216-2]

[356]   Martin DF, Kim YS. Long term peace river characteristics as a measure of a phosphate slime spill impact. Water Res 1977; 11(11): 963-70.
[http://dx.doi.org/10.1016/0043-1354(77)90153-1]

[357]   Phaup JD, Gannon J. Ecology of Sphaerotilus in an experimental outdoor channel. Water Res 1967; 1(7): 523-41.
[http://dx.doi.org/10.1016/0043-1354(67)90028-0]

[358]   Rice EL. 6 - Natural Ecosystems: Ecological Effects of Algal Allelopathy. In: Rice EL, Ed. Allelopathy. 2nd ed. San Diego: Academic Press 1984; pp. 189-205.
[http://dx.doi.org/10.1016/B978-0-08-092539-4.50010-1]

[359]   Sreenivasa MR, Nalewajko C. Phytoplankton Biomass and Species Composition in Northeastern Lake Ontario, April-November, 1965. J Great Lakes Res 1975; 1(1): 151-61.
[http://dx.doi.org/10.1016/S0380-1330(75)72342-0]

[360]   Viviani R. Eutrophication, marine biotoxins, human health. In: Vollenweider RA, Marchetti R, Viviani R, Eds. Marine Coastal Eutrophication. Amsterdam: Elsevier 1992; pp. 631-62.
[http://dx.doi.org/10.1016/B978-0-444-89990-3.50056-0]

[361]   Biological oceanography. Deep-sea Res, B, Oceanogr Lit Rev 1979; 26(4): 203-25.
[http://dx.doi.org/10.1016/0198-0254(79)94218-3]

[362]   Levitt J. 3 - Chilling Injury and Resistance. In: Levitt J, Ed. Chilling, Freezing, and High Temperature Stresses. 2nd ed. Academic Press 1980; pp. 23-64.
[http://dx.doi.org/10.1016/B978-0-12-445501-6.50008-7]

[363]   Singha Roy A, Pal R. Planktonic Cyanoprokaryota and Bacillariophyta of East Kolkata Wetlands Ecosystem, a Ramsar Site of India with reference to diversity and taxonomic study. J Algal Biomass Util 2015; 2015: 47-59.

[364]   Baresova M, Pivokonsky M, Novotna K, Naceradska J, Branyik T. An application of cellular organic matter to coagulation of cyanobacterial cells (Merismopedia tenuissima). Water Res 2017; 122: 70-7.
[http://dx.doi.org/10.1016/j.watres.2017.05.070] [PMID: 28591663]

[365]   dos Santos Severiano J, dos Santos Almeida-Melo VL, Bittencourt-Oliveira MC, Chia MA, do Nascimento Moura A. Effects of increased zooplankton biomass on phytoplankton and cyanotoxins: A tropical mesocosm study. Harmful Algae 2018; 71: 10-8.
[http://dx.doi.org/10.1016/j.hal.2017.11.003] [PMID: 29306392]

[366]   Fawzy MA. Phycoremediation and adsorption isotherms of cadmium and copper ions by Merismopedia tenuissima and their effect on growth and metabolism. Environ Toxicol Pharmacol 2016; 46: 116-21.
[http://dx.doi.org/10.1016/j.etap.2016.07.008] [PMID: 27458699]

[367]   Fawzy MA, Hifney AF, Adam MS, Al-Badaani AA. Biosorption of cobalt and its effect on growth and metabolites of Synechocystis pevalekii and Scenedesmus bernardii: Isothermal analysis. Environmental Technology & Innovation 2020; 19: 100953.
[http://dx.doi.org/10.1016/j.eti.2020.100953]

[368]   Issa A, Ali E, Abdel-Basset R, Awad MF, Ebied AM, Hassan SA. The impact of nitrogen

concentrations on production and quality of food and feed supplements from three cyanobacteria and potential application in biotechnology. Biocatal Agric Biotechnol 2020; 24: 101533.
[http://dx.doi.org/10.1016/j.bcab.2020.101533]

[369] Lucena-Silva D, Molozzi J, Severiano JS, Becker V, Lucena Barbosa JE. Removal efficiency of phosphorus, cyanobacteria and cyanotoxins by the "flock & sink" mitigation technique in semi-arid eutrophic waters. Water Res 2019; 159: 262-73.
[http://dx.doi.org/10.1016/j.watres.2019.04.057] [PMID: 31102855]

[370] Naceradska J, Novotna K, Cermakova L, Cajthaml T, Pivokonsky M. Investigating the coagulation of non-proteinaceous algal organic matter: Optimizing coagulation performance and identification of removal mechanisms. J Environ Sci (China) 2019; 79: 25-34.
[http://dx.doi.org/10.1016/j.jes.2018.09.024] [PMID: 30784448]

[371] Rojo C, Alvarez Cobales M. Taxonomy and ecology of phytoplankton in a hypertrophic gravel-pit lake. I. blue-green algae. Arch Protistenkd 1992; 142(1-2): 77-90.
[http://dx.doi.org/10.1016/S0003-9365(11)80103-X]

[372] Amorim CA, Dantas ÊW, Moura AN. Modeling cyanobacterial blooms in tropical reservoirs: The role of physicochemical variables and trophic interactions. Sci Total Environ 2020; 744: 140659.
[http://dx.doi.org/10.1016/j.scitotenv.2020.140659] [PMID: 32711303]

[373] Bao Q, Liu Z, Zhao M, *et al*. Primary productivity and seasonal dynamics of planktonic algae species composition in karst surface waters under different land uses. J Hydrol (Amst) 2020; 591: 125295.
[http://dx.doi.org/10.1016/j.jhydrol.2020.125295]

[374] Blomqvist P. Phytoplankton responses to biomanipulated grazing pressure and nutrient additions — enclosure studies in unlimed and limed Lake Njupfatet, central Sweden. Environ Pollut 2001; 111(2): 333-48.
[http://dx.doi.org/10.1016/S0269-7491(00)00037-3] [PMID: 11202737]

[375] Esqueda-Lara K, Sánchez AJ, Valdés-Lagunes G, Salcedo MÁ, Franco-Torres AE, Florido R. Fitoplancton en el humedal tropical Chaschoc en la cuenca baja del río Usumacinta. Rev Mex Biodivers 2016; 87(4): 1177-88.
[http://dx.doi.org/10.1016/j.rmb.2016.10.015]

[376] Karosienė J, Kasperovičienė J, Koreivienė J, Vitonytė I. Assessment of the vulnerability of Lithuanian lakes to expansion of Gonyostomum semen (Raphidophyceae). Limnologica 2014; 45: 7-15.
[http://dx.doi.org/10.1016/j.limno.2013.10.005]

[377] Nankabirwa A, De Crop W, Van der Meeren T, *et al*. Phytoplankton communities in the crater lakes of western Uganda, and their indicator species in relation to lake trophic status. Ecol Indic 2019; 107: 105563.
[http://dx.doi.org/10.1016/j.ecolind.2019.105563]

[378] Roshith CM, Meena DK, Manna RK, *et al*. Phytoplankton community structure of the Gangetic (Hooghly-Matla) estuary: Status and ecological implications in relation to eco-climatic variability. Flora (Jena) 2018; 240: 133-43.
[http://dx.doi.org/10.1016/j.flora.2018.01.001]

[379] Zhao CS, Shao NF, Yang ST, *et al*. Quantitative assessment of the effects of human activities on phytoplankton communities in lakes and reservoirs. Sci Total Environ 2019; 665: 213-25.
[http://dx.doi.org/10.1016/j.scitotenv.2019.02.117] [PMID: 30772551]

[380] Amorim CA, Moura AN. Ecological impacts of freshwater algal blooms on water quality, plankton biodiversity, structure, and ecosystem functioning. Sci Total Environ 2021; 758: 143605.
[http://dx.doi.org/10.1016/j.scitotenv.2020.143605] [PMID: 33248793]

[381] Daam MA, Rodrigues AMF, Van den Brink PJ, Nogueira AJA. Ecological effects of the herbicide linuron in tropical freshwater microcosms. Ecotoxicol Environ Saf 2009; 72(2): 410-23.
[http://dx.doi.org/10.1016/j.ecoenv.2008.07.009] [PMID: 18718661]

[382]   del Arco A, Álvarez-Manzaneda I, Funes A, Pérez-Martínez C, de Vicente I. Assessing the toxic effects of magnetic particles used for lake restoration on phytoplankton: A community-based approach. Ecotoxicol Environ Saf 2021; 207: 111288.
[http://dx.doi.org/10.1016/j.ecoenv.2020.111288] [PMID: 32949929]

[383]   Ioki M, Ohkoshi M, Nakajima N, Nakahira-Yanaka Y, Watanabe MM. Isolation of herbicide-resistant mutants of Botryococcus braunii. Bioresour Technol 2012; 109: 300-3.
[http://dx.doi.org/10.1016/j.biortech.2011.07.101] [PMID: 21906932]

[384]   Kasai F, Takamura N, Hatakeyama S. Effects of simetryne on growth of various freshwater algal taxa. Environ Pollut 1993; 79(1): 77-83.
[http://dx.doi.org/10.1016/0269-7491(93)90180-V] [PMID: 15091916]

[385]   Latinopoulos D, Ntislidou C, Kagalou I. Relationships of environmental conditions and phytoplankton functional groups in a new re-constructed shallow lentic system in draught conditions. Ecohydrol Hydrobiol 2020; 20(3): 369-81.
[http://dx.doi.org/10.1016/j.ecohyd.2020.04.003]

[386]   Liu C, Liu L, Shen H. Seasonal variations of phytoplankton community structure in relation to physico-chemical factors in Lake Baiyangdian, China. Procedia Environ Sci 2010; 2: 1622-31.
[http://dx.doi.org/10.1016/j.proenv.2010.10.173]

[387]   Liu Y, Liu F, Wang C, Quan Z, Li J. Effects of Bt-transgenic rice cultivation on planktonic communities in paddy fields and adjacent ditches. Sci Total Environ 2016; 565: 690-7.
[http://dx.doi.org/10.1016/j.scitotenv.2016.05.103] [PMID: 27219503]

[388]   Xiao F, Xiao P, Wang D. Influence of allochthonous organic matters on algae removal: Organic removal and floc characteristics. Colloids Surf A Physicochem Eng Asp 2019; 583: 123995.
[http://dx.doi.org/10.1016/j.colsurfa.2019.123995]

[389]   Calado SLM, Santos GS, Leite TPB, et al. Depuration time and sublethal effects of microcystins in a freshwater fish from water supply reservoir. Chemosphere 2018; 210: 805-15.
[http://dx.doi.org/10.1016/j.chemosphere.2018.07.075] [PMID: 30041158]

[390]   Lin Z, Li J, Luan Y, Dai W. Application of algae for heavy metal adsorption: A 20-year meta-analysis. Ecotoxicol Environ Saf 2020; 190: 110089.
[http://dx.doi.org/10.1016/j.ecoenv.2019.110089] [PMID: 31896472]

[391]   Nawrocka L, Kobos J. The trophic state of the Vistula Lagoon: An assessment based on selected biotic and abiotic parameters according to the Water Framework Directive**This work was supported by the Norway grants PNRF 82 A I. Oceanologia 2011; 53(3): 881-94.
[http://dx.doi.org/10.5697/oc.53-3.881]

[392]   Oberholster PJ, Myburgh JG, Ashton PJ, Botha AM. Responses of phytoplankton upon exposure to a mixture of acid mine drainage and high levels of nutrient pollution in Lake Loskop, South Africa. Ecotoxicol Environ Saf 2010; 73(3): 326-35.
[http://dx.doi.org/10.1016/j.ecoenv.2009.08.011] [PMID: 19896711]

[393]   Shcherbak V, Sherman I, Semeniuk N, Kutishchev P. Autotrophic communities' diversity in natural and artificial water-bodies of a river estuary — A case-study of the Dnieper–Bug Estuary, Ukraine. Ecohydrol Hydrobiol 2020; 20(1): 112-22.
[http://dx.doi.org/10.1016/j.ecohyd.2019.07.001]

[394]   Tang C, Yi Y, Yang Z, et al. Planktonic indicators of trophic states for a shallow lake (Baiyangdian Lake, China). Limnologica 2019; 78: 125712.
[http://dx.doi.org/10.1016/j.limno.2019.125712]

[395]   Zhao CS, Shao NF, Yang ST, et al. Predicting cyanobacteria bloom occurrence in lakes and reservoirs before blooms occur. Sci Total Environ 2019; 670: 837-48.
[http://dx.doi.org/10.1016/j.scitotenv.2019.03.161] [PMID: 30921717]

[396]   Aboulela H, Amin A, Lashin A, El Rayes A. Contribution of geothermal resources to the future of

renewable energy in Egypt: A case study, Gulf of Suez-Egypt. Renew Energy 2020.

[397] Kavitha S, Yukesh Kannah R, Rajesh Banu J, Kaliappan S, Johnson M. Biological disintegration of microalgae for biomethane recovery-prediction of biodegradability and computation of energy balance. Bioresour Technol 2017; 244(Pt 2): 1367-75.
[http://dx.doi.org/10.1016/j.biortech.2017.05.007] [PMID: 28522200]

[398] Lepšová-Skácelová O, Fibich P, Wild J, Lepš J. Trophic gradient is the main determinant of species and large taxonomic groups representation in phytoplankton of standing water bodies. Ecol Indic 2018; 85: 262-70.
[http://dx.doi.org/10.1016/j.ecolind.2017.10.034]

[399] Schaeffer BA, Kurtz JC, Hein MK. Phytoplankton community composition in nearshore coastal waters of Louisiana. Mar Pollut Bull 2012; 64(8): 1705-12.
[http://dx.doi.org/10.1016/j.marpolbul.2012.03.017] [PMID: 22498318]

[400] Schneider S, Lindstrøm EA. Bioindication in Norwegian rivers using non-diatomaceous benthic algae: The acidification index periphyton (AIP). Ecol Indic 2009; 9(6): 1206-11.
[http://dx.doi.org/10.1016/j.ecolind.2009.02.008]

[401] Sehnal L, Procházková T, Smutná M, Kohoutek J, Lepšová-Skácelová O, Hilscherová K. Widespread occurrence of retinoids in water bodies associated with cyanobacterial blooms dominated by diverse species. Water Res 2019; 156: 136-47.
[http://dx.doi.org/10.1016/j.watres.2019.03.009] [PMID: 30909126]

[402] Ülgüdür N, Ergüder TH, Demirer GN. Simultaneous dissolution and uptake of nutrients in microalgal treatment of the secondarily treated digestate. Algal Res 2019; 43: 101633.
[http://dx.doi.org/10.1016/j.algal.2019.101633]

[403] Zhang S, Zhang S, Li G. Acorus calamus root extracts to control harmful cyanobacteria blooms. Ecol Eng 2016; 94: 95-101.
[http://dx.doi.org/10.1016/j.ecoleng.2016.05.053]

[404] Aguilera A, Haakonsson S, Martin MV, Salerno GL, Echenique RO. Bloom-forming cyanobacteria and cyanotoxins in Argentina: A growing health and environmental concern. Limnologica 2018; 69: 103-14.
[http://dx.doi.org/10.1016/j.limno.2017.10.006]

[405] Andersson A, Höglander H, Karlsson C, Huseby S. Key role of phosphorus and nitrogen in regulating cyanobacterial community composition in the northern Baltic Sea. Estuar Coast Shelf Sci 2015; 164: 161-71.
[http://dx.doi.org/10.1016/j.ecss.2015.07.013]

[406] Brogueira MJ, Oliveira MR, Cabeçadas G. Phytoplankton community structure defined by key environmental variables in Tagus estuary, Portugal. Mar Environ Res 2007; 64(5): 616-28.
[http://dx.doi.org/10.1016/j.marenvres.2007.06.007] [PMID: 17884159]

[407] Chen R, Ao D, Ji J, *et al.* Insight into the risk of replenishing urban landscape ponds with reclaimed wastewater. J Hazard Mater 2017; 324(Pt B): 573-82.
[http://dx.doi.org/10.1016/j.jhazmat.2016.11.028] [PMID: 27856052]

[408] Gaysina LA, Saraf A, Singh P. Cyanobacteria in Diverse Habitats. In: Mishra AK, Tiwari DN, Rai AN, Eds. Cyanobacteria. Academic Press 2019; pp. 1-28.
[http://dx.doi.org/10.1016/B978-0-12-814667-5.00001-5]

[409] Li XY, Li B, Sun XL. Effects of a coastal power plant thermal discharge on phytoplankton community structure in Zhanjiang Bay, China. Mar Pollut Bull 2014; 81(1): 210-7.
[http://dx.doi.org/10.1016/j.marpolbul.2013.08.006] [PMID: 24635985]

[410] López-Mendoza Z, Tavera R, Novelo E. El fitoplancton de un canal de Xochimilco y la importancia de estudiar ecosistemas acuáticos urbanos. TIP 2015; 18(1): 13-28.
[http://dx.doi.org/10.1016/j.recqb.2015.05.002]

[411] Sinistro R, Sánchez ML, Marinone MC, Izaguirre I. Experimental study of the zooplankton impact on the trophic structure of phytoplankton and the microbial assemblages in a temperate wetland (Argentina). Limnologica 2007; 37(1): 88-99.
[http://dx.doi.org/10.1016/j.limno.2006.09.001]

[412] Sutherland DL, Turnbull MH, Craggs RJ. Environmental drivers that influence microalgal species in fullscale wastewater treatment high rate algal ponds. Water Res 2017; 124: 504-12.
[http://dx.doi.org/10.1016/j.watres.2017.08.012] [PMID: 28802135]

[413] Bakrač K, Ilijanić N, Miko S, Hasan O. Evidence of sapropel S1 formation from Holocene lacustrine sequences in Lake Vrana in Dalmatia (Croatia). Quat Int 2018; 494: 5-18.
[http://dx.doi.org/10.1016/j.quaint.2018.06.010]

[414] Hodoki Y, Ohbayashi K, Kobayashi Y, Okuda N, Nakano S. Detection and identification of potentially toxic cyanobacteria: Ubiquitous distribution of *Microcystis aeruginosa* and Cuspidothrix issatschenkoi in Japanese lakes. Harmful Algae 2012; 16: 49-57.
[http://dx.doi.org/10.1016/j.hal.2012.01.003]

[415] Larras F, Lambert AS, Pesce S, Rimet F, Bouchez A, Montuelle B. The effect of temperature and a herbicide mixture on freshwater periphytic algae. Ecotoxicol Environ Saf 2013; 98: 162-70.
[http://dx.doi.org/10.1016/j.ecoenv.2013.09.007] [PMID: 24119653]

[416] Marshall HG, Lacouture RV, Buchanan C, Johnson JM. Phytoplankton assemblages associated with water quality and salinity regions in Chesapeake Bay, USA. Estuar Coast Shelf Sci 2006; 69(1-2): 10-8.
[http://dx.doi.org/10.1016/j.ecss.2006.03.019]

[417] Mataloni G, Pose M. Non-marine algae from islands near Cierva Point, Antarctic Peninsula. Cryptogam, Algol 2001; 22(1): 41-64.
[http://dx.doi.org/10.1016/S0181-1568(00)01049-7]

[418] Mohamed ZA. Macrophytes-cyanobacteria allelopathic interactions and their implications for water resources management—A review. Limnologica 2017; 63: 122-32.
[http://dx.doi.org/10.1016/j.limno.2017.02.006]

[419] Novotna K, Cermakova L, Pivokonska L, Cajthaml T, Pivokonsky M. Microplastics in drinking water treatment – Current knowledge and research needs. Sci Total Environ 2019; 667: 730-40.
[http://dx.doi.org/10.1016/j.scitotenv.2019.02.431] [PMID: 30851606]

[420] Wiedner C, Rücker J, Fastner J, Chorus I, Nixdorf B. Seasonal dynamics of cylindrospermopsin and cyanobacteria in two German lakes. Toxicon 2008; 52(6): 677-86.
[http://dx.doi.org/10.1016/j.toxicon.2008.07.017] [PMID: 18725243]

[421] Papadimitriou T, Katsiapi M, Kormas KA, Moustaka-Gouni M, Kagalou I. Artificially-born "killer" lake: Phytoplankton based water quality and microcystin affected fish in a reconstructed lake. Sci Total Environ 2013; 452-453: 116-24.
[http://dx.doi.org/10.1016/j.scitotenv.2013.02.035] [PMID: 23500405]

[422] Qian X, Kim MK, Kumaraswamy GK, Agarwal A, Lun DS, Dismukes GC. Flux balance analysis of photoautotrophic metabolism: Uncovering new biological details of subsystems involved in cyanobacterial photosynthesis. Biochim Biophys Acta Bioenerg 2017; 1858(4): 276-87.
[http://dx.doi.org/10.1016/j.bbabio.2016.12.007] [PMID: 28012908]

[423] Tasić MB, Pinto LFR, Klein BC, Veljković VB, Filho RM. Botryococcus braunii for biodiesel production. Renew Sustain Energy Rev 2016; 64: 260-70.
[http://dx.doi.org/10.1016/j.rser.2016.06.009]

[424] Terry KL, Hirata J, Laws EA. Light, nitrogen, and phosphorus-limited growth of Phaeodactylum tricornutum Bohlin strain TFX-1: Chemical composition, carbon partitioning, and the diel periodicity of physiological processes. J Exp Mar Biol Ecol 1985; 86(1): 85-100.
[http://dx.doi.org/10.1016/0022-0981(85)90044-9]

[425] Bose R, Nandi C, Singha Roy A, Gorain P, Pal R. Floristic survey of microplanktonic cyanobacteria and chlorophyta from different ecological niches of West Bengal, India. Phytomorphology 2016; 66: 77-93.

[426] Gao Y, Sun L, Wu C, *et al.* Inter-annual and seasonal variations of phytoplankton community and its relation to water pollution in Futian Mangrove of Shenzhen, China. Cont Shelf Res 2018; 166: 138-47.
[http://dx.doi.org/10.1016/j.csr.2018.07.010]

[427] Pham DT, Everaert G, Janssens N, Alvarado A, Nopens I, Goethals PLM. Algal community analysis in a waste stabilisation pond. Ecol Eng 2014; 73: 302-6.
[http://dx.doi.org/10.1016/j.ecoleng.2014.09.046]

[428] Rai LC, Singh AK, Mallick N. Employment of CEPEX enclosures for monitoring toxicity of Hg and Zn on *in situ* structural and functional characteristics of algal communities of river ganga in Varanasi, India. Ecotoxicol Environ Saf 1990; 20(2): 211-21.
[http://dx.doi.org/10.1016/0147-6513(90)90060-I] [PMID: 1980458]

[429] Xiao R, Wang Q, Zhang M, Pan W, Wang JJ. Plankton distribution patterns and the relationship with environmental gradients and hydrological connectivity of wetlands in the Yellow River Delta. Ecohydrol Hydrobiol 2020; 20(4): 584-96.
[http://dx.doi.org/10.1016/j.ecohyd.2020.01.002]

[430] Yuan Y, Jiang M, Liu X, *et al.* Environmental variables influencing phytoplankton communities in hydrologically connected aquatic habitats in the Lake Xingkai basin. Ecol Indic 2018; 91: 1-12.
[http://dx.doi.org/10.1016/j.ecolind.2018.03.085]

[431] Chaparro G, Marinone MC, Lombardo RJ, Schiaffino MR, de Souza Guimarães A, O'Farrell I. Zooplankton succession during extraordinary drought–flood cycles: A case study in a South American floodplain lake. Limnologica 2011; 41(4): 371-81.
[http://dx.doi.org/10.1016/j.limno.2011.04.003]

[432] Drobac D, Tokodi N, Lujić J, *et al.* Cyanobacteria and cyanotoxins in fishponds and their effects on fish tissue. Harmful Algae 2016; 55: 66-76.
[http://dx.doi.org/10.1016/j.hal.2016.02.007] [PMID: 28073548]

[433] Randolph K, Wilson J, Tedesco L, Li L, Pascual DL, Soyeux E. Hyperspectral remote sensing of cyanobacteria in turbid productive water using optically active pigments, chlorophyll a and phycocyanin. Remote Sens Environ 2008; 112(11): 4009-19.
[http://dx.doi.org/10.1016/j.rse.2008.06.002]

[434] Mahar M. Ecology and Taxonomy of Plankton of Manchhar Lake (Distt: Dadu). Pakistan: Sindh 2020.

[435] Rosen B, Davis T, Gobler C, Kramer B, Loftin K. Cyanobacteria of the 2016 Lake Okeechobee and Okeechobee Waterway Harmful Algal Bloom. Anatomy of a Cyanobacteria Bloom 2017.
[http://dx.doi.org/10.3133/ofr20171054]

[436] Borah D, Rethinam G, Gopalakrishnan S, *et al.* Ozone enhanced production of potentially useful exopolymers from the cyanobacterium Nostoc muscorum. Polym Test 2020; 84: 106385.
[http://dx.doi.org/10.1016/j.polymertesting.2020.106385]

[437] Hirose Y, Chihong S, Watanabe M, *et al.* Diverse chromatic acclimation processes regulating phycoerythrocyanin and rod-shaped phycobilisome in cyanobacteria. Mol Plant 2019; 12(5): 715-25.
[http://dx.doi.org/10.1016/j.molp.2019.02.010] [PMID: 30818037]

[438] Jones F. The dynamics of suspended algal populations in the lower Wye catchment. Water Res 1984; 18(1): 25-35.
[http://dx.doi.org/10.1016/0043-1354(84)90044-7]

[439] Sykes J. Centrifugal Techniques for the Isolation and Characterisation of Sub-Cellular Components from Bacteria. In: Norris JR, Ribbons DW, Eds. Methods in Microbiology. Academic Press 1971; 5: pp. 55-207.
[http://dx.doi.org/10.1016/S0580-9517(08)70640-8]

[440] Tao Y, Yu J, Xue B, Yao S, Wang S. Precipitation and temperature drive seasonal variation in bioaccumulation of polycyclic aromatic hydrocarbons in the planktonic food webs of a subtropical shallow eutrophic lake in China. Sci Total Environ 2017; 583: 447-57.
[http://dx.doi.org/10.1016/j.scitotenv.2017.01.100] [PMID: 28110880]

[441] Umamaheswara Rao M, Mohanchand V. Water quality characteristics and phytoplankton of polluted Visakhapatnam Harbour. Mar Environ Res 1988; 25(1): 23-43.
[http://dx.doi.org/10.1016/0141-1136(88)90359-5]

[442] Bharathi MD, Sarma VVSS, Ramaneswari K. Intra-annual variations in phytoplankton biomass and its composition in the tropical estuary: Influence of river discharge. Mar Pollut Bull 2018; 129(1): 14-25.
[http://dx.doi.org/10.1016/j.marpolbul.2018.02.007] [PMID: 29680531]

[443] Sathicq MB, Gómez N, Bauer DE, Donadelli J. Use of phytoplankton assemblages to assess the quality of coastal waters of a transitional ecosystem: Río de la Plata estuary. Cont Shelf Res 2017; 150: 10-7.
[http://dx.doi.org/10.1016/j.csr.2016.08.009]

[444] Tang H, Vanderploeg HA, Johengen TH, Liebig JR. Quagga mussel (Dreissena rostriformis bugensis) selective feeding of phytoplankton in Saginaw Bay. J Great Lakes Res 2014; 40: 83-94.
[http://dx.doi.org/10.1016/j.jglr.2013.11.011]

[445] Xie P. Gut contents of silver carp, Hypophthalmichthys molitrix, and the disruption of a centric diatom, Cyclotella, on passage through the esophagus and intestine. Aquaculture 1999; 180(3-4): 295-305.
[http://dx.doi.org/10.1016/S0044-8486(99)00205-7]

[446] Xie P. Gut contents of bighead carp (Aristichthys nobilis) and the processing and digestion of algal cells in the alimentary canal. Aquaculture 2001; 195(1-2): 149-61.
[http://dx.doi.org/10.1016/S0044-8486(00)00549-4]

[447] Zhang D, Yang Y, Hu J, Ran Y, Mao J. Occurrence of aliphatic biopolymer in chlorophyceae algae and cyanobacteria-rich phytoplankton. Org Geochem 2019; 135: 1-10.
[http://dx.doi.org/10.1016/j.orggeochem.2019.04.008]

[448] Zhang X, Liu Z, Jeppesen E, Taylor WD. Effects of deposit-feeding tubificid worms and filter-feeding bivalves on benthic-pelagic coupling: Implications for the restoration of eutrophic shallow lakes. Water Res 2014; 50: 135-46.
[http://dx.doi.org/10.1016/j.watres.2013.12.003] [PMID: 24370657]

[449] Asmus R. Field measurements on seasonal variation of the activity of primary producers on a sandy tidal flat in the northern Wadden sea. Neth J Sea Res 1982; 16: 389-402.
[http://dx.doi.org/10.1016/0077-7579(82)90045-X]

[450] Gürevin C, Erturk A, Albay M. Predicting the effects of sediment based internal nutrient loads on eutrophication in Küçükçekmece Lagoon for rehabilitation planning. Int J Sediment Res 2017; 32(4): 527-54.
[http://dx.doi.org/10.1016/j.ijsrc.2016.08.002]

[451] Luo S, Berges JA, He Z, Young EB. Algal-microbial community collaboration for energy recovery and nutrient remediation from wastewater in integrated photobioelectrochemical systems. Algal Res 2017; 24: 527-39.
[http://dx.doi.org/10.1016/j.algal.2016.10.006]

[452] Spencer DF, Nichols LH. Free nickel ion inhibits growth of two species of green algae. Environ Pollut A 1983; 31(2): 97-104.
[http://dx.doi.org/10.1016/0143-1471(83)90002-8]

[453] Walter J, Selim KA, Leganés F, et al. A novel Ca^{2+}-binding protein influences photosynthetic electron transport in Anabaena sp. PCC 7120. Biochim Biophys Acta Bioenerg 2019; 1860(6): 519-32.
[http://dx.doi.org/10.1016/j.bbabio.2019.04.007] [PMID: 31034800]

[454] Hamels I, Sabbe K, Muylaert K, *et al.* Organisation of microbenthic communities in intertidal Estuarine flats, a case study from the molenplaat (Westerschelde estuary, The Netherlands). Eur J Protistol 1998; 34(3): 308-20.
[http://dx.doi.org/10.1016/S0932-4739(98)80058-8]

[455] Herrera-Silveira JA, Morales-Ojeda SM. Evaluation of the health status of a coastal ecosystem in southeast Mexico: Assessment of water quality, phytoplankton and submerged aquatic vegetation. Mar Pollut Bull 2009; 59(1-3): 72-86.
[http://dx.doi.org/10.1016/j.marpolbul.2008.11.017] [PMID: 19157464]

[456] Mendoza-Vera JM, Kâ S, Cuoc C, Bouvy M, Pagano M. Decline of Pseudodiaptomus hessei (Copepoda, Calanoida) in two water bodies located in the Senegal River hydrosystem (West Africa): Hypotheses and perspectives. Estuar Coast Shelf Sci 2008; 79(4): 740-50.
[http://dx.doi.org/10.1016/j.ecss.2008.07.002]

[457] Xyländer M, Braune W. Influence of nickel on the green alga *Haematococcus lacustris* Rostafinski in phases of its life cycle. J Plant Physiol 1994; 144(1): 86-93.
[http://dx.doi.org/10.1016/S0176-1617(11)80998-3]

[458] Aktan Y, Balkıs N, Balkıs N. Seasonal variations of epipelic algal community in relation to environmental factors in the Istanbul Strait (the Bosphorus), Turkey. Mar Pollut Bull 2014; 81(1): 268-75.
[http://dx.doi.org/10.1016/j.marpolbul.2014.01.027] [PMID: 24467854]

[459] Arumugam S, Sigamani S, Samikannu M, Perumal M. Assemblages of phytoplankton diversity in different zonation of Muthupet mangroves. Reg Stud Mar Sci 2016; 3: 234-41.
[http://dx.doi.org/10.1016/j.rsma.2015.11.005]

[460] Beltram FL, Lamb RW, Smith F, Witman JD. Rapid proliferation and impacts of cyanobacterial mats on Galapagos rocky reefs during the 2014–2017 El Niño Southern Oscillation. J Exp Mar Biol Ecol 2019; 514-515: 18-26.
[http://dx.doi.org/10.1016/j.jembe.2019.03.007]

[461] Zhao C, Liu C, Xia J, Zhang Y, Yu Q, Eamus D. Recognition of key regions for restoration of phytoplankton communities in the Huai River basin, China. J Hydrol (Amst) 2012; 420-421: 292-300.
[http://dx.doi.org/10.1016/j.jhydrol.2011.12.016]

[462] Żyszka-Haberecht B, Poliwoda A, Lipok J. 'Structural constraints in cyanobacteria-mediated whole-cell biotransformation of methoxylated and methylated derivatives of 2'-hydroxychalcone. J Biotechnol 2019; 293: 36-46.
[http://dx.doi.org/10.1016/j.jbiotec.2019.01.005] [PMID: 30690100]

[463] George B, Nirmal Kumar JI, Kumar RN. Study on the influence of hydro-chemical parameters on phytoplankton distribution along Tapi estuarine area of Gulf of Khambhat, India. Egypt J Aquat Res 2012; 38(3): 157-70.
[http://dx.doi.org/10.1016/j.ejar.2012.12.010]

[464] Sathicq Á, Paola A, Pérez M, *et al.* Furylchalcones as new potential marine antifoulants. Int Biodeterior Biodegradation 2019; 143: 104730.
[http://dx.doi.org/10.1016/j.ibiod.2019.104730]

[465] Sorokovikova E, Belykh O, Krasnopeev A, *et al.* First data on cyanobacterial biodiversity in benthic biofilms during mass mortality of endemic sponges in Lake Baikal. J Great Lakes Res 2020; 46(1): 75-84.
[http://dx.doi.org/10.1016/j.jglr.2019.10.017]

[466] Wang H, Ki JS. Molecular identification, differential expression and protective roles of iron/manganese superoxide dismutases in the green algae Closterium ehrenbergii against metal stress. Eur J Protistol 2020; 74: 125689.
[http://dx.doi.org/10.1016/j.ejop.2020.125689] [PMID: 32193064]

[467] Xiao L, Young EB, Grothjan JJ, Lyon S, Zhang H, He Z. Wastewater treatment and microbial communities in an integrated photo-bioelectrochemical system affected by different wastewater algal inocula. Algal Res 2015; 12: 446-54.
[http://dx.doi.org/10.1016/j.algal.2015.10.008]

[468] Walley WJ, Grbović J, Džeroski S. A reappraisal of saprobic values and indicator weights based on slovenian river quality data. Water Res 2001; 35(18): 4285-92.
[http://dx.doi.org/10.1016/S0043-1354(01)00162-2] [PMID: 11763029]

[469] Amaral LM, Carolina de Almeida Castilho M, Henry R, Ferragut C. Epipelon, phytoplankton and zooplankton responses to the experimental oligotrophication in a eutrophic shallow reservoir. Environ Pollut 2020; 263(Pt A): 114603.
[http://dx.doi.org/10.1016/j.envpol.2020.114603] [PMID: 33618459]

[470] Barros MUG, Wilson AE, Leitão JIR, *et al.* Environmental factors associated with toxic cyanobacterial blooms across 20 drinking water reservoirs in a semi-arid region of Brazil. Harmful Algae 2019; 86: 128-37.
[http://dx.doi.org/10.1016/j.hal.2019.05.006] [PMID: 31358272]

[471] Fishman DB, Adlerstein SA, Vanderploeg HA, Fahnenstiel GL, Scavia D. Phytoplankton community composition of Saginaw Bay, Lake Huron, during the zebra mussel (Dreissena polymorpha) invasion: A multivariate analysis. J Great Lakes Res 2010; 36(1): 9-19.
[http://dx.doi.org/10.1016/j.jglr.2009.10.004]

[472] Hawkins PR, Holliday J, Kathuria A, Bowling L. Change in cyanobacterial biovolume due to preservation by Lugol's Iodine. Harmful Algae 2005; 4(6): 1033-43.
[http://dx.doi.org/10.1016/j.hal.2005.03.001]

[473] Kumari A. Phytorestoration of abandoned ash-ponds by native algal strains. In: Bauddh K, Korstad J, Sharma P, Eds. Phytorestoration of Abandoned Mining and Oil Drilling Sites. Elsevier 2021; pp. 105-14.
[http://dx.doi.org/10.1016/B978-0-12-821200-4.00005-4]

[474] Napiórkowska-Krzebietke A, Dunalska JA, Zębek E. Taxa-specific eco-sensitivity in relation to phytoplankton bloom stability and ecologically relevant lake state. Acta Oecol 2017; 81: 10-21.
[http://dx.doi.org/10.1016/j.actao.2017.04.002]

[475] Šmarda J, Šmajs D, Komrska J, Krzyžánek V. S-layers on cell walls of cyanobacteria. Micron 2002; 33(3): 257-77.
[http://dx.doi.org/10.1016/S0968-4328(01)00031-2] [PMID: 11742749]

[476] Gągała I, Izydorczyk K, Skowron A, *et al.* Appearance of toxigenic cyanobacteria in two Polish lakes dominated by *Microcystis aeruginosa* and Planktothrix agardhii and environmental factors influence. Ecohydrol Hydrobiol 2010; 10(1): 25-34.
[http://dx.doi.org/10.2478/v10104-009-0045-5]

[477] Kim HG, Hong S, Kim DK, Joo GJ. Drivers shaping episodic and gradual changes in phytoplankton community succession: Taxonomic *versus* functional groups. Sci Total Environ 2020; 734: 138940.
[http://dx.doi.org/10.1016/j.scitotenv.2020.138940] [PMID: 32460064]

[478] Messineo V, Bogialli S, Melchiorre S, *et al.* Cyanobacterial toxins in Italian freshwaters. Limnologica 2009; 39(2): 95-106.
[http://dx.doi.org/10.1016/j.limno.2008.09.001]

[479] Pełechaty M, Pełechata A, Pukacz A, Burchardt L. Interrelationships between macrophytes (including charophytes) and phytoplankton and the ecological state of lakes. Ecohydrol Hydrobiol 2006; 6(1-4): 79-88.
[http://dx.doi.org/10.1016/S1642-3593(06)70129-X]

[480] Rachel R, Pum D, Šmarda J, *et al.* Fine structure of S-layers1This review is part of a series of reviews dealing with different aspects of bacterial S-layers; all these reviews appeared in Volume 20/1-2 (June

1997) of FEMS Microbiology Reviews, thematic issue devoted to bacterial S-layers.1. FEMS Microbiol Rev 1997; 20(1): 13-23.
[http://dx.doi.org/10.1111/j.1574-6976.1997.tb00302.x]

[481] Rangel LM, Soares MCS, Paiva R, Silva LHS. Morphology-based functional groups as effective indicators of phytoplankton dynamics in a tropical cyanobacteria-dominated transitional river–reservoir system. Ecol Indic 2016; 64: 217-27.
[http://dx.doi.org/10.1016/j.ecolind.2015.12.041]

[482] Fishman DB, Adlerstein SA, Vanderploeg HA, Fahnenstiel GL, Scavia D. Causes of phytoplankton changes in Saginaw Bay, Lake Huron, during the zebra mussel invasion. J Great Lakes Res 2009; 35(4): 482-95.
[http://dx.doi.org/10.1016/j.jglr.2009.08.003]

[483] Kutovaya OA, Watson SB. Development and application of a molecular assay to detect and monitor geosmin-producing cyanobacteria and actinomycetes in the Great Lakes. J Great Lakes Res 2014; 40(2): 404-14.
[http://dx.doi.org/10.1016/j.jglr.2014.03.016]

[484] Das SK, Maurya ON. Floristic survey of algae in Vikramsila Gangetic Dolphin Sanctuary, Bihar (India). Nelumbo 2015; 57: 124-34.

[485] Dondajewska R, Kozak A, Budzyńska A, Kowalczewska-Madura K, Gołdyn R. Nature-based solutions for protection and restoration of degraded Bielsko Lake. Ecohydrol Hydrobiol 2018; 18(4): 401-11.
[http://dx.doi.org/10.1016/j.ecohyd.2018.04.001]

[486] Dong F, Lin Q, Li C, Wang L, García A. UV/chlorination process of algal-laden water: Algal inactivation and disinfection byproducts attenuation. Separ Purif Tech 2021; 257: 117896.
[http://dx.doi.org/10.1016/j.seppur.2020.117896]

[487] Dong F, Liu J, Li C, *et al.* Ferrate(VI) pre-treatment and subsequent chlorination of blue-green algae: Quantification of disinfection byproducts. Environ Int 2019; 133(Pt B): 105195.
[http://dx.doi.org/10.1016/j.envint.2019.105195] [PMID: 31654918]

[488] Kakimoto M, Ishikawa T, Miyagi A, *et al.* Culture temperature affects gene expression and metabolic pathways in the 2-methylisoborneol-producing cyanobacterium Pseudanabaena galeata. J Plant Physiol 2014; 171(3-4): 292-300.
[http://dx.doi.org/10.1016/j.jplph.2013.09.005] [PMID: 24140001]

[489] Rosińska J, Kozak A, Dondajewska R, Gołdyn R. Cyanobacteria blooms before and during the restoration process of a shallow urban lake. J Environ Manage 2017; 198(Pt 1): 340-7.
[http://dx.doi.org/10.1016/j.jenvman.2017.04.091] [PMID: 28494422]

[490] Rosińska J, Kozak A, Dondajewska R, Kowalczewska-Madura K, Gołdyn R. Water quality response to sustainable restoration measures – Case study of urban Swarzędzkie Lake. Ecol Indic 2018; 84: 437-49.
[http://dx.doi.org/10.1016/j.ecolind.2017.09.009]

[491] Smith SM. Chlorophyll fluorescence responses to additions of Pinus rigida L. pollen in freshwater kettle ponds of Cape Cod National Seashore (Massachusetts, USA). Limnologica 2020; 83: 125781.
[http://dx.doi.org/10.1016/j.limno.2020.125781]

[492] Zhao H, Wang Y, Yang L, Yuan L, Peng D. Relationship between phytoplankton and environmental factors in landscape water supplemented with reclaimed water. Ecol Indic 2015; 58: 113-21.
[http://dx.doi.org/10.1016/j.ecolind.2015.03.033]

[493] Bormans M, Amzil Z, Mineaud E, *et al.* Demonstrated transfer of cyanobacteria and cyanotoxins along a freshwater-marine continuum in France. Harmful Algae 2019; 87: 101639.
[http://dx.doi.org/10.1016/j.hal.2019.101639] [PMID: 31349891]

[494] Dondajewska R, Kozak A, Rosińska J, Gołdyn R. Water quality and phytoplankton structure changes

under the influence of effective microorganisms (EM) and barley straw – Lake restoration case study. Sci Total Environ 2019; 660: 1355-66.
[http://dx.doi.org/10.1016/j.scitotenv.2019.01.071] [PMID: 30743930]

[495]   Lu X, Tian C, Pei H, Hu W, Xie J. Environmental factors influencing cyanobacteria community structure in Dongping Lake, China. J Environ Sci (China) 2013; 25(11): 2196-206.
[http://dx.doi.org/10.1016/S1001-0742(12)60297-6] [PMID: 24552047]

[496]   Watson SB, Monis P, Baker P, Giglio S. Biochemistry and genetics of taste- and odor-producing cyanobacteria. Harmful Algae 2016; 54: 112-27.
[http://dx.doi.org/10.1016/j.hal.2015.11.008] [PMID: 28073471]

[497]   Mihaljević M, Špoljarić D, Stević F, Cvijanović V, Hackenberger Kutuzović B. The influence of extreme floods from the River Danube in 2006 on phytoplankton communities in a floodplain lake: Shift to a clear state. Limnologica - Ecology and Management of Inland Waters 2010; 40(3): 260-8.

[498]   Mischke U. Cyanobacteria associations in shallow polytrophic lakes: Influence of environmental factors. Acta Oecol 2003; 24: S11-23.
[http://dx.doi.org/10.1016/S1146-609X(03)00003-1]

[499]   Muhid P, Davis TW, Bunn SE, Burford MA. Effects of inorganic nutrients in recycled water on freshwater phytoplankton biomass and composition. Water Res 2013; 47(1): 384-94.
[http://dx.doi.org/10.1016/j.watres.2012.10.015] [PMID: 23131396]

[500]   Singh Y, Gulati A, Singh DP, Khattar JIS. Cyanobacterial community structure in hot water springs of Indian North-Western Himalayas: A morphological, molecular and ecological approach. Algal Res 2018; 29: 179-92.
[http://dx.doi.org/10.1016/j.algal.2017.11.023]

[501]   Zhang S, Xu H, Zhang Y, Li Y, Wei J, Pei H. Variation of phytoplankton communities and their driving factors along a disturbed temperate river-to-sea ecosystem. Ecol Indic 2020; 118: 106776.
[http://dx.doi.org/10.1016/j.ecolind.2020.106776]

[502]   Gavrilović BR, Petrović TG, Radovanović TB, *et al.* Hepatic oxidative stress and neurotoxicity in Pelophylax kl. esculentus frogs: Influence of long-term exposure to a cyanobacterial bloom. Sci Total Environ 2021; 750: 141569.
[http://dx.doi.org/10.1016/j.scitotenv.2020.141569] [PMID: 32853936]

[503]   Gavrilović BR, Prokić MD, Petrović TG, *et al.* Biochemical parameters in skin and muscle of Pelophylax kl. esculentus frogs: Influence of a cyanobacterial bloom in situ. Aquat Toxicol 2020; 220: 105399.
[http://dx.doi.org/10.1016/j.aquatox.2019.105399] [PMID: 31896464]

[504]   Ji X, Liu C, Zhang M, Yin Y, Pan G. Mitigation of methylmercury production in eutrophic waters by interfacial oxygen nanobubbles. Water Res 2020; 173: 115563.
[http://dx.doi.org/10.1016/j.watres.2020.115563] [PMID: 32059129]

[505]   Petrou M, Karas PA, Vasileiadis S, *et al.* Irrigation of radish (Raphanus sativus L.) with microcystin-enriched water holds low risk for plants and their associated rhizopheric and epiphytic microbiome. Environ Pollut 2020; 266(Pt 1): 115208.
[http://dx.doi.org/10.1016/j.envpol.2020.115208] [PMID: 32683235]

[506]   Posadas E, García-Encina PA, Domínguez A, *et al.* Enclosed tubular and open algal–bacterial biofilm photobioreactors for carbon and nutrient removal from domestic wastewater. Ecol Eng 2014; 67: 156-64.
[http://dx.doi.org/10.1016/j.ecoleng.2014.03.007]

[507]   Coring E, Bäthe J. Effects of reduced salt concentrations on plant communities in the River Werra (Germany). Limnologica 2011; 41(2): 134-42.
[http://dx.doi.org/10.1016/j.limno.2010.08.004]

[508]   Izaguirre G, Taylor WD. A pseudanabaena species from Castaic Lake, California, that produces 2-

methylisoborneol. Water Res 1998; 32(5): 1673-7.
[http://dx.doi.org/10.1016/S0043-1354(97)00379-5]

[509] Rosińska J, Rybak M, Gołdyn R. Patterns of macrophyte community recovery as a result of the restoration of a shallow urban lake. Aquat Bot 2017; 138: 45-52.
[http://dx.doi.org/10.1016/j.aquabot.2016.12.005]

[510] Sidelev S, Koksharova O, Babanazarova O, Fastner J, Chernova E, Gusev E. Phylogeographic, toxicological and ecological evidence for the global distribution of *Raphidiopsis raciborskii* and its northernmost presence in Lake Nero, Central Western Russia. Harmful Algae 2020; 98: 101889.
[http://dx.doi.org/10.1016/j.hal.2020.101889] [PMID: 33129449]

[511] Tamm M, Laas P, Freiberg R, Nõges P, Nõges T. Parallel assessment of marine autotrophic picoplankton using flow cytometry and chemotaxonomy. Sci Total Environ 2018; 625: 185-93.
[http://dx.doi.org/10.1016/j.scitotenv.2017.12.234] [PMID: 29289004]

[512] Tamm M, Ligi M, Panksep K, *et al.* Boosting the monitoring of phytoplankton in optically complex coastal waters by combining pigment-based chemotaxonomy and in situ radiometry. Ecol Indic 2019; 97: 329-40.
[http://dx.doi.org/10.1016/j.ecolind.2018.10.024]

[513] Watson SB, Zastepa A, Boyer GL, Matthews E. Algal bloom response and risk management: On-site response tools. Toxicon 2017; 129: 144-52.
[http://dx.doi.org/10.1016/j.toxicon.2017.02.005] [PMID: 28209478]

[514] Posadas E, Alcántara C, García-Encina PA, *et al.* 3 - Microalgae cultivation in wastewater. In: Gonzalez-Fernandez C, Muñoz R, Eds. Microalgae-Based Biofuels and Bioproducts. Woodhead Publishing 2017; pp. 67-91.
[http://dx.doi.org/10.1016/B978-0-08-101023-5.00003-0]

[515] Sarma MK, Kaushik S, Goswami P. Cyanobacteria: A metabolic power house for harvesting solar energy to produce bio-electricity and biofuels. Biomass Bioenergy 2016; 90: 187-201.
[http://dx.doi.org/10.1016/j.biombioe.2016.03.043]

[516] Pham TL, Dao TS, Shimizu K, Lan-Chi DH, Utsumi M. Isolation and characterization of microcystin-producing cyanobacteria from Dau Tieng Reservoir, Vietnam. Nova Hedwigia 2015; 101(1-2): 3-20.
[http://dx.doi.org/10.1127/nova_hedwigia/2014/0243]

[517] Mann FD, Steinke TD. Biological nitrogen fixation (acetylene reduction) associated with blue-green algal (cyanobacterial) communities in the Beachwood Mangrove Nature Reserve II. Seasonal variation in acetylene reduction activity. S Afr J Bot 1993; 59(1): 1-8.
[http://dx.doi.org/10.1016/S0254-6299(16)30767-0]

[518] Skulberg OM. Blue-green Algae in lake mjøsa and other norwegian lakes. In: Jenkins SH, Ed. Eutrophication of Deep Lakes. Pergamon 1980; pp. 121-40.
[http://dx.doi.org/10.1016/B978-0-08-026024-2.50012-2]

[519] Kossalbayev BD, Tomo T, Zayadan BK, *et al.* Determination of the potential of cyanobacterial strains for hydrogen production. Int J Hydrogen Energy 2020; 45(4): 2627-39.
[http://dx.doi.org/10.1016/j.ijhydene.2019.11.164]

[520] Shi XY, Yu HQ. Simultaneous metabolism of benzoate and photobiological hydrogen production by Lyngbya sp. Renew Energy 2016; 95: 474-7.
[http://dx.doi.org/10.1016/j.renene.2016.04.051]

[521] Tiwari A, Pandey A. Cyanobacterial hydrogen production – A step towards clean environment. Int J Hydrogen Energy 2012; 37(1): 139-50.
[http://dx.doi.org/10.1016/j.ijhydene.2011.09.100]

[522] Villbrandt M, Stal LJ, Krumbein WE. Interactions between nitrogen fixation and oxegenic photosynthesis in a marine cyanobacterial mat. FEMS Microbiol Lett 1990; 74(1): 59-71.
[http://dx.doi.org/10.1111/j.1574-6968.1990.tb04052.x]

[523]   Felisberto SA, Souza DBS, Silva D. Characteristics and diversity of cyanobacteria in periphyton from lentic tropical ecosystem, Brazil. Adv Microbiol 2014; 4(15): 1076-87.
[http://dx.doi.org/10.4236/aim.2014.415118]

[524]   Barco M, Flores C, Rivera J, Caixach J. Determination of microcystin variants and related peptides present in a water bloom of Planktothrix (Oscillatoria) rubescens in a Spanish drinking water reservoir by LC/ESI-MS. Toxicon 2004; 44(8): 881-6.
[http://dx.doi.org/10.1016/j.toxicon.2004.08.011] [PMID: 15530970]

[525]   Barroin G, Feuillade M. Hydrogen peroxide as a potential algicide for oscillatoria rubescens D.C. Water Res 1986; 20(5): 619-23.
[http://dx.doi.org/10.1016/0043-1354(86)90026-6]

[526]   Bourdier G, Bohatier J, Feuillade M, Feuillade J. Amino acid incorporation by a natural population of *Oscillatoria rubescens*. A microautoradiographic study. FEMS Microbiol Lett 1989; 62(3): 185-90.
[http://dx.doi.org/10.1111/j.1574-6968.1989.tb03692.x]

[527]   Bruno M, Gucci PMB, Pierdominici E, *et al.* Microcystin-like toxins in different freshwater species of Oscillatoria. Toxicon 1992; 30(10): 1307-11.
[http://dx.doi.org/10.1016/0041-0101(92)90448-E] [PMID: 1440636]

[528]   El-Sheekh MM, Gharieb MM, Abou-El-Souod GW. Biodegradation of dyes by some green algae and cyanobacteria. Int Biodeterior Biodegradation 2009; 63(6): 699-704.
[http://dx.doi.org/10.1016/j.ibiod.2009.04.010]

[529]   Hertzberg S, Liaaen Jensen S. The carotenoids of blue-green algae—I. Phytochemistry 1966; 5(4): 557-63.
[http://dx.doi.org/10.1016/S0031-9422(00)83633-X]

[530]   Jann-Para G, Schwob I, Feuillade M. Occurrence of toxic Planktothrix rubescens blooms in lake Nantua, France. Toxicon 2004; 43(3): 279-85.
[http://dx.doi.org/10.1016/j.toxicon.2003.12.005] [PMID: 15033326]

[531]   Mur LR, Schreurs H. Light as a selective factor in the distribution of phytoplankton species. Water Sci Technol 1995; 32(4): 25-34.
[http://dx.doi.org/10.2166/wst.1995.0161]

[532]   Omlin M, Reichert P, Forster R. Biogeochemical model of Lake Zürich: model equations and results. Ecol Modell 2001; 141(1-3): 77-103.
[http://dx.doi.org/10.1016/S0304-3800(01)00256-3]

[533]   Rønneberg H, Foss P, Ramdahl T, Borch GM, Skulberg O, Liaaen-Jensen S. Occurrence and chirality of oscillaxanthin. Phytochemistry 1980; 19(10): 2167-70.
[http://dx.doi.org/10.1016/S0031-9422(00)82216-5]

[534]   List of Taxa. Ecology and Management of Inland Waters. Los Huertos M, Ed. Elsevier 2020; pp. 479-83.
[http://dx.doi.org/10.1016/B978-0-12-814266-0.00031-3]

[535]   Button DK, Robertson BR, Jüttner F. Microflora of a subalpine lake: Bacterial populations, size and DNA distributions, and their dependence on phosphate. FEMS Microbiol Ecol 1996; 21(2): 87-101.
[http://dx.doi.org/10.1111/j.1574-6941.1996.tb00336.x]

[536]   Chalansonnet S, Largeau C, Casadevall E, Berkaloff C, Peniguel G, Couderc R. Cyanobacterial resistant biopolymers. Geochemical implications of the properties of Schizothrix sp. resistant material. Org Geochem 1988; 13(4-6): 1003-10.
[http://dx.doi.org/10.1016/0146-6380(88)90282-3]

[537]   Clasen J, Bernhardt H. The use of algal assays for determining the effect of iron and phosphorus compounds on the growth of various algal species. Water Res 1974; 8(1): 31-44.
[http://dx.doi.org/10.1016/0043-1354(74)90006-2]

[538] Husain S, Afreen S, Hemlata , Yasin D, Afzal B, Fatma T. Cyanobacteria as a bioreactor for synthesis of silver nanoparticles-an effect of different reaction conditions on the size of nanoparticles and their dye decolorization ability. J Microbiol Methods 2019; 162: 77-82.
[http://dx.doi.org/10.1016/j.mimet.2019.05.011] [PMID: 31132377]

[539] Messineo V, Mattei D, Melchiorre S, *et al.* Microcystin diversity in a Planktothrix rubescens population from Lake Albano (Central Italy). Toxicon 2006; 48(2): 160-74.
[http://dx.doi.org/10.1016/j.toxicon.2006.04.006] [PMID: 16828137]

[540] Omlin M, Brun R, Reichert P. Biogeochemical model of Lake Zürich: Sensitivity, identifiability and uncertainty analysis. Ecol Modell 2001; 141(1-3): 105-23.
[http://dx.doi.org/10.1016/S0304-3800(01)00257-5]

[541] Piorreck M, Baasch KH, Pohl P. Biomass production, total protein, chlorophylls, lipids and fatty acids of freshwater green and blue-green algae under different nitrogen regimes. Phytochemistry 1984; 23(2): 207-16.
[http://dx.doi.org/10.1016/S0031-9422(00)80304-0]

[542] Rai J, Kumar D, Pandey LK, Yadav A, Gaur JP. Potential of cyanobacterial biofilms in phosphate removal and biomass production. J Environ Manage 2016; 177: 138-44.
[http://dx.doi.org/10.1016/j.jenvman.2016.04.010] [PMID: 27088210]

[543] Walsby AE, Avery A. Measurement of filamentous cyanobacteria by image analysis. J Microbiol Methods 1996; 26(1-2): 11-20.
[http://dx.doi.org/10.1016/0167-7012(96)00816-0]

[544] Collos Y. A physiological basis for estimating inorganic carbon release during photosynthesis by natural phytoplankton. Ecol Modell 1997; 96(1-3): 285-92.
[http://dx.doi.org/10.1016/S0304-3800(96)00070-1]

[545] Dale VH, Beyeler SC. Challenges in the development and use of ecological indicators. Ecol Indic 2001; 1(1): 3-10.
[http://dx.doi.org/10.1016/S1470-160X(01)00003-6]

[546] Fastner J, Heinze R, Chorus I. Microcystin-content, hepatotoxicity and cytotoxicity of cyanobacteria in some german water bodies. Water Sci Technol 1995; 32(4): 165-70.
[http://dx.doi.org/10.2166/wst.1995.0179]

[547] Konopka A. Accumulation and utilization of polysaccharide by hot-spring phototrophs during a light-dark transition. FEMS Microbiol Lett 1992; 102(1): 27-32.
[http://dx.doi.org/10.1111/j.1574-6968.1992.tb05792.x]

[548] Kreitlow S, Mundt S, Lindequist U. Cyanobacteria—a potential source of new biologically active substances. J Biotechnol 1999; 70(1-3): 61-3.
[http://dx.doi.org/10.1016/S0168-1656(99)00058-9] [PMID: 10412206]

[549] Kreitlow S, Mundt S, Lindequist U. Cyanobacteria — a potential source of new biologically active substances. In: Osinga R, Tramper J, Burgess JG, Wijffels RH, Eds. Progress in Industrial Microbiology. Elsevier 1999; Vol. 35: pp. 61-3.

[550] Kumar L, Bharadvaja N. Microorganisms: A remedial source for dye pollution. In: Shah MP, Ed. Removal of Toxic Pollutants Through Microbiological and Tertiary Treatment. Elsevier 2020; pp. 309-33.

[551] Los Huertos M. Water quality and catchments. In: Los Huertos M, Ed. Ecology and Management of Inland Waters. Elsevier 2020; pp. 315-58.
[http://dx.doi.org/10.1016/B978-0-12-814266-0.00024-6]

[552] Schanz F, Burri J. Phytoplankton photoadaptation during the spring period in Lake Zürich. Water Sci Technol 1995; 32(4): 59-62.
[http://dx.doi.org/10.2166/wst.1995.0166]

[553] Wang B, Song Q, Long J, Song G, Mi W, Bi Y. Optimization method for Microcystis bloom mitigation by hydrogen peroxide and its stimulative effects on growth of chlorophytes. Chemosphere 2019; 228: 503-12.
[http://dx.doi.org/10.1016/j.chemosphere.2019.04.138] [PMID: 31051353]

[554] Adey WH, Loveland K. Freshwater ecosystem models. In: Adey WH, Loveland K, Eds. Dynamic Aquaria. 3rd ed.. London: Academic Press 2007; pp. 443-53.

[555] Erdmann N, Schiewer U. 14CO2 Fixation Pattern of Cyanobacteria. Biochem Physiol Pflanz 1985; 180(7): 515-32.
[http://dx.doi.org/10.1016/S0015-3796(85)80100-1]

[556] Fastner J, Flieger I, Neumann U. Optimised extraction of microcystins from field samples — a comparison of different solvents and procedures. Water Res 1998; 32(10): 3177-81.
[http://dx.doi.org/10.1016/S0043-1354(98)00073-6]

[557] Gleason FH, Carney LT, Lilje O, Glockling SL. Ecological potentials of species of Rozella (Cryptomycota). Fungal Ecol 2012; 5(6): 651-6.
[http://dx.doi.org/10.1016/j.funeco.2012.05.003]

[558] Heynig VH. Massenentwicklung eines Vertreters der Oscillatoria agardhii/rubescens-Gruppe in einigen Teichen des Bezirkes Halle (DDR). Arch Protistenkd 1986; 131(1-2): 171-6.
[http://dx.doi.org/10.1016/S0003-9365(86)80076-8]

[559] Khataee AR, Zarei M, Dehghan G, Ebadi E, Pourhassan M. Biotreatment of a triphenylmethane dye solution using a Xanthophyta alga: Modeling of key factors by neural network. J Taiwan Inst Chem Eng 2011; 42(3): 380-6.
[http://dx.doi.org/10.1016/j.jtice.2010.08.006]

[560] Parsaeimehr A, Tabatabaei M, Parra-Saldivar R. Approaches to improve the quality of microalgae biodiesel: Challenges and future prospects. In: Hosseini M, Ed. Advances in Feedstock Conversion Technologies for Alternative Fuels and Bioproducts. Woodhead Publishing 2019; pp. 89-103.
[http://dx.doi.org/10.1016/B978-0-12-817937-6.00005-9]

[561] Sarkar P, Dey A. Phycoremediation – An emerging technique for dye abatement: An overview. Process Saf Environ Prot 2021; 147: 214-25.
[http://dx.doi.org/10.1016/j.psep.2020.09.031]

[562] Wang Y, Ho SH, Cheng CL, et al. Perspectives on the feasibility of using microalgae for industrial wastewater treatment. Bioresour Technol 2016; 222: 485-97.
[http://dx.doi.org/10.1016/j.biortech.2016.09.106] [PMID: 27765375]

[563] Weisse T, Bergkemper V. Rapid detection and quantification of the potentially toxic cyanobacterium Planktothrix rubescens by in-vivo fluorometry and flow cytometry. Water Res 2018; 138: 234-40.
[http://dx.doi.org/10.1016/j.watres.2018.03.052] [PMID: 29604575]

[564] Codd GA, Brooks WP, Lawton LA, Beattie KA. Cyanobacterial toxins in European waters: Occurrence, properties, problems and requirements. In: Wheeler D, Richardson ML, Bridges J, Eds. Watershed 89. Pergamon 1989; pp. 211-20.

[565] Dugan NR, Williams DJ. Cyanobacteria passage through drinking water filters during perturbation episodes as a function of cell morphology, coagulant and initial filter loading rate. Harmful Algae 2006; 5(1): 26-35.
[http://dx.doi.org/10.1016/j.hal.2005.04.001]

[566] hem L, Efraimsen H. Isolation of natural organic matter? The influence on the assimilable organic carbon. Environ Int 1999; 25(2-3): 367-71.
[http://dx.doi.org/10.1016/S0160-4120(98)00113-5]

[567] Janani S, Kumar SS. Performance analysis of different textile effluent treatment processes involving marine diatom Odontella aurita. Environmental Technology & Innovation 2018; 11: 153-64.
[http://dx.doi.org/10.1016/j.eti.2018.06.002]

[568] Kessel M, Klink F. Identification and comparison of eighteen archaebacteria by means of the diphtheria toxin reaction. Zentralbl Bakteriol Mikrobiol Hyg Ser B 1982; 3(1): 140-8.

[569] Khan R, Patel V, Khan Z. Bioremediation of dyes from textile and dye manufacturing industry effluent. In: Singh P, Kumar A, Borthakur A, Eds. Abatement of Environmental Pollutants. Elsevier 2020; pp. 107-25.
[http://dx.doi.org/10.1016/B978-0-12-818095-2.00005-9]

[570] Lankoff A, Kołątaj A. Influence of microcystine–YR and nodularin on the activity of some glucosidases in mouse liver. Toxicology 2000; 146(2-3): 177-85.
[http://dx.doi.org/10.1016/S0300-483X(00)00173-6] [PMID: 10814850]

[571] Lukavský J. Long-Term Preservation of Algal Strains by Immobilization. Arch Protistenkd 1988; 135(1-4): 65-8.
[http://dx.doi.org/10.1016/S0003-9365(88)80054-X]

[572] Abdel-Mawgoud AM, Stephanopoulos G. Simple glycolipids of microbes: Chemistry, biological activity and metabolic engineering. Synth Syst Biotechnol 2018; 3(1): 3-19.
[http://dx.doi.org/10.1016/j.synbio.2017.12.001] [PMID: 29911195]

[573] Dodds WK, Whiles MR. Trophic state and eutrophication. In: Dodds WK, Whiles MR, Eds. Freshwater Ecology. 3rd ed.. Academic Press 2020; pp. 537-81.

[574] French DP, Furnas MJ, Smayda TJ. Diel changes in nitrite concentration in the chlorophyll maximum in the Gulf of Mexico. Deep-Sea Res A, Oceanogr Res Pap 1983; 30(7): 707-22.
[http://dx.doi.org/10.1016/0198-0149(83)90018-3]

[575] Gibson MT. Developing biocidal compounds as freshwater algicides. International Biodeterioration 1990; 26(2-4): 149-58.
[http://dx.doi.org/10.1016/0265-3036(90)90055-C]

[576] Hayat K, Menhas S, Bundschuh J, Chaudhary HJ. Microbial biotechnology as an emerging industrial wastewater treatment process for arsenic mitigation: A critical review. J Clean Prod 2017; 151: 427-38.
[http://dx.doi.org/10.1016/j.jclepro.2017.03.084]

[577] Honegger R. Zytologie der Blaualgen-Hornmoos-Symbiose bei Anthoceros laevis aus Island. Flora (Jena) 1980; 170(4): 290-302.
[http://dx.doi.org/10.1016/S0367-2530(17)31218-5]

[578] Kouřil R, Ilík P, Tomek P, Nauš J, Poulíčková A. Chlorophyll fluorescence temperature curve on Klebsormidium flaccidum cultivated at different temperature regimes. J Plant Physiol 2001; 158(9): 1131-6.
[http://dx.doi.org/10.1078/0176-1617-00281]

[579] Matern U, Oberer L, Erhard M, Herdman M, Weckesser J. Hofmannolin, a cyanopeptolin from Scytonema hofmanni PCC 7110. Phytochemistry 2003; 64(6): 1061-7.
[http://dx.doi.org/10.1016/S0031-9422(03)00467-9] [PMID: 14568072]

[580] Naeher S, Gilli A, North RP, Hamann Y, Schubert CJ. Tracing bottom water oxygenation with sedimentary Mn/Fe ratios in Lake Zurich, Switzerland. Chem Geol 2013; 352: 125-33.
[http://dx.doi.org/10.1016/j.chemgeo.2013.06.006]

[581] Sangolkar LN, Maske SS, Muthal PL, Kashyap SM, Chakrabarti T. Isolation and characterization of microcystin producing Microcystis from a Central Indian water bloom. Harmful Algae 2009; 8(5): 674-84.
[http://dx.doi.org/10.1016/j.hal.2008.12.003]

[582] De Groot WT. Modelling the multiple nutrient limitation of algal growth. Ecol Modell 1983; 18(2): 99-119.
[http://dx.doi.org/10.1016/0304-3800(83)90049-2]

[583]   Fogg GE, Stewart WDP, Fay P, Walsby AE. Cellular organization of blue-green algae. In: Fogg GE, Stewart WDP, Fay P, Walsby AE, Eds. The Blue-green Algae. Academic Press 1973; pp. 35-77.
[http://dx.doi.org/10.1016/B978-0-12-261650-1.50007-1]

[584]   Guilizzoni P, Lami A, Marchetto A, Jones V, Manca M, Bettinetti R. Palaeoproductivity and environmental changes during the Holocene in central Italy as recorded in two crater lakes (Albano and Nemi). Quat Int 2002; 88(1): 57-68.
[http://dx.doi.org/10.1016/S1040-6182(01)00073-8]

[585]   Ihsanullah I, Jamal A, Ilyas M, Zubair M, Khan G, Atieh MA. Bioremediation of dyes: Current status and prospects. J Water Process Eng 2020; 38: 101680.
[http://dx.doi.org/10.1016/j.jwpe.2020.101680]

[586]   Ribau Teixeira M, Rosa MJ. Neurotoxic and hepatotoxic cyanotoxins removal by nanofiltration. Water Res 2006; 40(15): 2837-46.
[http://dx.doi.org/10.1016/j.watres.2006.05.035] [PMID: 16839584]

[587]   Skowroński T, Szubińska S, Jakubowski M, Pawlik B. Cadmium availability to the cyanobacterium Synechocystis aquatilis in solutions containing chloride. Environ Pollut 1992; 76(2): 163-7.
[http://dx.doi.org/10.1016/0269-7491(92)90104-I] [PMID: 15091998]

[588]   Feroci LT, Margheri MC, Pelosi E. Ultrastructure of Spirulina in comparison with Oscillatoria. Zentralbl Bakteriol Parasitenkd Infektionskr Hyg 1976; 131(7): 592-601.
[PMID: 827871]

[589]   Zainith S, Saxena G, Kishor R, Bharagava RN. Application of microalgae in industrial effluent treatment, contaminants removal, and biodiesel production: Opportunities, challenges, and future prospects. In: Saxena G, Kumar V, Shah MP, Eds. Bioremediation for Environmental Sustainability. Elsevier 2021; pp. 481-517.

[590]   Ho S-H, Chen Y-D, Qu W-Y, Liu F-Y, Wang Y. Algal culture and biofuel production using wastewater. In: Pandey A, Chang J-S, Soccol CR, Lee D-J, Chisti Y, Eds. Biofuels from Algae. 2nd ed.. Elsevier 2019; pp. 167-98.

[591]   Sajjadi B, Chen WY, Raman AAA, Ibrahim S. Microalgae lipid and biomass for biofuel production: A comprehensive review on lipid enhancement strategies and their effects on fatty acid composition. Renew Sustain Energy Rev 2018; 97: 200-32.
[http://dx.doi.org/10.1016/j.rser.2018.07.050]

[592]   Ciraolo G, la Loggia G, Maltese A. Sprectroradiometric characteristics of inland water bodies infestated by Oscillatoria rubescens algae. Proceedings of SPIE. 7824.
[http://dx.doi.org/10.1117/12.864674]

[593]   Çelekli A, Külköylüoğlu O. On the relationship between ecology and phytoplankton composition in a karstic spring (Çepni, Bolu). Ecol Indic 2007; 7(2): 497-503.
[http://dx.doi.org/10.1016/j.ecolind.2006.02.006]

[594]   Chia WY, Ying Tang DY, Khoo KS, Kay Lup AN, Chew KW. Nature's fight against plastic pollution: Algae for plastic biodegradation and bioplastics production. Environmental Science and Ecotechnology 2020; 4: 100065.
[http://dx.doi.org/10.1016/j.ese.2020.100065]

[595]   Chrapusta E, Węgrzyn M, Zabaglo K, et al. Microcystins and anatoxin-a in Arctic biocrust cyanobacterial communities. Toxicon 2015; 101: 35-40.
[http://dx.doi.org/10.1016/j.toxicon.2015.04.016] [PMID: 25937338]

[596]   Gammoudi S, Athmouni K, Nasri A, et al. Optimization, isolation, characterization and hepatoprotective effect of a novel pigment-protein complex (phycocyanin) producing microalga: Phormidium versicolorNCC-466 using response surface methodology. Int J Biol Macromol 2019; 137: 647-56.
[http://dx.doi.org/10.1016/j.ijbiomac.2019.06.237] [PMID: 31265852]

[597] Li L, Li J, Zhu C, Yu S. Study of the binding regularity and corresponding mechanism of drinking water odorous compound 2-MIB with coexisting dissolved organic matter. Chem Eng J 2020; 395: 125015.
[http://dx.doi.org/10.1016/j.cej.2020.125015]

[598] Mundt S, Kreitlow S, Nowotny A, Effmert U. Biochemical and pharmacological investigations of selected cyanobacteria. Int J Hyg Environ Health 2001; 203(4): 327-34.
[http://dx.doi.org/10.1078/1438-4639-00045] [PMID: 11434213]

[599] Xie L, Hagar J, Rediske RR, *et al.* The influence of environmental conditions and hydrologic connectivity on cyanobacteria assemblages in two drowned river mouth lakes. J Great Lakes Res 2011; 37(3): 470-9.
[http://dx.doi.org/10.1016/j.jglr.2011.05.002]

[600] Zimmerman WJ, Soliman CM, Rosen BH. Growth and 2-methylisoborneol production by the cyanobacterium Phormidium LM689. Water Sci Technol 1995; 31(11): 181-6.
[http://dx.doi.org/10.2166/wst.1995.0433]

[601] Chernova E, Sidelev S, Russkikh I, *et al.* Spatial distribution of cyanotoxins and ratios of microcystin to biomass indicators in the reservoirs of the Volga, Kama and Don Rivers, the European part of Russia. Limnologica 2020; 84: 125819.
[http://dx.doi.org/10.1016/j.limno.2020.125819]

[602] Gautam K, Rajvanshi M, Chugh N, *et al.* Microalgal applications toward agricultural sustainability: Recent trends and future prospects. In: Galanakis CM, Ed. Microalgae. Academic Press 2021; pp. 339-79.
[http://dx.doi.org/10.1016/B978-0-12-821218-9.00011-6]

[603] Kannaujiya VK, Kumar D, Singh V, Sinha RP. Advances in phycobiliproteins research: innovations and commercialization. In: Sinha Rp, Häder D-P, Eds. Natural Bioactive Compounds. Academic Press 2021; pp. 57-81.
[http://dx.doi.org/10.1016/B978-0-12-820655-3.00004-5]

[604] Papp ZG. Off-flavour problems in farmed fish. In: Lie Ø, Ed. Improving Farmed Fish Quality and Safety. Woodhead Publishing 2008; pp. 471-93.
[http://dx.doi.org/10.1533/9781845694920.2.471]

[605] Podduturi R, Petersen MA, Vestergaard M, Jørgensen NOG. Geosmin fluctuations and potential hotspots for elevated levels in recirculated aquaculture system (RAS): A case study from pikeperch (Stizostedion lucioperca) production in Denmark. Aquaculture 2020; 514: 734501.
[http://dx.doi.org/10.1016/j.aquaculture.2019.734501]

[606] Varol M, Balcı M. Characteristics of effluents from trout farms and their impact on water quality and benthic algal assemblages of the receiving stream. Environ Pollut 2020; 266(Pt 1): 115101.
[http://dx.doi.org/10.1016/j.envpol.2020.115101] [PMID: 32623272]

[607] Wang Z, Li R. Effects of light and temperature on the odor production of 2-methylisoborneo--producing Pseudanabaena sp. and geosmin-producing Anabaena ucrainica (cyanobacteria). Biochem Syst Ecol 2015; 58: 219-26.
[http://dx.doi.org/10.1016/j.bse.2014.12.013]

[608] Brittain S, Mohamed ZA, Wang J, Lehmann VKB, Carmichael WW, Rinehart KL. Isolation and characterization of microcystins from a River Nile strain of Oscillatoria tenuis Agardh ex Gomont. Toxicon 2000; 38(12): 1759-71.
[http://dx.doi.org/10.1016/S0041-0101(00)00105-7] [PMID: 10858515]

[609] Corry MJ, Payne PI, Dyer TA. A sequence analysis of 5 S rRNA from the blue-green alga *Oscillatoria tenuis* and a comparison of blue-green alga 5 S rRNA with those of bacterial and eukaryotic origin. FEBS Lett 1974; 46(1-2): 67-70.
[http://dx.doi.org/10.1016/0014-5793(74)80336-4] [PMID: 4214272]

[610] Donkor VA, Amewowor DHAK, Häder D-P. Effects of tropical solar radiation on the motility of filamentous cyanobacteria. FEMS Microbiol Ecol 1993; 12(2): 143-8.
[http://dx.doi.org/10.1111/j.1574-6941.1993.tb00026.x]

[611] Donkor VA, Häner DP. Protective Strategies of Several Cyanobacteria against Solar Radiation. J Plant Physiol 1995; 145(5-6): 750-5.
[http://dx.doi.org/10.1016/S0176-1617(11)81291-5]

[612] El Herry S, Fathalli A, Rejeb AJB, Bouaïcha N. Seasonal occurrence and toxicity of Microcystis spp. and Oscillatoria tenuis in the Lebna Dam, Tunisia. Water Res 2008; 42(4-5): 1263-73.
[http://dx.doi.org/10.1016/j.watres.2007.09.019] [PMID: 17936328]

[613] Ghobrial MG, Nassr HS, Kamil AW. Bioactivity effect of two macrophyte extracts on growth performance of two bloom-forming cyanophytes. Egypt J Aquat Res 2015; 41(1): 69-81.
[http://dx.doi.org/10.1016/j.ejar.2015.01.001]

[614] Higashi Y, Seki H. Application of an *in situ* gradostat for a natural phytoplankton community in a eutrophic environment. Environ Pollut 1999; 105(1): 101-9.
[http://dx.doi.org/10.1016/S0269-7491(98)00185-7]

[615] Nakanishi M, Hoson T, Inoue Y, Yagi M. Relationship between the maximum standing crop of musty-odor producing algae and nutrient concentrations in the southern basin water of Lake Biwa. Water Sci Technol 1999; 40(6): 179-84.
[http://dx.doi.org/10.2166/wst.1999.0293]

[616] Shunyu S, Yongding L, Yinwu S, Genbao L, Dunhai L. Lysis of Aphanizomenon flos-aquae (Cyanobacterium) by a bacterium Bacillus cereus. Biol Control 2006; 39(3): 345-51.
[http://dx.doi.org/10.1016/j.biocontrol.2006.06.011]

[617] Thangam R, Suresh V, Asenath Princy W, *et al.* C-Phycocyanin from Oscillatoria tenuis exhibited an antioxidant and *in vitro* antiproliferative activity through induction of apoptosis and G0/G1 cell cycle arrest. Food Chem 2013; 140(1-2): 262-72.
[http://dx.doi.org/10.1016/j.foodchem.2013.02.060] [PMID: 23578642]

[618] Akanmu RT, Onyema IC. Phytoplankton composition and dynamics off the coast of Lagos south-west, Nigeria. Reg Stud Mar Sci 2020; 37: 101356.
[http://dx.doi.org/10.1016/j.rsma.2020.101356]

[619] Alam MGM, Jahan N, Thalib L, Wei B, Maekawa T. Effects of environmental factors on the seasonally change of phytoplankton populations in a closed freshwater pond. Environ Int 2001; 27(5): 363-71.
[http://dx.doi.org/10.1016/S0160-4120(01)00087-3] [PMID: 11757850]

[620] Çelekli A, Bozkurt H. Assessing biochemical responses of filamentous algae integrated with surface waters in Yavuzeli-Araban catchment. Ecol Eng 2020; •••: 106126.

[621] Dwivedi S, Srivastava S, Mishra S, *et al.* Characterization of native microalgal strains for their chromium bioaccumulation potential: Phytoplankton response in polluted habitats. J Hazard Mater 2010; 173(1-3): 95-101.
[http://dx.doi.org/10.1016/j.jhazmat.2009.08.053] [PMID: 19744773]

[622] Hartley WR, Weiss CM. Light intensity and the vertical distribution of algae in tertiary oxidation ponds. Water Res 1970; 4(11): 751-63.
[http://dx.doi.org/10.1016/0043-1354(70)90011-4]

[623] Ram NM, Morris JC. Environmental significance of nitrogenous organic compounds in aquatic sources. Environ Int 1980; 4(5-6): 397-405.
[http://dx.doi.org/10.1016/0160-4120(80)90018-5]

[624] Sepehr A, Hassanzadeh M, Rodriguez-Caballero E. The protective role of cyanobacteria on soil stability in two Aridisols in northeastern Iran. Geoderma Reg 2019; 16: e00201.
[http://dx.doi.org/10.1016/j.geodrs.2018.e00201]

[625] Terauchi N, Ohtani T, Yamanaka K, Tsuji T, Sudou T, Ito K. Studies on a biological filter for musty odor removal in drinking water treatment processes. Water Sci Technol 1995; 31(11): 229-35.
[http://dx.doi.org/10.2166/wst.1995.0440]

[626] Zhuang LL, Li M, Hao Ngo H. Non-suspended microalgae cultivation for wastewater refinery and biomass production. Bioresour Technol 2020; 308: 123320.
[http://dx.doi.org/10.1016/j.biortech.2020.123320] [PMID: 32284252]

[627] Kajino M, Sakamoto K. The relationship between musty-odor-causing organisms and water quality in Lake Biwa. Water Sci Technol 1995; 31(11): 153-8.
[http://dx.doi.org/10.2166/wst.1995.0426]

[628] Perendeci NA, Yılmaz V, Ertit Taştan B, *et al.* Correlations between biochemical composition and biogas production during anaerobic digestion of microalgae and cyanobacteria isolated from different sources of Turkey. Bioresour Technol 2019; 281: 209-16.
[http://dx.doi.org/10.1016/j.biortech.2019.02.086] [PMID: 30822642]

[629] Salem Z, Ghobara M, El Nahrawy AA. Spatio-temporal evaluation of the surface water quality in the middle Nile Delta using Palmer's algal pollution index. Egyptian Journal of Basic and Applied Sciences 2017; 4(3): 219-26.
[http://dx.doi.org/10.1016/j.ejbas.2017.05.003]

[630] Zhang X, Mei X. Effects of benthic algae on release of soluble reactive phosphorus from sediments: A radioisotope tracing study. Water Sci Eng 2015; 8(2): 127-31.
[http://dx.doi.org/10.1016/j.wse.2015.04.008]

[631] Chakraborty S, Verma E, Singh SS. Cyanobacterial siderophores: Ecological and biotechnological significance. In: Mishra AK, Tiwari DN, Rai AN, Eds. Cyanobacteria. Academic Press 2019; pp. 383-97.
[http://dx.doi.org/10.1016/B978-0-12-814667-5.00019-2]

[632] Hosaka M, Murata K, Iikura Y, Oshimi A, Udagawa T. Off-flavor problem in drinking water of Tokyo arising from the occurrence of musty odor in a downstream tributary. Water Sci Technol 1995; 31(11): 29-34.
[http://dx.doi.org/10.2166/wst.1995.0394]

[633] Mohamed ZA. Potentially harmful microalgae and algal blooms in the Red Sea: Current knowledge and research needs. Mar Environ Res 2018; 140: 234-42.
[http://dx.doi.org/10.1016/j.marenvres.2018.06.019] [PMID: 29970250]

[634] Nwankwegu AS, Li Y, Huang Y, *et al.* Nutrient addition bioassay and phytoplankton community structure monitored during autumn in Xiangxi Bay of Three Gorges Reservoir, China. Chemosphere 2020; 247: 125960.
[http://dx.doi.org/10.1016/j.chemosphere.2020.125960] [PMID: 32069727]

[635] Sinang SC, Daud N, Kamaruddin N, Poh KB. Potential growth inhibition of freshwater algae by herbaceous plant extracts. Acta Ecol Sin 2019; 39(3): 229-33.
[http://dx.doi.org/10.1016/j.chnaes.2018.12.005]

[636] Zhang Q, Li X, Guo D, *et al.* Operation of a vertical algal biofilm enhanced raceway pond for nutrient removal and microalgae-based byproducts production under different wastewater loadings. Bioresour Technol 2018; 253: 323-32.
[http://dx.doi.org/10.1016/j.biortech.2018.01.014] [PMID: 29367158]

[637] Chittora D, Meena M, Barupal T, Swapnil P, Sharma K. Cyanobacteria as a source of biofertilizers for sustainable agriculture. Biochem Biophys Rep 2020; 22: 100737.
[http://dx.doi.org/10.1016/j.bbrep.2020.100737] [PMID: 32083191]

[638] Davoodbasha M, Edachery B, Nooruddin T, Lee SY, Kim JW. An evidence of C16 fatty acid methyl esters extracted from microalga for effective antimicrobial and antioxidant property. Microb Pathog 2018; 115: 233-8.

[http://dx.doi.org/10.1016/j.micpath.2017.12.049] [PMID: 29277474]

[639] Laraib N, Hussain A, Javid A, *et al.* Recent advancements in microalgal-induced remediation of wastewaters. In: Chowdhary P, Raj A, Verma D, Akhter Y, Eds. Microorganisms for Sustainable Environment and Health. Elsevier 2020; pp. 205-17.

[640] Pandey SN, Verma I, Kumar M. Cyanobacteria: potential source of biofertilizer and synthesizer of metallic nanoparticles. In: Singh PK, Kumar A, Singh VK, Shrivastava AK, Eds. Advances in Cyanobacterial Biology. Academic Press 2020; pp. 351-67.
[http://dx.doi.org/10.1016/B978-0-12-819311-2.00023-1]

[641] Reissig M, Trochine C, Queimaliños C, Balseiro E, Modenutti B. Impact of fish introduction on planktonic food webs in lakes of the Patagonian Plateau. Biol Conserv 2006; 132(4): 437-47.
[http://dx.doi.org/10.1016/j.biocon.2006.04.036]

[642] Shinfuku Y, Takanashi H, Nakajima T, Ogura A, Kitamura H, Akiba M. Exploration of an odorous aldehydes and ketones produced by Uroglena americana using high resolution mass spectrometry, GC-Olfactometry, and multivariate analysis. Chemosphere 2020; 257: 127174.
[http://dx.doi.org/10.1016/j.chemosphere.2020.127174] [PMID: 32497839]

[643] Wang Y, Wu M, Yu J, *et al.* Differences in growth, pigment composition and photosynthetic rates of two phenotypes Microcystis aeruginosa strains under high and low iron conditions. Biochem Syst Ecol 2014; 55: 112-7.
[http://dx.doi.org/10.1016/j.bse.2014.02.019]

[644] Belokda W, Khalil K, Loudiki M, Aziz F, Elkalay K. First assessment of phytoplankton diversity in a Marrocan shallow reservoir (Sidi Abderrahmane). Saudi J Biol Sci 2019; 26(3): 431-8.
[http://dx.doi.org/10.1016/j.sjbs.2017.11.047] [PMID: 30899154]

[645] Krztoń W, Kosiba J. Variations in zooplankton functional groups density in freshwater ecosystems exposed to cyanobacterial blooms. Sci Total Environ 2020; 730: 139044.
[http://dx.doi.org/10.1016/j.scitotenv.2020.139044] [PMID: 32402967]

[646] Li W, Su HN, Pu Y, *et al.* Phycobiliproteins: Molecular structure, production, applications, and prospects. Biotechnol Adv 2019; 37(2): 340-53.
[http://dx.doi.org/10.1016/j.biotechadv.2019.01.008] [PMID: 30685481]

[647] Oberholster PJ, De Klerk AR, De Klerk L, Chamier J, Botha AM. Algal assemblage responses to acid mine drainage and steel plant wastewater effluent up and downstream of pre and post wetland rehabilitation. Ecol Indic 2016; 62: 106-16.
[http://dx.doi.org/10.1016/j.ecolind.2015.11.025]

[648] Chinnasamy S, Bhatnagar A, Hunt RW, Das KC. Microalgae cultivation in a wastewater dominated by carpet mill effluents for biofuel applications. Bioresour Technol 2010; 101(9): 3097-105.
[http://dx.doi.org/10.1016/j.biortech.2009.12.026] [PMID: 20053551]

[649] Fazal T, Mushtaq A, Rehman F, *et al.* Bioremediation of textile wastewater and successive biodiesel production using microalgae. Renew Sustain Energy Rev 2018; 82: 3107-26.
[http://dx.doi.org/10.1016/j.rser.2017.10.029]

[650] Peng W, Wu Q, Tu P, Zhao N. Pyrolytic characteristics of microalgae as renewable energy source determined by thermogravimetric analysis. Bioresour Technol 2001; 80(1): 1-7.
[http://dx.doi.org/10.1016/S0960-8524(01)00072-4] [PMID: 11554596]

[651] Rawat I, Ranjith Kumar R, Mutanda T, Bux F. Biodiesel from microalgae: A critical evaluation from laboratory to large scale production. Appl Energy 2013; 103: 444-67.
[http://dx.doi.org/10.1016/j.apenergy.2012.10.004]

[652] Solé-Bundó M, Passos F, Romero-Güiza MS, Ferrer I, Astals S. Co-digestion strategies to enhance microalgae anaerobic digestion: A review. Renew Sustain Energy Rev 2019; 112: 471-82.
[http://dx.doi.org/10.1016/j.rser.2019.05.036]

[653] Gerasimenko LM, Mikhodyuk OS. Halophilic algal-bacterial and cyanobacterial communities and

their role in carbonate precipitation. Paleontol J 2009; 43(8): 940-57.
[http://dx.doi.org/10.1134/S0031030109080127]

[654] Aakermann T, Skulberg OM, Liaaen-Jensen S. A comparison of the carotenoids of strains of Oscillatoria and Spirulina (Cyanobacteria). Biochem Syst Ecol 1992; 20(8): 761-9.
[http://dx.doi.org/10.1016/0305-1978(92)90035-C]

[655] Belkin S, Padan E. Sulfide-dependent hydrogen evolution in the cyanobacterium *Oscillatoria limnetica*. FEBS Lett 1978; 94(2): 291-4.
[http://dx.doi.org/10.1016/0014-5793(78)80959-4] [PMID: 3111888]

[656] Daly MG, Bishop RH. Ferredoxin from the cyanobacterium Oscillatoria agardhii. Phytochemistry 1984; 23(11): 2669-70.
[http://dx.doi.org/10.1016/S0031-9422(00)84122-9]

[657] Leboulanger C, Rimet F, Hème de Lacotte M, Bérard A. Effects of atrazine and nicosulfuron on freshwater microalgae. Environ Int 2001; 26(3): 131-5.
[http://dx.doi.org/10.1016/S0160-4120(00)00100-8] [PMID: 11341696]

[658] Oikawa E, Ishibashi Y. Activity of geranyl pyrophosphate synthase in musty odor producing cyanobacteria. Water Sci Technol 1999; 40(6): 195-202.
[http://dx.doi.org/10.2166/wst.1999.0297]

[659] Schopf JW, Kudryavtsev AB, Osterhout JT, *et al.* An anaerobic □3400 Ma shallow-water microbial consortium: Presumptive evidence of Earth's Paleoarchean anoxic atmosphere. Precambrian Res 2017; 299: 309-18.
[http://dx.doi.org/10.1016/j.precamres.2017.07.021]

[660] Shabana EF, Ali GH. Phytoplankton activities in hypersaline, anoxic conditions. II-Photosynthetic activity of some sulphide adapted cyanobacterial strains isolated from Solar Lake, Taba, Egypt. Water Sci Technol 1999; 40(7): 127-32.
[http://dx.doi.org/10.2166/wst.1999.0344]

[661] Shahak Y, Arieli B, Binder B, Padan E. Sulfide-dependent photosynthetic electron flow coupled to proton translocation in thylakoids of the cyanobacterium Oscillatoria limnetica. Arch Biochem Biophys 1987; 259(2): 605-15.
[http://dx.doi.org/10.1016/0003-9861(87)90527-3] [PMID: 2827581]

[662] Slooten L, De Smet M, Sybesma C. Sulfide-dependent electron transport in thylakoids from the cyanobacterium Oscillatoria limnetica. Biochim Biophys Acta Bioenerg 1989; 973(2): 272-80.
[http://dx.doi.org/10.1016/S0005-2728(89)80432-3]

[663] Tien CJ. Biosorption of metal ions by freshwater algae with different surface characteristics. Process Biochem 2002; 38(4): 605-13.
[http://dx.doi.org/10.1016/S0032-9592(02)00183-8]

[664] Al-Homaidan AA, Arif IA. Ecology and bloom-forming algae of a semi-permanent rain-fed pool at Al-Kharj, Saudi Arabia. J Arid Environ 1998; 38(1): 15-25.
[http://dx.doi.org/10.1006/jare.1997.0319]

[665] Belkin S, Siderer Y, Shahak Y, Arieli B, Padan E. 2,3-Dimercaptopropan-1-ol (BAL). An aerobic electron-transport inhibitor, but an anaerobic photosynthetic electron donor. Biochim Biophys Acta Bioenerg 1984; 766(3): 563-9.
[http://dx.doi.org/10.1016/0005-2728(84)90115-4]

[666] Bramburger AJ, Reavie ED. A comparison of phytoplankton communities of the deep chlorophyll layers and epilimnia of the Laurentian Great Lakes. J Great Lakes Res 2016; 42(5): 1016-25.
[http://dx.doi.org/10.1016/j.jglr.2016.07.004]

[667] Hussein RA, Salama AAA, El Naggar ME, Ali GH. Medicinal impact of microalgae collected from high rate algal ponds; phytochemical and pharmacological studies of microalgae and its application in medicated bandages. Biocatal Agric Biotechnol 2019; 20: 101237.

[http://dx.doi.org/10.1016/j.bcab.2019.101237]

[668] Mishra M, Singh SK, Singh M, *et al.* Environmental contaminants and their management using microorganisms. In: Kumar A, Singh VK, Singh P, Mishra VK, Eds. Microbe Mediated Remediation of Environmental Contaminants. Woodhead Publishing 2021; pp. 37-45.
[http://dx.doi.org/10.1016/B978-0-12-821199-1.00004-3]

[669] Mohamed ZA. Toxic effect of norharmane on a freshwater plankton community. Ecohydrol Hydrobiol 2013; 13(3): 226-32.
[http://dx.doi.org/10.1016/j.ecohyd.2013.08.002]

[670] Sousa FM, Pereira JG, Marreiros BC, Pereira MM. Taxonomic distribution, structure/function relationship and metabolic context of the two families of sulfide dehydrogenases: SQR and FCSD. Biochim Biophys Acta Bioenerg 2018; 1859(9): 742-53.
[http://dx.doi.org/10.1016/j.bbabio.2018.04.004] [PMID: 29684324]

[671] Ali GH. Phytoplankton activities in hypersaline, anoxic conditions. I—Dynamics of phytoplankton succession in the solar lake (Taba, Egypt). Water Sci Technol 1999; 40(7): 117-25.
[http://dx.doi.org/10.2166/wst.1999.0342]

[672] Belkin S, Shahak Y, Padan E. Anoxygenic photosynthetic electron transport. Methods in Enzymology. Academic Press 1988; Vol. 167: pp. 380-6. [40]

[673] Jacquet S, Briand JF, Leboulanger C, *et al.* The proliferation of the toxic cyanobacterium Planktothrix rubescens following restoration of the largest natural French lake (Lac du Bourget). Harmful Algae 2005; 4(4): 651-72.
[http://dx.doi.org/10.1016/j.hal.2003.12.006]

[674] Kashyap A, Pandey KD, Sarkar S. Enhanced hydrogen photoproduction by non-heterocystous cyanobacterium Plectonema boryanum. Int J Hydrogen Energy 1996; 21(2): 107-9.
[http://dx.doi.org/10.1016/0360-3199(95)00050-X]

[675] Mansour EA, Abo El-Enin SA, Hamouda AS, Mahmoud HM. Efficacy of extraction techniques and solvent polarity on lipid recovery from domestic wastewater microalgae. Environ Nanotechnol Monit Manag 2019; 12: 100271.
[http://dx.doi.org/10.1016/j.enmm.2019.100271]

[676] Moezelaar R, Mattos MJT, Stal LJ. Lactate dehydrogenase in the cyanobacterium *Microcystis* PCC7806. FEMS Microbiol Lett 1995; 127(1-2): 47-50.
[http://dx.doi.org/10.1111/j.1574-6968.1995.tb07448.x] [PMID: 7557309]

[677] Schmidt A. Sulfur metabolism in cyanobacteria. Methods in Enzymology. Academic Press 1988; Vol. 167: pp. 572-83. [62]

[678] Allen JF, Thake B, Martin WF. Nitrogenase Inhibition Limited Oxygenation of Earth's Proterozoic Atmosphere. Trends Plant Sci 2019; 24(11): 1022-31.
[http://dx.doi.org/10.1016/j.tplants.2019.07.007] [PMID: 31447302]

[679] Boon JJ, De Leeuw JW. Amino acid sequence information in proteins and complex proteinaceous material revealed by pyrolysis-capillary gas chromatography-low and high resolution mass spectrometry. J Anal Appl Pyrolysis 1987; 11: 313-27.
[http://dx.doi.org/10.1016/0165-2370(87)85038-6]

[680] Camacho A, Garcia-Pichel F, Vicente E, Castenholz RW. Adaptation to sulfide and to the underwater light field in three cyanobacterial isolates from Lake Arcas (Spain). FEMS Microbiol Ecol 1996; 21(4): 293-301.
[http://dx.doi.org/10.1111/j.1574-6941.1996.tb00126.x]

[681] Khan S, Shamshad I, Waqas M, Nawab J, Ming L. Remediating industrial wastewater containing potentially toxic elements with four freshwater algae. Ecol Eng 2017; 102: 536-41.
[http://dx.doi.org/10.1016/j.ecoleng.2017.02.038]

[682] Momper L, Hu E, Moore KR, *et al.* Metabolic versatility in a modern lineage of cyanobacteria from

terrestrial hot springs. Free Radic Biol Med 2019; 140: 224-32.
[http://dx.doi.org/10.1016/j.freeradbiomed.2019.05.036] [PMID: 31163257]

[683]   Salem MZM, EL-Hefny M, Nasser RA, Ali HM, El-Shanhorey NA, Elansary HO. Medicinal and biological values of Callistemon viminalis extracts: History, current situation and prospects. Asian Pac J Trop Med 2017; 10(3): 229-37.
[http://dx.doi.org/10.1016/j.apjtm.2017.03.015] [PMID: 28442106]

[684]   Seguin F, Le Bihan F, Leboulanger C, Bérard A. A risk assessment of pollution: induction of atrazine tolerance in phytoplankton communities in freshwater outdoor mesocosms, using chlorophyll fluorescence as an endpoint. Water Res 2002; 36(13): 3227-36.
[http://dx.doi.org/10.1016/S0043-1354(02)00013-1] [PMID: 12188119]

[685]   Alamri SA, Mohamed ZA. Selective inhibition of toxic cyanobacteria by β-carboline-containing bacterium Bacillus flexus isolated from Saudi freshwaters. Saudi J Biol Sci 2013; 20(4): 357-63.
[http://dx.doi.org/10.1016/j.sjbs.2013.04.002] [PMID: 24235872]

[686]   Allen JF. A redox switch hypothesis for the origin of two light reactions in photosynthesis. FEBS Lett 2005; 579(5): 963-8.
[http://dx.doi.org/10.1016/j.febslet.2005.01.015] [PMID: 15710376]

[687]   Alonso-Rodríguez R, Páez-Osuna F. Nutrients, phytoplankton and harmful algal blooms in shrimp ponds: A review with special reference to the situation in the Gulf of California. Aquaculture 2003; 219(1-4): 317-36.
[http://dx.doi.org/10.1016/S0044-8486(02)00509-4]

[688]   Kalin M, Cao Y, Smith M, Olaveson MM. Development of the phytoplankton community in a pit-lake in relation to water quality changes. Water Res 2001; 35(13): 3215-25.
[http://dx.doi.org/10.1016/S0043-1354(01)00016-1] [PMID: 11487119]

[689]   Mohamed ZA, El-Sharouny HM, Ali WSM. Microcystin production in benthic mats of cyanobacteria in the Nile River and irrigation canals, Egypt. Toxicon 2006; 47(5): 584-90.
[http://dx.doi.org/10.1016/j.toxicon.2006.01.029] [PMID: 16564062]

[690]   Munawar M, Mudroch A, Munawar IF, Thomas RL. The Impact of Sediment-Associated Contaminants from the Niagara River Mouth on Various Size Assemblages of Phytoplankton. J Great Lakes Res 1983; 9(2): 303-13.
[http://dx.doi.org/10.1016/S0380-1330(83)71899-X]

[691]   Shibata H, Takahashi M, Yamaguchi I, Kobayashi S. Sulfide oxidation by gene expressions of sulfide-quinone oxidoreductase and ubiquinone-8 biosynthase in Escherichia coli. J Biosci Bioeng 1999; 88(3): 244-9.
[http://dx.doi.org/10.1016/S1389-1723(00)80004-3] [PMID: 16232606]

[692]   Friedrich CG. Physiology and genetics of sulfur-oxidizing bacteria. In: Poole RK, Ed. Advances in Microbial Physiology. Academic Press 1997; 39: pp. 235-89.
[http://dx.doi.org/10.1016/S0065-2911(08)60018-1]

[693]   Harel A, Karkar S, Cheng S, Falkowski PG, Bhattacharya D. Deciphering primordial cyanobacterial genome functions from protein network analysis. Curr Biol 2015; 25(5): 628-34.
[http://dx.doi.org/10.1016/j.cub.2014.12.061] [PMID: 25683807]

[694]   Mowe MAD, Song Y, Sim DZH, *et al.* Comparative study of six emergent macrophyte species for controlling cyanobacterial blooms in a tropical reservoir. Ecol Eng 2019; 129: 11-21.
[http://dx.doi.org/10.1016/j.ecoleng.2018.12.026]

[695]   Orsini S, Duce C, Bonaduce I. Analytical pyrolysis of ovalbumin. J Anal Appl Pyrolysis 2018; 130: 62-71.
[http://dx.doi.org/10.1016/j.jaap.2018.01.026]

[696]   Shibata H, Takahashi M, Yamaguchi I, Kobayashi S. Efficient removal of sulfide following integration of multiple copies of the sulfide-quinone oxidoreductase gene (sqr) into the Escherichia

coli chromosome. J Biosci Bioeng 2001; 91(5): 493-9.
[http://dx.doi.org/10.1016/S1389-1723(01)80279-6] [PMID: 16233028]

[697] Stal LJ. Poly(hydroxyalkanoate) in cyanobacteria: An overview. FEMS Microbiol Lett 1992; 103(2-4): 169-80.
[http://dx.doi.org/10.1111/j.1574-6968.1992.tb05835.x] [PMID: 1537546]

[698] Walsh BJC, Brito JA, Giedroc DP. Hydrogen sulfide signaling and enzymology. In: Liu H-W, Begley TP, Eds. Comprehensive Natural Products III. Oxford: Elsevier 2020; pp. 430-73.
[http://dx.doi.org/10.1016/B978-0-12-409547-2.14699-2]

[699] Wentzky VC, Frassl MA, Rinke K, Boehrer B. Metalimnetic oxygen minimum and the presence of Planktothrix rubescens in a low-nutrient drinking water reservoir. Water Res 2019; 148: 208-18.
[http://dx.doi.org/10.1016/j.watres.2018.10.047] [PMID: 30388522]

[700] Acosta Pomar L, Bruni V, Decembrini F, Giuffré G, Maugeri TL. Distribution and activity of picophytoplankton in a brackish environment. Prog Oceanogr 1988; 21(2): 129-38.
[http://dx.doi.org/10.1016/0079-6611(88)90031-6]

[701] Batchu NK, Khater S, Patil S, *et al.* Whole genome sequence analysis of Geitlerinema sp. FC II unveils competitive edge of the strain in marine cultivation system for biofuel production. Genomics 2019; 111(3): 465-72.
[http://dx.doi.org/10.1016/j.ygeno.2018.03.004] [PMID: 29518464]

[702] Cho HJ, Baek K, Jeon JK, Park SH, Suh DJ, Park YK. Removal characteristics of copper by marine macro-algae-derived chars. Chem Eng J 2013; 217: 205-11.
[http://dx.doi.org/10.1016/j.cej.2012.11.123]

[703] El-Kassas HY, Mohamed LA. Bioremediation of the textile waste effluent by Chlorella vulgaris. Egypt J Aquat Res 2014; 40(3): 301-8.
[http://dx.doi.org/10.1016/j.ejar.2014.08.003]

[704] Friedberg D, Seijffers J. Plasmids in two cyanobacterial strains. FEBS Lett 1979; 107(1): 165-8.
[http://dx.doi.org/10.1016/0014-5793(79)80487-1] [PMID: 115717]

[705] Kim MJ, Bai SJ, Youn JR, Song YS. Anomalous power enhancement of biophotovoltaic cell. J Power Sources 2019; 412: 301-10.
[http://dx.doi.org/10.1016/j.jpowsour.2018.11.056]

[706] Nitschmann WH, Schmetterer G, Muchl R, Peschek GA. Active sodium extrusion reduces net efficiencies of oxidative phosphorylation in the strictly photoautotrophic cyanobacterium Anacystis nidulans. Biochim Biophys Acta Bioenerg 1982; 682(2): 293-6.
[http://dx.doi.org/10.1016/0005-2728(82)90111-6]

[707] Rao KK, Hall DO. Hydrogenases: Isolation and assay. Methods in Enzymology. Academic Press 1988; Vol. 167: pp. 501-9. [53]

[708] Camargo JA. Fluoride toxicity to aquatic organisms: A review. Chemosphere 2003; 50(3): 251-64.
[http://dx.doi.org/10.1016/S0045-6535(02)00498-8] [PMID: 12656244]

[709] Izaguirre G, Taylor WD, Pasek J. Off-flavor problems in two reservoirs, associated with planktonic Pseudanabaena species. Water Sci Technol 1999; 40(6): 85-90.
[http://dx.doi.org/10.2166/wst.1999.0268]

[710] Roeselers G, Stal LJ, van Loosdrecht MCM, Muyzer G. Development of a PCR for the detection and identification of cyanobacterial nifD genes. J Microbiol Methods 2007; 70(3): 550-6.
[http://dx.doi.org/10.1016/j.mimet.2007.06.011] [PMID: 17681622]

[711] van Gemerden H. Microbial mats: A joint venture. Mar Geol 1993; 113(1-2): 3-25.
[http://dx.doi.org/10.1016/0025-3227(93)90146-M]

[712] Zwart G, Hiorns WD, Methé BA, *et al.* Nearly identical 16S rRNA sequences recovered from lakes in North America and Europe indicate the existence of clades of globally distributed freshwater bacteria.

Syst Appl Microbiol 1998; 21(4): 546-56.
[http://dx.doi.org/10.1016/S0723-2020(98)80067-2] [PMID: 9924823]

[713] de Farias Silva CE, Bertucco A. Bioethanol from microalgae and cyanobacteria: A review and technological outlook. Process Biochem 2016; 51(11): 1833-42.
[http://dx.doi.org/10.1016/j.procbio.2016.02.016]

[714] Phwan CK, Ong HC, Chen WH, Ling TC, Ng EP, Show PL. Overview: Comparison of pretreatment technologies and fermentation processes of bioethanol from microalgae. Energy Convers Manage 2018; 173: 81-94.
[http://dx.doi.org/10.1016/j.enconman.2018.07.054]

[715] Ueno Y, Kurano N, Miyachi S. Ethanol production by dark fermentation in the marine green alga, Chlorococcum littorale. J Ferment Bioeng 1998; 86(1): 38-43.
[http://dx.doi.org/10.1016/S0922-338X(98)80031-7]

[716] Karan T, Erenler R, Altuner Z. Isolation and Molecular Identification of Some Blue-Green Algae (Cyanobacteria) from Freshwater Sites in Tokat Province of Turkey. Turkish J Agri Food Sci Technol (Campinas) 2017; 5: 1371.

[717] Bakeeva LE, Chumakov KM, Drachev AL, Metlina AL, Skulachev VP. The sodium cycle. III. Vibrio alginolyticus resembles Vibrio cholerae and some other vibriones by flagellar motor and ribosomal 5S-RNA structures. Biochim Biophys Acta Bioenerg 1986; 850(3): 466-72.
[http://dx.doi.org/10.1016/0005-2728(86)90115-5] [PMID: 3730372]

[718] Hemens J, Mason MH. Sewage nutrient removal by a shallow algal stream. Water Res 1968; 2(4): 277-87.
[http://dx.doi.org/10.1016/0043-1354(68)90020-1]

[719] Jüttner F. Physiology and biochemistry of odorous compounds from freshwater cyanobacteria and algae. Water Sci Technol 1995; 31(11): 69-78.
[http://dx.doi.org/10.2166/wst.1995.0405]

[720] Ndlela LL, Oberholster PJ, Van Wyk JH, Cheng PH. An overview of cyanobacterial bloom occurrences and research in Africa over the last decade. Harmful Algae 2016; 60: 11-26.
[http://dx.doi.org/10.1016/j.hal.2016.10.001] [PMID: 28073554]

[721] Reyes-Caballero H, Campanello GC, Giedroc DP. Metalloregulatory proteins: Metal selectivity and allosteric switching. Biophys Chem 2011; 156(2-3): 103-14.
[http://dx.doi.org/10.1016/j.bpc.2011.03.010] [PMID: 21511390]

[722] Sadvakasova AK, Kossalbayev BD, Zayadan BK, *et al.* Bioprocesses of hydrogen production by cyanobacteria cells and possible ways to increase their productivity. Renew Sustain Energy Rev 2020; 133: 110054.
[http://dx.doi.org/10.1016/j.rser.2020.110054]

[723] Simons J. Vaucheria species from estuarine areas in the Netherlands. Neth J Sea Res 1975; 9(1): 1-23.
[http://dx.doi.org/10.1016/0077-7579(75)90020-4]

[724] Smith PT. Toxic effects of blooms of marine species of Oscillatoriales on farmed prawns (Penaeus monodon, Penaeus japonicus) and brine shrimp (Artemia salina). Toxicon 1996; 34(8): 857-69.
[http://dx.doi.org/10.1016/0041-0101(96)00048-7] [PMID: 8875773]

[725] Subashchandrabose SR, Ramakrishnan B, Megharaj M, Venkateswarlu K, Naidu R. Mixotrophic cyanobacteria and microalgae as distinctive biological agents for organic pollutant degradation. Environ Int 2013; 51: 59-72.
[http://dx.doi.org/10.1016/j.envint.2012.10.007] [PMID: 23201778]

[726] Benemann JR, Miyamoto K, Hallenbeck PC. Bioengineering aspects of biophotolysis. Enzyme Microb Technol 1980; 2(2): 103-11.
[http://dx.doi.org/10.1016/0141-0229(80)90064-2]

[727] Défarge C, Trichet J, Maurin A, Hucher M. Kopara in Polynesian atolls: early stages of formation of

calcareous stromatolites. Sediment Geol 1994; 89(1-2): 9-23.
[http://dx.doi.org/10.1016/0037-0738(94)90080-9]

[728] Dibrov PA, Kostyrko VA, Lazarova RL, Skulachev VP, Smirnova IA. The sodium cycle. I. Na+-dependent motility and modes of membrane energization in the marine alkalotolerant Vibrio alginolyticus. Biochim Biophys Acta Bioenerg 1986; 850(3): 449-57.
[http://dx.doi.org/10.1016/0005-2728(86)90113-1] [PMID: 2425848]

[729] Gantt E. Structure and function of phycobilisomes: Light harvesting pigment complexes in red and blue-green algae. In: Bourne GH, Danielli JF, Jeon KW, Eds. International Review of Cytology. Academic Press 1980; Vol. 66: pp. 45-80.

[730] Hemens J, Stander GJ. Nutrient removal from sewage effluents by algal activity. In: Jenkins SH, Ed. Advances in Water Pollution Research. Pergamon 1969; pp. 701-11.

[731] Metting B, Pyne JW. Biologically active compounds from microalgae. Enzyme Microb Technol 1986; 8(7): 386-94.
[http://dx.doi.org/10.1016/0141-0229(86)90144-4]

[732] Skulachev VP. Bacterial sodium transport: Bioenergetic functions of sodium ions. In: Rosen BP, Silver S, Eds. Ion Transport in Prokaryotes. Academic Press 1987; pp. 131-64.
[http://dx.doi.org/10.1016/B978-0-12-596935-2.50006-2]

[733] Skulachev VP. Bacterial Na$^+$ energetics. FEBS Lett 1989; 250(1): 106-14.
[http://dx.doi.org/10.1016/0014-5793(89)80693-3] [PMID: 2544457]

[734] Suffet IHM, Khiari D, Bruchet A. The drinking water taste and odor wheel for the millennium: Beyond geosmin and 2-methylisoborneol. Water Sci Technol 1999; 40(6): 1-13.
[http://dx.doi.org/10.2166/wst.1999.0248]

[735] Tian X, Chen L, Wang J, Qiao J, Zhang W. Quantitative proteomics reveals dynamic responses of Synechocystis sp. PCC 6803 to next-generation biofuel butanol. J Proteomics 2013; 78: 326-45.
[http://dx.doi.org/10.1016/j.jprot.2012.10.002] [PMID: 23079071]

[736] Botello-Morte L, González A, Bes MT, Peleato ML, Fillat MF. Functional genomics of metalloregulators in cyanobacteria. In: Chauvat F, Cassier-Chauvat C, Eds. Advances in Botanical Research. Academic Press 2013; Vol. 65: pp. 107-56.

[737] Brown II, Fadeyev SI, Kirik II, Severina II, Skulachev VP, Light-dependent Δ. Light-dependent Δ{smNa-generation and utilization in the marine cyanobacterium *Oscillatoria brevis*. FEBS Lett 1990; 270(1-2): 203-6.
[http://dx.doi.org/10.1016/0014-5793(90)81268-S] [PMID: 2171990]

[738] Friedmann EI, Galun M. Desert algae, lichens, and fungi. In: Brown GW, Ed. Desert Biology. Academic Press 1974; pp. 165-212.

[739] Hirose K, Ezaki B, Liu T, Nakashima S. Diamide stress induces a metallothionein BmtA through a repressor BxmR and is modulated by Zn-inducible BmtA in the cyanobacterium Oscillatoria brevis. Toxicol Lett 2006; 163(3): 250-6.
[http://dx.doi.org/10.1016/j.toxlet.2005.11.008] [PMID: 16359831]

[740] Hube AE, Heyduck-Söller B, Fischer U. Phylogenetic classification of heterotrophic bacteria associated with filamentous marine cyanobacteria in culture. Syst Appl Microbiol 2009; 32(4): 256-65.
[http://dx.doi.org/10.1016/j.syapm.2009.03.001] [PMID: 19423262]

[741] le Campion-Alsumard T, Plante-Cuny MR, Vacelet E, Mille G. Evolution des hydrocarbures et des populations bactériennes et microphytiques dans les sédiments des marais maritimes de l'ile grande pollués par l'Amoco cadiz: 2—Evolution des peuplements microphytiques. Mar Environ Res 1984; 11(4): 275-303.
[http://dx.doi.org/10.1016/0141-1136(84)90039-4]

[742] Liu T, Nakashima S, Hirose K, *et al.* A metallothionein and CPx-ATPase handle heavy-metal tolerance in the filamentous cyanobacterium *Oscillatoria brevis*[1]. FEBS Lett 2003; 542(1-3): 159-63.

[http://dx.doi.org/10.1016/S0014-5793(03)00370-3] [PMID: 12729917]

[743] McCleary JA. The biology of desert plants. In: Brown GW, Ed. Desert Biology. Academic Press 1968; pp. 141-94.

[744] Nies DH. Efflux-mediated heavy metal resistance in prokaryotes. FEMS Microbiol Rev 2003; 27(2-3): 313-39.
[http://dx.doi.org/10.1016/S0168-6445(03)00048-2] [PMID: 12829273]

[745] Weaver PF, Lien S, Seibert M. Photobiological production of hydrogen. Sol Energy 1980; 24(1): 3-45.
[http://dx.doi.org/10.1016/0038-092X(80)90018-3]

[746] Arias DM, García J, Uggetti E. Production of polymers by cyanobacteria grown in wastewater: Current status, challenges and future perspectives. N Biotechnol 2020; 55: 46-57.
[http://dx.doi.org/10.1016/j.nbt.2019.09.001] [PMID: 31541716]

[747] Dionigi CP, Greene DA, Millie DF, Johnsen PB. Mixed function oxidase inhibitors affect production of the off-flavor microbial metabolite geosmin. Pestic Biochem Physiol 1990; 38(1): 76-80.
[http://dx.doi.org/10.1016/0048-3575(90)90151-Q]

[748] Kraus MP. Resistance of blue-green algae to 60Co gamma radiation. Radiat Bot 1969; 9(6): 481-9.
[http://dx.doi.org/10.1016/S0033-7560(69)80054-2]

[749] Lambert GR, Smith GD. Hydrogen metabolism by filamentous cyanobacteria. Arch Biochem Biophys 1980; 205(1): 36-50.
[http://dx.doi.org/10.1016/0003-9861(80)90081-8] [PMID: 6778397]

[750] Lopespinto F, Troshina O, Lindblad P. A brief look at three decades of research on cyanobacterial hydrogen evolution. Int J Hydrogen Energy 2002; 27(11-12): 1209-15.
[http://dx.doi.org/10.1016/S0360-3199(02)00089-7]

[751] Moazeni M, Ahmadi A, Mootabi Alavi A. A new method for laboratory rearing of Galba truncatula, the intermediate host of Fasciola hepatica. Vet Parasitol 2018; 253: 12-5.
[http://dx.doi.org/10.1016/j.vetpar.2018.02.008] [PMID: 29604994]

[752] Talaiekhozani A, Rezania S. Application of photosynthetic bacteria for removal of heavy metals, macro-pollutants and dye from wastewater: A review. J Water Process Eng 2017; 19: 312-21.
[http://dx.doi.org/10.1016/j.jwpe.2017.09.004]

[753] Buck D, Smith GD. Evidence for a Na+/H+ electrogenic antiporter in an alkaliphilic cyanobacterium Synechocystis. FEMS Microbiol Lett 1995; 128(3): 315-20.
[http://dx.doi.org/10.1016/0378-1097(95)00126-P]

[754] Gallo G, Antonucci I, Pirone L, *et al.* A physicochemical investigation on the metal binding properties of TtSmtB, a thermophilic member of the ArsR/SmtB transcription factor family. Int J Biol Macromol 2019; 138: 1056-63.
[http://dx.doi.org/10.1016/j.ijbiomac.2019.07.174] [PMID: 31356933]

[755] Lambert GR, Smith GD. Hydrogen formation by marine blue-green algae. FEBS Lett 1977; 83(1): 159-62.
[http://dx.doi.org/10.1016/0014-5793(77)80664-9] [PMID: 411686]

[756] Li J, Chen Q, Bao B, *et al.* RNA-seq analysis reveals the significant effects of different light conditions on oil degradation by marine Chlorella vulgaris. Mar Pollut Bull 2018; 137: 267-76.
[http://dx.doi.org/10.1016/j.marpolbul.2018.10.021] [PMID: 30503435]

[757] Saadoun IMK, Schrader KK, Blevins WT. Environmental and nutritional factors affecting geosmin synthesis by Anabaena SP. Water Res 2001; 35(5): 1209-18.
[http://dx.doi.org/10.1016/S0043-1354(00)00381-X] [PMID: 11268841]

[758] Stoyanov JV, Magnani D, Solioz M. Measurement of cytoplasmic copper, silver, and gold with a *lux* biosensor shows copper and silver, but not gold, efflux by the CopA ATPase of *Escherichia coli.* FEBS Lett 2003; 546(2-3): 391-4.

[http://dx.doi.org/10.1016/S0014-5793(03)00640-9] [PMID: 12832075]

[759]   Tranchida MC, Riccillo PM, Rodriguero MS, García JJ, Micieli MV. Isolation and identification of α-proteobacteria from Culex pipiens (Diptera Culicidae) larvae. J Invertebr Pathol 2012; 109(1): 143-7.
[http://dx.doi.org/10.1016/j.jip.2011.10.007] [PMID: 22036984]

[760]   Volk RB. Screening of microalgae for species excreting norharmane, a manifold biologically active indole alkaloid. Microbiol Res 2008; 163(3): 307-13.
[http://dx.doi.org/10.1016/j.micres.2006.06.002] [PMID: 16872816]

[761]   Gómez N, Sierra MV, Cortelezzi A, Rodrigues Capítulo A. Effects of discharges from the textile industry on the biotic integrity of benthic assemblages. Ecotoxicol Environ Saf 2008; 69(3): 472-9.
[http://dx.doi.org/10.1016/j.ecoenv.2007.03.007] [PMID: 17490744]

[762]   Izaguirre G, Taylor WD. Geosmin and 2-methylisoborneol production in a major aqueduct system. Water Sci Technol 1995; 31(11): 41-8.
[http://dx.doi.org/10.2166/wst.1995.0398]

[763]   Kidd SP, Brown NL, Zcc R. ZccR—a MerR-like regulator from Bordetella pertussis which responds to zinc, cadmium, and cobalt. Biochem Biophys Res Commun 2003; 302(4): 697-702.
[http://dx.doi.org/10.1016/S0006-291X(03)00249-3] [PMID: 12646225]

[764]   Li Z, Hobson P, An W, Burch MD, House J, Yang M. Earthy odor compounds production and loss in three cyanobacterial cultures. Water Res 2012; 46(16): 5165-73.
[http://dx.doi.org/10.1016/j.watres.2012.06.008] [PMID: 22818951]

[765]   Naes H. Geosmin production. Methods in Enzymology. Academic Press 1988; Vol. 167: pp. 605-8.
[66]

[766]   Nainangu P, Antonyraj APM, Subramanian K, *et al. In vitro* screening of antimicrobial, antioxidant, cytotoxic activities, and characterization of bioactive substances from freshwater cyanobacteria Oscillatoria sp. SSCM01 and Phormidium sp. SSCM02. Biocatal Agric Biotechnol 2020; 29: 101772.
[http://dx.doi.org/10.1016/j.bcab.2020.101772]

[767]   Tranchida MC, Pelizza SA, Bisaro V, Beltrán C, García JJ, Micieli MV. Use of the neotropical fish Cnesterodon decemmaculatus for long-term control of Culex pipiens L. in Argentina. Biol Control 2010; 53(2): 183-7.
[http://dx.doi.org/10.1016/j.biocontrol.2009.11.006]

[768]   van Veenhuyzen B, Hörstmann C, Peens J, Brink HG. Bioremediation and biorecovery of aqueous lead by local lead-resistant organisms. In: Shah MP, Rodriguez-Couto S, Şengör SS, Eds. Emerging Technologies in Environmental Bioremediation. Elsevier 2020; pp. 407-24.

[769]   Blevins WT, Schrader KK, Saadoun I. Comparative physiology of geosmin production by Streptomyces halstedii and Anabaena sp. Water Sci Technol 1995; 31(11): 127-33.
[http://dx.doi.org/10.2166/wst.1995.0419]

[770]   Jones GJ, Korth W. *In situ* production of volatile odour compounds by river and reservoir phytoplankton populations in Australia. Water Sci Technol 1995; 31(11): 145-51.
[http://dx.doi.org/10.2166/wst.1995.0424]

[771]   Naik MM, Dubey SK. Lead resistant bacteria: Lead resistance mechanisms, their applications in lead bioremediation and biomonitoring. Ecotoxicol Environ Saf 2013; 98: 1-7.
[http://dx.doi.org/10.1016/j.ecoenv.2013.09.039] [PMID: 24144999]

[772]   Robin J, Cravedi JP, Hillenweck A, Deshayes C, Vallod D. Off flavor characterization and origin in French trout farming. Aquaculture 2006; 260(1-4): 128-38.
[http://dx.doi.org/10.1016/j.aquaculture.2006.05.058]

[773]   Roy U, Sengupta S, Banerjee P, Das P, Bhowal A, Datta S. Assessment on the decolourization of textile dye (Reactive Yellow) using Pseudomonas sp. immobilized on fly ash: Response surface methodology optimization and toxicity evaluation. J Environ Manage 2018; 223: 185-95.
[http://dx.doi.org/10.1016/j.jenvman.2018.06.026] [PMID: 29929074]

[774] Verret F, Gravot A, Auroy P, *et al.* Heavy metal transport by AtHMA4 involves the N-terminal degenerated metal binding domain and the C-terminal His $_{11}$ stretch. FEBS Lett 2005; 579(6): 1515-22.
[http://dx.doi.org/10.1016/j.febslet.2005.01.065] [PMID: 15733866]

[775] Wu JW, Shi Y, Zhu YX, Wang YC, Gong HJ. Mechanisms of Enhanced Heavy Metal Tolerance in Plants by Silicon: A Review. Pedosphere 2013; 23(6): 815-25.
[http://dx.doi.org/10.1016/S1002-0160(13)60073-9]

[776] Marrez D, Naguib M M, Sultan Y, Daw Z Y, Zaher S S, Higazy A M. Phytoplankton profile and toxicity assessment of dominant algal species from different Egyptian aquatic ecosystems. 2016; 7: 1453-61.

[777] Gaignard C, Laroche C, Pierre G, *et al.* Screening of marine microalgae: Investigation of new exopolysaccharide producers. Algal Res 2019; 44: 101711.
[http://dx.doi.org/10.1016/j.algal.2019.101711]

[778] Itoh N, Tsujita M, Ando T, Hisatomi G, Higashi T. Formation and emission of monohalomethanes from marine algae. Phytochemistry 1997; 45(1): 67-73.
[http://dx.doi.org/10.1016/S0031-9422(96)00786-8]

[779] Kabir F, Gulfraz M, Raja GK, *et al.* Screening of native hyper-lipid producing microalgae strains for biomass and lipid production. Renew Energy 2020; 160: 1295-307.
[http://dx.doi.org/10.1016/j.renene.2020.07.004]

[780] Kisand V, NÃµges T. Abiotic and biotic factors regulating dynamics of bacterioplankton in a large shallow lake. FEMS Microbiol Ecol 2004; 50(1): 51-62.
[http://dx.doi.org/10.1016/j.femsec.2004.05.009] [PMID: 19712376]

[781] Metcalf JS, Beattie KA, Purdie EL, Bryant JA, Irvine LM, Codd GA. Analysis of microcystins and microcystin genes in 60–170-year-old dried herbarium specimens of cyanobacteria. Harmful Algae 2012; 15: 47-52.
[http://dx.doi.org/10.1016/j.hal.2011.11.004]

[782] Ozaki K, Ohta A, Iwata C, *et al.* Lysis of cyanobacteria with volatile organic compounds. Chemosphere 2008; 71(8): 1531-8.
[http://dx.doi.org/10.1016/j.chemosphere.2007.11.052] [PMID: 18179811]

[783] Smith JL, Boyer GL, Zimba PV. A review of cyanobacterial odorous and bioactive metabolites: Impacts and management alternatives in aquaculture. Aquaculture 2008; 280(1-4): 5-20.
[http://dx.doi.org/10.1016/j.aquaculture.2008.05.007]

[784] Wang M, Chen S, Zhou W, Yuan W, Wang D. Algal cell lysis by bacteria: A review and comparison to conventional methods. Algal Res 2020; 46: 101794.
[http://dx.doi.org/10.1016/j.algal.2020.101794]

[785] Allen MB. Utilization of thermal energy by living organisms. In: Florkin M, Mason HS, Eds. Comparative Biochemistry. Academic Press 1960; pp. 487-514.
[http://dx.doi.org/10.1016/B978-0-12-395542-5.50021-6]

[786] De Philippis R, Vincenzini M. Exocellular polysaccharides from cyanobacteria and their possible applications. FEMS Microbiol Rev 1998; 22(3): 151-75.
[http://dx.doi.org/10.1016/S0168-6445(98)00012-6]

[787] Fogg GE, Stewart WDP, Fay P, Walsby AE. Pathogens of blue-green algae. In: Fogg GE, Stewart WDP, Fay P, Walsby AE, Eds. The Blue-green Algae. Academic Press 1973; pp. 281-97.
[http://dx.doi.org/10.1016/B978-0-12-261650-1.50017-4]

[788] Fogg GE, Stewart WDP, Fay P, Walsby AE. Marine blue-green algae. In: Fogg GE, Stewart WDP, Fay P, Walsby AE, Eds. The Blue-green Algae. Academic Press 1973; pp. 298-310.
[http://dx.doi.org/10.1016/B978-0-12-261650-1.50018-6]

[789] Komárek J, Johansen JR. Filamentous cyanobacteria. In: Wehr JD, Sheath RG, Kociolek JP, Eds.

Freshwater Algae of North America. 2nd ed.. Boston: Academic Press 2015; pp. 135-235.

[790] Komárek J, Kling H, Komárková J. Filamentous cyanobacteria. In: Wehr JD, Sheath RG, Eds. Freshwater Algae of North America. Burlington: Academic Press 2003; pp. 117-96.
[http://dx.doi.org/10.1016/B978-012741550-5/50005-2]

[791] Pankow H, Haendel D, Richter W, Wand U. Algologische Beobachtungen in der Schirmacher- und Unterseeoase (Dronning -Maud -Land, Ostantarktika). Arch Protistenkd 1987; 134(1): 59-82.
[http://dx.doi.org/10.1016/S0003-9365(87)80029-5]

[792] Tidgewell K, Clark BR, Gerwick WH. The natural products chemistry of cyanobacteria. In: Liu H-W, Mander L, Eds. Comprehensive Natural Products II. Oxford: Elsevier 2010; pp. 141-88.
[http://dx.doi.org/10.1016/B978-008045382-8.00041-1]

[793] Verbist J-F, Sallenave C, Pouchus Y-F. Marine fungal substances. In: Atta ur R, Ed. Studies in Natural Products Chemistry. Elsevier 2000; 24: pp. 979-1092.
[http://dx.doi.org/10.1016/S1572-5995(00)80059-9]

[794] Kobayashi Ji, Ishibashi M. Marine natural products and marine chemical ecology. In: Barton SD, Nakanishi K, Meth-Cohn O, Eds. Comprehensive Natural Products Chemistry. Oxford: Pergamon 1999; pp. 415-649.
[http://dx.doi.org/10.1016/B978-0-08-091283-7.00055-2]

[795] Kumar P, Vijayan D, Antony L, Kumar M. Thajuddin, Phylogenetic diversity of cultivable bacteria associated with filamentous non-hetrocystous marine cyanobacteria. J Algal Biomass Util 2009; 1: 87-102.

[796] Forest HS, Wade JQ, Maxwell TF. The limnology of conesus lake. In: Bloomfield JA, Ed. Lakes of New York State. Academic Press 1978; pp. 121-224.
[http://dx.doi.org/10.1016/B978-0-12-107301-5.50009-0]

[797] Kuznetsov SI. Trends in the development of ecological microbiology. In: Droop MR, Jannasch HW, Eds. Advances in Aquatic Microbiology. San Diego: Academic Press 1977; pp. 1-48.
[http://dx.doi.org/10.1016/B978-0-12-003001-9.50005-6]

[798] Schaffner WR, Oglesby RT. Limnology of eight finger lakes: Hemlock, canadice, honeoye, keuka, seneca, owasco, skaneateles, and otisco. In: Bloomfield JA, Ed. Lakes of New York State. Academic Press 1978; pp. 313-470.
[http://dx.doi.org/10.1016/B978-0-12-107301-5.50011-9]

[799] Stewart WDP, Daft MJ. Microbial pathogens of cyanophycean blooms.Advances in Aquatic Microbiology. San Diego: Academic Press 1977; pp. 177-218.
[http://dx.doi.org/10.1016/B978-0-12-003001-9.50008-1]

[800] Akazawa T. Starch, inulin, and other reserve polysaccharides.Plant Biochemistry. Academic Press 1965; pp. 258-97.
[http://dx.doi.org/10.1016/B978-1-4832-3243-0.50016-2]

[801] Fogg GE, Stewart WDP, Fay P, Walsby AE. Movements.The Blue-green Algae. Academic Press 1973; pp. 111-28.
[http://dx.doi.org/10.1016/B978-0-12-261650-1.50010-1]

[802] Fogg GE, Stewart WDP, Fay P, Walsby AE. Cell biology.The Blue-green Algae. Academic Press 1973; pp. 78-92.
[http://dx.doi.org/10.1016/B978-0-12-261650-1.50008-3]

[803] Fredrick JF. Evolution of isozymes forming storage polyglucans.Algae In Isozymes. Academic Press 1975; pp. 307-21.
[http://dx.doi.org/10.1016/B978-0-12-472704-5.50025-2]

[804] Lackie JM. The Dictionary of Cell & Molecular Biology. 5th ed. Boston: Academic Press 2013; pp. 464-76.
[http://dx.doi.org/10.1016/B978-0-12-384931-1.00015-5]

[805] Painter TJ. Algal Polysaccharides.The Polysaccharides. Academic Press 1983; pp. 195-285.
[http://dx.doi.org/10.1016/B978-0-12-065602-8.50009-1]

[806] Sarmah P, Rout J. Role of algae and cyanobacteria in bioremediation: prospects in polyethylene biodegradation.Advances in Cyanobacterial Biology. Academic Press 2020; pp. 333-49.
[http://dx.doi.org/10.1016/B978-0-12-819311-2.00022-X]

[807] Trichet J. Étude des premiers stades d'évolution de la matière organique.des Mares En Milieu Récifal (Polynésie française) In Advances in Organic Geochemistry. Pergamon 1970; pp. 265-84.

[808] Liaaen-Jensen S. Marine carotenoids.Marine Natural Products. Academic Press 1978; pp. 1-73.
[http://dx.doi.org/10.1016/B978-0-12-624002-3.50008-5]

[809] Kumar D, Kumar H. Hydrogen production by several cyanobacteria. Int J Hydrogen Energy 1992; 17(11): 847-52.
[http://dx.doi.org/10.1016/0360-3199(92)90034-T]

[810] Vyas D, Kumar HD. Nitrogen fixation and hydrogen uptake in four cyanobacteria. Int J Hydrogen Energy 1995; 20(2): 163-8.
[http://dx.doi.org/10.1016/0360-3199(94)E0008-M]

[811] Abdel-Raouf N, Ibraheem I. Antibiotic activity of two Anabaena species against four fish pathogenic Aeromonas species. Afr J Biotechnol 2010; •••: 7.

[812] Frost TM. Selection in sponge feeding processes. In: Smith DC, Tiffon Y, Eds. Nutrition in the Lower Metazoa. Pergamon 1980; pp. 33-44.
[http://dx.doi.org/10.1016/B978-0-08-025904-8.50006-7]

[813] Hammerton D. The nile river – A case history.River Ecology and Man. Academic Press 1972; pp. 171-214.
[http://dx.doi.org/10.1016/B978-0-12-524450-3.50017-4]

[814] Oglesby RT. The limnology of cayuga lake In: Bloomfield JA, Ed. Lakes of New York State. Academic Press 1978; pp. 1-120.

[815] Kushwaha D, Srivastava N, Prasad D, Mishra PK, Upadhyay SN. Biobutanol production from hydrolysates of cyanobacteria Lyngbya limnetica and Oscillatoria obscura. Fuel 2020; 271: 117583.
[http://dx.doi.org/10.1016/j.fuel.2020.117583]

[816] Zayadan BK, Sadvakasova AK, Usserbayeva AA, *et al.* Waste-free technology of wastewater treatment to obtain microalgal biomass for biodiesel production. Int J Hydrogen Energy 2017; 42(12): 8586-91.
[http://dx.doi.org/10.1016/j.ijhydene.2016.12.058]

[817] Smock LA, Wright AB, Benke AC. Atlantic coast rivers of The Southeastern United States.Rivers of North America. Burlington: Academic Press 2005; pp. 72-122.
[http://dx.doi.org/10.1016/B978-012088253-3/50006-7]

[818] Tuchman NC. The role of heterotrophy in algae.Algal Ecology. San Diego: Academic Press 1996; pp. 299-319.
[http://dx.doi.org/10.1016/B978-012668450-6/50039-4]

[819] Wehr JD, Sheath RG. Habitats of freshwater algae. In: Wehr JD, Sheath RG, Kociolek JP, Eds. Freshwater Algae of North America. 2nd ed.. Boston: Academic Press 2015; pp. 13-74.

[820] Sushchik NN, Kuchkina AY, Gladyshev MI. Fatty acid content and composition of sediments from Siberian eutrophic water bodies: Implications for biodiesel production. Water Res 2013; 47(9): 3192-200.
[http://dx.doi.org/10.1016/j.watres.2013.03.031] [PMID: 23561504]

[821] Veit S, Takeda K, Tsunoyama Y, *et al.* Structural and functional characterisation of the cyanobacterial PetC3 Rieske protein family. Biochim Biophys Acta Bioenerg 2016; 1857(12): 1879-91.
[http://dx.doi.org/10.1016/j.bbabio.2016.09.007] [PMID: 27663073]

[822] Humbert JF. Toxins of cyanobacteria.Handbook of Toxicology of Chemical Warfare Agents. San Diego: Academic Press 2009; pp. 371-9.
[http://dx.doi.org/10.1016/B978-012374484-5.00027-4]

[823] Kulal DK, Loni PC, Dcosta C, Some S, Kalambate PK. Cyanobacteria: As a promising candidate for heavy-metals removal.Advances in Cyanobacterial Biology. Academic Press 2020; pp. 291-300.
[http://dx.doi.org/10.1016/B978-0-12-819311-2.00019-X]

[824] Luckas B, Krüger T, Röder K. Phycotoxins and food safety.Chemical Contaminants and Residues in Food. Woodhead Publishing 2012; pp. 342-93.
[http://dx.doi.org/10.1533/9780857095794.2.342]

[825] Puschner B. Cyanobacterial (blue-green algae) toxins.Veterinary Toxicology. 3rd ed. Academic Press 2018; pp. 763-77.

[826] Watson SB, Whitton BA, Higgins SN, Paerl HW, Brooks BW, Wehr JD. Harmful algal blooms. In: Wehr JD, Sheath RG, Kociolek JP, Eds. Freshwater Algae of North America. 2nd ed.. Boston: Academic Press 2015; pp. 873-920.

[827] Geada P, Gkelis S, Teixeira J, Vasconcelos V, Vicente AA, Fernandes B. Cyanobacterial toxins as a high value-added product. In: Gonzalez-Fernandez C, Muñoz R, Eds. Microalgae-Based Biofuels and Bioproducts. Woodhead Publishing 2017; pp. 401-28.
[http://dx.doi.org/10.1016/B978-0-08-101023-5.00017-0]

[828] Hudon C, De Sève M, Cattaneo A. Increasing occurrence of the benthic filamentous cyanobacterium *Lyngbya wollei* : A symptom of freshwater ecosystem degradation. Freshw Sci 2014; 33(2): 606-18.
[http://dx.doi.org/10.1086/675932]

[829] Haugen J-E, Oehme M, Müller MD. Enantiomer-specific analysis of homoanatoxin-a, A cyanophyte neurotoxin.Detection Methods for Cynobacterial Toxins. Woodhead Publishing 1994; pp. 40-4.
[http://dx.doi.org/10.1533/9781845698164.1.40]

[830] Malik JK, Bharti VK, Rahal A, Kumar D, Gupta RC. Cyanobacterial (blue-green algae) toxins.Handbook of Toxicology of Chemical Warfare Agents. 3rd ed. Boston: Academic Press 2020; pp. 467-78.
[http://dx.doi.org/10.1016/B978-0-12-819090-6.00031-3]

[831] Skulberg OM, Carmichael WW, Codd GA, Skulberg R. Taxonomy of toxic cyanophyceae (Cyanobacteria).Algal Toxins in Seafood and Drinking Water. San Diego: Academic Press 1993; pp. 145-64.
[http://dx.doi.org/10.1016/B978-0-08-091811-2.50014-9]

[832] van der Merwe D. Freshwater cyanotoxins.Biomarkers in Toxicology. Boston: Academic Press 2014; pp. 539-48.
[http://dx.doi.org/10.1016/B978-0-12-404630-6.00031-2]

[833] van der Merwe D. Cyanobacterial (blue-green algae) toxins.Handbook of Toxicology of Chemical Warfare Agents. 2nd ed. Boston: Academic Press 2015; pp. 421-9.
[http://dx.doi.org/10.1016/B978-0-12-800159-2.00031-2]

[834] Anahas AMP, Muralitharan G. Characterization of heterocystous cyanobacterial strains for biodiesel production based on fatty acid content analysis and hydrocarbon production. Energy Convers Manage 2018; 157: 423-37.
[http://dx.doi.org/10.1016/j.enconman.2017.12.012]

[835] Jindal N, Singh DP, Khattar JIS. Kinetics and physico-chemical characterization of exopolysaccharides produced by the cyanobacterium Oscillatoria formosa. World J Microbiol Biotechnol 2011; 27(9): 2139-46.
[http://dx.doi.org/10.1007/s11274-011-0678-6]

[836] Genuário DB, Vaz MGMV, Santos SN, Kavamura VN, Melo IS. Cyanobacteria from Brazilian extreme environments: Toward functional exploitation.Microbial Diversity in the Genomic Era.

Academic Press 2019; pp. 265-84.
[http://dx.doi.org/10.1016/B978-0-12-814849-5.00016-2]

[837] Joshi H, Shourie A, Singh A. Cyanobacteria as a source of biofertilizers for sustainable agriculture.Advances in Cyanobacterial Biology. Academic Press 2020; pp. 385-96.
[http://dx.doi.org/10.1016/B978-0-12-819311-2.00025-5]

[838] Krishnamoorthy S, Manickam P. Phycoremediation of industrial wastewater: challenges and prospects. In: Kumar V, Saxena G, Shah MP, Eds. Bioremediation for Environmental Sustainability. Elsevier 2021; pp. 99-123.

[839] Asada Y, Kawamura S. Screening for cyanobacteria that evolve molecular hydrogen under dark and anaerobic conditions. J Ferment Technol 1986; 64(6): 553-6.
[http://dx.doi.org/10.1016/0385-6380(86)90081-6]

[840] Chi Z, O'Fallon JV, Chen S. Bicarbonate produced from carbon capture for algae culture. Trends Biotechnol 2011; 29(11): 537-41.
[http://dx.doi.org/10.1016/j.tibtech.2011.06.006] [PMID: 21775005]

[841] Chamizo S, Adessi A, Mugnai G, Simiani A, De Philippis R. Soil Type and Cyanobacteria Species Influence the Macromolecular and Chemical Characteristics of the Polysaccharidic Matrix in Induced Biocrusts. Microb Ecol 2019; 78(2): 482-93.
[http://dx.doi.org/10.1007/s00248-018-1305-y] [PMID: 30535915]

[842] Istvánovics V. Eutrophication of lakes and reservoirs.Encyclopedia of Inland Waters. Oxford: Academic Press 2009; pp. 157-65.
[http://dx.doi.org/10.1016/B978-012370626-3.00141-1]

[843] Schagerl M, Müller B. Acclimation of chlorophyll a and carotenoid levels to different irradiances in four freshwater cyanobacteria. J Plant Physiol 2006; 163(7): 709-16.
[http://dx.doi.org/10.1016/j.jplph.2005.09.015] [PMID: 16325961]

[844] Singh SP, Singh P. Effect of temperature and light on the growth of algae species: A review. Renew Sustain Energy Rev 2015; 50: 431-44.
[http://dx.doi.org/10.1016/j.rser.2015.05.024]

[845] Santos KRDS, Jacinavicius FR, Sant'Anna CL. Tese Tudo Kleber02.Julho.2013. 2014.

[846] Delia TJ, Hurst DT. Bicyclic 5-6 systems: Three heteroatoms 1:2.Comprehensive Heterocyclic Chemistry II. Oxford: Pergamon 1996; pp. 229-81.
[http://dx.doi.org/10.1016/B978-008096518-5.00151-9]

[847] Filatova D, Picardo M, Núñez O, Farré M. Analysis, levels and seasonal variation of cyanotoxins in freshwater ecosystems. Trends in Environmental Analytical Chemistry 2020; 26: e00091.
[http://dx.doi.org/10.1016/j.teac.2020.e00091]

[848] Kosek K, Polkowska Ż, Żyszka B, Lipok J. Phytoplankton communities of polar regions–Diversity depending on environmental conditions and chemical anthropopressure. J Environ Manage 2016; 171: 243-59.
[http://dx.doi.org/10.1016/j.jenvman.2016.01.026] [PMID: 26846983]

[849] Leong YK, Chang JS. Bioremediation of heavy metals using microalgae: Recent advances and mechanisms. Bioresour Technol 2020; 303: 122886.
[http://dx.doi.org/10.1016/j.biortech.2020.122886] [PMID: 32046940]

[850] Li X, Dreher TW, Li R. An overview of diversity, occurrence, genetics and toxin production of bloom-forming Dolichospermum (Anabaena) species. Harmful Algae 2016; 54: 54-68.
[http://dx.doi.org/10.1016/j.hal.2015.10.015] [PMID: 28073482]

[851] Steinhoff FS, Karlberg M, Graeve M, Wulff A. Cyanobacteria in Scandinavian coastal waters — A potential source for biofuels and fatty acids? Algal Res 2014; 5: 42-51.
[http://dx.doi.org/10.1016/j.algal.2014.05.005]

[852] Sun J, Cheng J, Yang Z, Li K, Zhou J, Cen K. Microstructures and functional groups of Nannochloropsis sp. cells with arsenic adsorption and lipid accumulation. Bioresour Technol 2015; 194: 305-11.
[http://dx.doi.org/10.1016/j.biortech.2015.07.041] [PMID: 26210144]

[853] Blinkova M, Zdravevski N, Krstic S. Blue-green Algae, the Janus of Nature. 2012.

[854] Gorham PR, Carmichael WW. Toxic substances from freshwater algae.Eutrophication of Deep Lakes. Pergamon 1980; pp. 189-98.
[http://dx.doi.org/10.1016/B978-0-08-026024-2.50016-X]

[855] Masukawa H, Nakamura K, Mochimaru M, Sakurai H. Photobiological hydrogen production and nitrogenase activity in some heterocystous cyanobacteria.Biohydrogen II. Oxford: Pergamon 2001; pp. 63-6.
[http://dx.doi.org/10.1016/B978-008043947-1/50005-3]

[856] Singh VK, Singh SK, Singh PK, *et al.* Impact of pesticides applications on the growth and function of cyanobacteria.Advances in Cyanobacterial Biology. Academic Press 2020; pp. 151-62.
[http://dx.doi.org/10.1016/B978-0-12-819311-2.00010-3]

[857] Gassanova LG, Netrusov AI, Teplyakov VV, Modigell M. Fuel gases from organic wastes using membrane bioreactors. Desalination 2006; 198(1-3): 56-66.
[http://dx.doi.org/10.1016/j.desal.2006.09.009]

[858] Guo Y, Song W, Lu J, Ma Q, Xu D, Wang S. Hydrothermal liquefaction of Cyanophyta: Evaluation of potential bio-crude oil production and component analysis. Algal Res 2015; 11: 242-7.
[http://dx.doi.org/10.1016/j.algal.2015.06.025]

[859] Nalley JO, O'Donnell DR, Litchman E. Temperature effects on growth rates and fatty acid content in freshwater algae and cyanobacteria. Algal Res 2018; 35: 500-7.
[http://dx.doi.org/10.1016/j.algal.2018.09.018]

[860] Rasul I, Azeem F, Siddique MH, *et al.* Algae biotechnology: A green light for engineered algae. In: Zia KM, Zuber M, Ali M, Eds. Algae Based Polymers, Blends, and Composites. Elsevier 2017; pp. 301-34.
[http://dx.doi.org/10.1016/B978-0-12-812360-7.00008-2]

[861] Taylor RW, Sistani KR, Floyd M. Algal biomass production, N uptake and N2 fixation in a synthetic medium. Biomass 1988; 15(4): 249-57.
[http://dx.doi.org/10.1016/0144-4565(88)90060-1]

[862] Trivedi J, Aila M, Bangwal DP, Kaul S, Garg MO. Algae based biorefinery—How to make sense? Renew Sustain Energy Rev 2015; 47: 295-307.
[http://dx.doi.org/10.1016/j.rser.2015.03.052]

[863] Spence W, Hoto P, Blem C. Ospreys in the flathead lake catchment basin smoke on the water: Flathead lake remains clear while impact of fires is hazy. 2020.

[864] Fresenborg LS, Graf J, Schätzle H, Schleiff E. Iron homeostasis of cyanobacteria: Advancements in siderophores and metal transporters.Advances in Cyanobacterial Biology. Academic Press 2020; pp. 85-117.
[http://dx.doi.org/10.1016/B978-0-12-819311-2.00007-3]

[865] Jacob C. Computer models of developmental programs.Illustrating Evolutionary Computation with Mathematica. San Diego: Academic Press 2001; pp. 439-70.
[http://dx.doi.org/10.1016/B978-155860637-1/50022-9]

[866] Jean RV. A new approach to a problem of plant growth.Systems Research in Health Care, Biocybernetics and Ecology. Pergamon 1981; pp. 1906-10.
[http://dx.doi.org/10.1016/B978-0-08-027201-6.50055-8]

[867] Kranzler C, Rudolf M, Keren N, Schleiff E. Iron in cyanobacteria.Advances in Botanical Research.

Academic Press 2013; Vol. 65: pp. 57-105.

[868]   Subashchandrabose SR, Ramakrishnan B, Megharaj M, Venkateswarlu K, Naidu R. Consortia of cyanobacteria/microalgae and bacteria: Biotechnological potential. Biotechnol Adv 2011; 29(6): 896-907.
[http://dx.doi.org/10.1016/j.biotechadv.2011.07.009] [PMID: 21801829]

[869]   Prusinkiewicz P, Hammel M, Mjolsness E. Animation of plant development. 1993; 27: pp. 351-60.

[870]   Ali Z, Waheed H, Kazi AG, Hayat A, Ahmad M. Duckweed: An efficient hyperaccumulator of heavy metals in water bodies.Plant Metal Interaction. Elsevier 2016; pp. 411-29.
[http://dx.doi.org/10.1016/B978-0-12-803158-2.00016-3]

[871]   Koller   M.   "Bioplastics   from   microalgae"—Polyhydroxyalkanoate   production   by cyanobacteria.Handbook of Microalgae-Based Processes and Products. Academic Press 2020; pp. 597-645.
[http://dx.doi.org/10.1016/B978-0-12-818536-0.00022-1]

[872]   Wan D, Wu Q, Kuča K. Spirulina.Nutraceuticals. Boston: Academic Press 2016; pp. 569-83.
[http://dx.doi.org/10.1016/B978-0-12-802147-7.00042-5]

[873]   Zohoorian H, Ahmadzadeh H, Molazadeh M, Shourian M, Lyon S. Microalgal bioremediation of heavy metals and dyes.Handbook of Algal Science, Technology and Medicine. Academic Press 2020; pp. 659-74.
[http://dx.doi.org/10.1016/B978-0-12-818305-2.00041-3]

[874]   Ramanathan G, Sugumar R, Jeevarathinam A, Rajarathinam K. Studies of Cyanobacterial distribution in estuary region of Southeastern Coast of Tamil Nadu, India. J Algal Biomass Util 2013; 4: 26-34.

[875]   Borowitzka MA. Microalgae in medicine and human health: A historical perspective.Microalgae in Health and Disease Prevention. Academic Press 2018; pp. 195-210.
[http://dx.doi.org/10.1016/B978-0-12-811405-6.00009-8]

[876]   Cepoi   L.   Environmental   and   technological   stresses   and   their   management   in cyanobacteria.Cyanobacteria. Academic Press 2019; pp. 217-44.
[http://dx.doi.org/10.1016/B978-0-12-814667-5.00011-8]

[877]   Fleurence J, Levine IA. Antiallergic and allergic properties.Microalgae in Health and Disease Prevention. Academic Press 2018; pp. 307-15.
[http://dx.doi.org/10.1016/B978-0-12-811405-6.00014-1]

[878]   Hashemian M, Ahmadzadeh H, Hosseini M, Lyon S, Pourianfar HR. Production of microalgae-derived high-protein biomass to enhance food for animal feedstock and human consumption.Advanced Bioprocessing for Alternative Fuels, Biobased Chemicals, and Bioproducts. Woodhead Publishing 2019; pp. 393-405.
[http://dx.doi.org/10.1016/B978-0-12-817941-3.00020-6]

[879]   Malwade CR, Roda-Serrat MC, Christensen KV, Fretté X, Christensen LP. Kinetics of phycocyanobilin cleavage from C-phycocyanin by methanolysis.Computer Aided Chemical Engineering. Elsevier 2016; Vol. 38: pp. 61-6.

[880]   Soreanu G, Cretescu I, Diaconu M, *et al.* A model microalga for addressing air treatment in spacecrafts. In: Soreanu G, Dumont E, Eds. From Biofiltration to Promising Options in Gaseous Fluxes Biotreatment. Elsevier 2020; pp. 397-417.

[881]   Vieira RC, Medeiros JA, do Nascimento MAA, de Souza Abud AK, Raymundo A, de Farias Silva CE. Microalgae as sustainable food: incorporation as strategy in the formulation of functional food.New and Future Developments in Microbial Biotechnology and Bioengineering. Elsevier 2020; pp. 19-30.
[http://dx.doi.org/10.1016/B978-0-12-820528-0.00003-X]

[882]   Ge D, Zhang L, Long Z, Chi C, Liu H. A novel biomarker for marine environmental pollution: A metallothionein from Mytilus coruscus. Aquacult Rep 2020; 17: 100364.
[http://dx.doi.org/10.1016/j.aqrep.2020.100364]

[883]   Grahl S, Strack M, Mensching A, Mörlein D. Alternative protein sources in Western diets: Food product development and consumer acceptance of spirulina-filled pasta. Food Qual Prefer 2020; 84: 103933.
[http://dx.doi.org/10.1016/j.foodqual.2020.103933]

[884]   Liang Y, Bao Y, Gao X, *et al.* Effects of spirulina supplementation on lipid metabolism disorder, oxidative stress caused by high-energy dietary in Hu sheep. Meat Sci 2020; 164: 108094.
[http://dx.doi.org/10.1016/j.meatsci.2020.108094] [PMID: 32146297]

[885]   Lu J, Takeuchi T, Satoh H. Ingestion and assimilation of three species of freshwater algae by larval tilapia Oreochromis niloticus. Aquaculture 2004; 238(1-4): 437-49.
[http://dx.doi.org/10.1016/j.aquaculture.2004.05.002]

[886]   Marcus JB. Weight management: Finding the healthy balance: Practical applications for nutrition, food science and culinary professionals.Culinary Nutrition. San Diego: Academic Press 2013; pp. 431-73.
[http://dx.doi.org/10.1016/B978-0-12-391882-6.00010-8]

[887]   Mehrabadi A, Craggs R, Farid MM. Wastewater treatment high rate algal ponds (WWT HRAP) for low-cost biofuel production. Bioresour Technol 2015; 184: 202-14.
[http://dx.doi.org/10.1016/j.biortech.2014.11.004] [PMID: 25465780]

[888]   Paredes-Carbajal MC, Torres-Durán PV, Díaz-Zagoya JC, Mascher D, Juárez-Oropeza MA. EFFECTS OF DIETARY Spirulina maxima ON ENDOTHELIUM DEPENDENT VASOMOTOR RESPONSES OF RAT AORTIC RINGS. Life Sci 1997; 61(15): PL211-9.
[http://dx.doi.org/10.1016/S0024-3205(97)00715-7] [PMID: 9328235]

[889]   Serban C, Sahebkar A, Dragan S, Andrica F, Urosniu S, Banach M. A systematic review and meta-analysis of the impact of spirulina supplementation on plasma lipid concentrations. Atherosclerosis 2015; 241(1): e191.
[http://dx.doi.org/10.1016/j.atherosclerosis.2015.04.935] [PMID: 26433766]

[890]   Serban MC, Sahebkar A, Dragan S, *et al.* A systematic review and meta-analysis of the impact of Spirulina supplementation on plasma lipid concentrations. Clin Nutr 2016; 35(4): 842-51.
[http://dx.doi.org/10.1016/j.clnu.2015.09.007] [PMID: 26433766]

[891]   Vidé J, Romain C, Feillet-Coudray C, *et al.* Assessment of potential toxicological aspects of dietary exposure to silicon-rich spirulina in rats. Food Chem Toxicol 2015; 80: 108-13.
[http://dx.doi.org/10.1016/j.fct.2015.02.021] [PMID: 25778349]

[892]   Fox JA, Booth SJ, Martin EL. Cyanophage SM-2: A new blue-green algal virus. Virology 1976; 73(2): 557-60.
[http://dx.doi.org/10.1016/0042-6822(76)90420-7] [PMID: 8869]

[893]   Goyal D. A simplified method for screening and characterization of plasmid DNA in cyanobacteria. J Microbiol Methods 1992; 15(1): 7-15.
[http://dx.doi.org/10.1016/0167-7012(92)90063-A]

[894]   Rippka R. Recognition and identification of cyanobacteria.Methods in Enzymology. Academic Press 1988; Vol. 167: pp. 28-67. [2]

[895]   Smith KM, Brown RM Jr, Walne PL, Goldstein DA. Electron microscopy of the infection process of the blue-green alga virus. Virology 1966; 30(2): 182-92.
[http://dx.doi.org/10.1016/0042-6822(66)90094-8] [PMID: 5919226]

[896]   Vis C, Hudon C, Cattaneo A, Pinel-Alloul B. Periphyton as an indicator of water quality in the St Lawrence River (Québec, Canada). Environ Pollut 1998; 101(1): 13-24.
[http://dx.doi.org/10.1016/S0269-7491(98)00042-6] [PMID: 15093094]

[897]   Alekseenko VA, Maximovich NG, Alekseenko AV. Geochemical barriers for soil protection in mining areas.Assessment, Restoration and Reclamation of Mining Influenced Soils. Academic Press 2017; pp. 255-74.
[http://dx.doi.org/10.1016/B978-0-12-809588-1.00009-8]

[898] Boudouresque CF, Blanfuné A, Ruitton S, Thibaut T. Macroalgae as a tool for coastal management in the Mediterranean Sea.Handbook of Algal Science, Technology and Medicine. Academic Press 2020; pp. 277-90.
[http://dx.doi.org/10.1016/B978-0-12-818305-2.00017-6]

[899] Coelho JP. Arsenic speciation in algae: Case studies in Europe.Comprehensive Analytical Chemistry. Elsevier 2019; Vol. 85: pp. 179-98.

[900] de Oliveira AC, Morocho-Jácome AL, de Castro Lima CR, *et al.* Cosmetics applications. In: Galanakis CM, Ed. Microalgae. Academic Press 2021; pp. 313-38.
[http://dx.doi.org/10.1016/B978-0-12-821218-9.00010-4]

[901] Díaz-Reinoso B. Concentration and purification of seaweed extracts using membrane technologies.Sustainable Seaweed Technologies. Elsevier 2020; pp. 371-90.
[http://dx.doi.org/10.1016/B978-0-12-817943-7.00014-7]

[902] Dikmen BY, Filazi A. Nutraceuticals: Turkish perspective.Nutraceuticals. Boston: Academic Press 2016; pp. 971-81.
[http://dx.doi.org/10.1016/B978-0-12-802147-7.00069-3]

[903] Genter RB. Ecotoxicology of inorganic chemical stress to algae.Algal Ecology. San Diego: Academic Press 1996; pp. 403-68.
[http://dx.doi.org/10.1016/B978-012668450-6/50043-6]

[904] Koutsaviti A, Ioannou E, Roussis V. Bioactive seaweed substances.Bioactive Seaweeds for Food Applications. Academic Press 2018; pp. 25-52.
[http://dx.doi.org/10.1016/B978-0-12-813312-5.00002-9]

[905] Munekata PES, Pateiro M, Barba FJ, Dominguéz R, Gagaoua M, Lorenzo JM. Development of new food and pharmaceutical products: Nutraceuticals and food additives. In: Lorenzo JM, Barba FJ, Eds. Advances in Food and Nutrition Research. Academic Press 2020; Vol. 92: pp. 53-96.

[906] Shahidi F, Zhong Y. Antioxidants from marine by-products. In: Shahidi F, Ed. Maximising the Value of Marine By-Products. Woodhead Publishing 2007; pp. 397-412.
[http://dx.doi.org/10.1533/9781845692087.2.397]

[907] Gavrilescu M. Biomass—a resource for environmental bioremediation and bioenergy.Recent Developments in Bioenergy Research. Elsevier 2020; pp. 19-63.
[http://dx.doi.org/10.1016/B978-0-12-819597-0.00002-7]

[908] Galatchi M, Nenciu M, Niţă V, Nicolaev S, Tania Z, Valodia M. Viata in Marea Neagra 2014.

[909] Konur O. The pioneering research on the bioethanol production from green macroalgae.Handbook of Algal Science, Technology and Medicine. Academic Press 2020; pp. 385-401.
[http://dx.doi.org/10.1016/B978-0-12-818305-2.00025-5]

[910] Milledge JJ, Maneein S. Storage of seaweed for biofuel production: Ensilage.Sustainable Seaweed Technologies. Elsevier 2020; pp. 155-67.
[http://dx.doi.org/10.1016/B978-0-12-817943-7.00005-6]

[911] Padam BS, Chye FY. Seaweed components, properties, and applications.Sustainable Seaweed Technologies. Elsevier 2020; pp. 33-87.
[http://dx.doi.org/10.1016/B978-0-12-817943-7.00002-0]

[912] Pugnetti A, Viaroli P, Ferrari I. Processes leading to dystrophy in a Po River Delta lagoon (Sacca Di Goro): phytoplankton–macroalgae interactions.Marine Coastal Eutrophication. Amsterdam: Elsevier 1992; pp. 445-56.
[http://dx.doi.org/10.1016/B978-0-444-89990-3.50042-0]

[913] Viaroli P, Fumagalli I, Cavalca M. Chemical composition and decomposition of Ulva rigida in a coastal lagoon (Sacca di Goro, Po River Delta).Marine Coastal Eutrophication. Amsterdam: Elsevier 1992; pp. 471-4.

[http://dx.doi.org/10.1016/B978-0-444-89990-3.50044-4]

[914] Cecchi F, Pavan P, Mata-Alvarez J. Anaerobic co-digestion of sewage sludge: Application to the macroalgae from the Venice lagoon. Resour Conserv Recycling 1996; 17(1): 57-66.
[http://dx.doi.org/10.1016/0921-3449(96)88182-1]

[915] Chemodanov A, Robin A, Golberg A. Design of marine macroalgae photobioreactor integrated into building to support seagriculture for biorefinery and bioeconomy. Bioresour Technol 2017; 241: 1084-93.
[http://dx.doi.org/10.1016/j.biortech.2017.06.061] [PMID: 28651325]

[916] Dave N, Selvaraj R, Varadavenkatesan T, Vinayagam R. A critical review on production of bioethanol from macroalgal biomass. Algal Res 2019; 42: 101606.
[http://dx.doi.org/10.1016/j.algal.2019.101606]

[917] El Harchi M, Fakihi Kachkach FZ, El Mtili N. Optimization of thermal acid hydrolysis for bioethanol production from Ulva rigida with yeast Pachysolen tannophilus. S Afr J Bot 2018; 115: 161-9.
[http://dx.doi.org/10.1016/j.sajb.2018.01.021]

[918] Gohain M, Hasin M, Eldiehy KSH, *et al.* Bio-ethanol production: A route to sustainability of fuels using bio-based heterogeneous catalyst derived from waste. Process Saf Environ Prot 2021; 146: 190-200.
[http://dx.doi.org/10.1016/j.psep.2020.08.046]

[919] Gordillo FJL, Jiménez C, Goutx M, Niell X. Effects of CO2 and nitrogen supply on the biochemical composition of Ulva rigida with especial emphasis on lipid class analysis. J Plant Physiol 2001; 158(3): 367-73.
[http://dx.doi.org/10.1078/0176-1617-00209]

[920] Karray R, Hamza M, Sayadi S. Evaluation of ultrasonic, acid, thermo-alkaline and enzymatic pre-treatments on anaerobic digestion of Ulva rigida for biogas production. Bioresour Technol 2015; 187: 205-13.
[http://dx.doi.org/10.1016/j.biortech.2015.03.108] [PMID: 25855526]

[921] Karray R, Hamza M, Sayadi S. Production and characterization of enzymatic cocktail produced by Aspergillus niger using green macroalgae as nitrogen source and its application in the pre-treatment for biogas production from Ulva rigida. Bioresour Technol 2016; 216: 622-8.
[http://dx.doi.org/10.1016/j.biortech.2016.05.067] [PMID: 27285578]

[922] Karray R, Karray F, Loukil S, Mhiri N, Sayadi S. Anaerobic co-digestion of Tunisian green macroalgae Ulva rigida with sugar industry wastewater for biogas and methane production enhancement. Waste Manag 2017; 61: 171-8.
[http://dx.doi.org/10.1016/j.wasman.2016.11.042] [PMID: 28038905]

[923] Sivaprakash G, Mohanrasu K, Ananthi V, *et al.* Biodiesel production from Ulva linza, Ulva tubulosa, Ulva fasciata, Ulva rigida, Ulva reticulate by using Mn2ZnO4 heterogenous nanocatalysts. Fuel 2019; 255: 115744.
[http://dx.doi.org/10.1016/j.fuel.2019.115744]

[924] Allen E, Browne J, Hynes S, Murphy JD. The potential of algae blooms to produce renewable gaseous fuel. Waste Manag 2013; 33(11): 2425-33.
[http://dx.doi.org/10.1016/j.wasman.2013.06.017] [PMID: 23850117]

[925] del Río PG, Gomes-Dias JS, Rocha CMR, Romaní A, Garrote G, Domingues L. Recent trends on seaweed fractionation for liquid biofuels production. Bioresour Technol 2020; 299: 122613.
[http://dx.doi.org/10.1016/j.biortech.2019.122613] [PMID: 31870706]

[926] Filote C, Ungureanu G, Boaventura R, Santos S, Volf I, Botelho C. Green macroalgae from the Romanian coast of Black Sea: Physico-chemical characterization and future perspectives on their use as metal anions biosorbents. Process Saf Environ Prot 2017; 108: 34-43.
[http://dx.doi.org/10.1016/j.psep.2016.06.002]

[927] Hong Y, Wu YR. Acidolysis as a biorefinery approach to producing advanced bioenergy from macroalgal biomass: A state-of-the-art review. Bioresour Technol 2020; 318: 124080.
[http://dx.doi.org/10.1016/j.biortech.2020.124080] [PMID: 32927316]

[928] Maceiras R, Rodrı́guez M, Cancela A, Urréjola S, Sánchez A. Macroalgae: Raw material for biodiesel production. Appl Energy 2011; 88(10): 3318-23.
[http://dx.doi.org/10.1016/j.apenergy.2010.11.027]

[929] Parvathy Eswari A, Anu Alias Meena R, Yukesh Kannah R, Sakthinathan G, Karthikeyan OP, Rajesh Banu J. Bioconversion of marine waste biomass for biofuel and value-added products recovery.Refining Biomass Residues for Sustainable Energy and Bioproducts. Academic Press 2020; pp. 481-507.
[http://dx.doi.org/10.1016/B978-0-12-818996-2.00022-3]

[930] Qarri A, Israel A. Seasonal biomass production, fermentable saccharification and potential ethanol yields in the marine macroalga Ulva sp. (Chlorophyta). Renew Energy 2020; 145: 2101-7.
[http://dx.doi.org/10.1016/j.renene.2019.07.155]

[931] Aparicio E, Rodríguez-Jasso RM, Lara A, *et al.* Biofuels production of third generation biorefinery from macroalgal biomass in the Mexican context: An overview.Sustainable Seaweed Technologies. Elsevier 2020; pp. 393-446.
[http://dx.doi.org/10.1016/B978-0-12-817943-7.00015-9]

[932] Bardhan P, Gupta K, Mandal M. Microbes as bio-resource for sustainable production of biofuels and other bioenergy products.New and Future Developments in Microbial Biotechnology and Bioengineering. Elsevier 2019; pp. 205-22.
[http://dx.doi.org/10.1016/B978-0-444-64191-5.00015-8]

[933] Fernand F, Israel A, Skjermo J, Wichard T, Timmermans KR, Golberg A. Offshore macroalgae biomass for bioenergy production: Environmental aspects, technological achievements and challenges. Renew Sustain Energy Rev 2017; 75: 35-45.
[http://dx.doi.org/10.1016/j.rser.2016.10.046]

[934] Ganesh Saratale R, Kumar G, Banu R, Xia A, Periyasamy S, Dattatraya Saratale G. A critical review on anaerobic digestion of microalgae and macroalgae and co-digestion of biomass for enhanced methane generation. Bioresour Technol 2018; 262: 319-32.
[http://dx.doi.org/10.1016/j.biortech.2018.03.030] [PMID: 29576518]

[935] Han W, Clarke W, Pratt S. Composting of waste algae: A review. Waste Manag 2014; 34(7): 1148-55.
[http://dx.doi.org/10.1016/j.wasman.2014.01.019] [PMID: 24602833]

[936] Kumar M, Vivekanand V, Pareek N. Enzymatic degradation of lignocellulosic waste: bioremediation and industrial implementation. In: Saxena G, Kumar V, Shah MP, Eds. Bioremediation for Environmental Sustainability. Elsevier 2021; pp. 163-91.

[937] McKennedy J, Sherlock O. Anaerobic digestion of marine macroalgae: A review. Renew Sustain Energy Rev 2015; 52: 1781-90.
[http://dx.doi.org/10.1016/j.rser.2015.07.101]

[938] Tan IS, Lam MK, Foo HCY, Lim S, Lee KT. Advances of macroalgae biomass for the third generation of bioethanol production. Chin J Chem Eng 2020; 28(2): 502-17.
[http://dx.doi.org/10.1016/j.cjche.2019.05.012]

[939] Terme N, Hardouin K, Cortès HP, *et al.* Emerging seaweed extraction techniques: Enzyme-assisted extraction a key step of seaweed biorefinery?Sustainable Seaweed Technologies. Elsevier 2020; pp. 225-56.
[http://dx.doi.org/10.1016/B978-0-12-817943-7.00009-3]

[940] Wall DM, McDonagh S, Murphy JD. Cascading biomethane energy systems for sustainable green gas production in a circular economy. Bioresour Technol 2017; 243: 1207-15.
[http://dx.doi.org/10.1016/j.biortech.2017.07.115] [PMID: 28803063]

[941] Ben Yahmed N, Jmel MA, Ben Alaya M, Bouallagui H, Marzouki MN, Smaali I. A biorefinery concept using the green macroalgae Chaetomorpha linum for the coproduction of bioethanol and biogas. Energy Convers Manage 2016; 119: 257-65.
[http://dx.doi.org/10.1016/j.enconman.2016.04.046]

[942] Hosseini Koupaie E, Dahadha S, Bazyar Lakeh AA, Azizi A, Elbeshbishy E. Enzymatic pretreatment of lignocellulosic biomass for enhanced biomethane production-A review. J Environ Manage 2019; 233: 774-84.
[http://dx.doi.org/10.1016/j.jenvman.2018.09.106] [PMID: 30314871]

[943] Hou X, Hansen JH, Bjerre AB. Integrated bioethanol and protein production from brown seaweed Laminaria digitata. Bioresour Technol 2015; 197: 310-7.
[http://dx.doi.org/10.1016/j.biortech.2015.08.091] [PMID: 26342344]

[944] Lara A, Rodríguez-Jasso RM, Loredo-Treviño A, Aguilar CN, Meyer AS, Ruiz HA. Enzymes in the third generation biorefinery for macroalgae biomass.Biomass, Biofuels, Biochemicals. Elsevier 2020; pp. 363-96.
[http://dx.doi.org/10.1016/B978-0-12-819820-9.00017-X]

[945] Polikovsky M, Gillis A, Steinbruch E, *et al.* Biorefinery for the co-production of protein, hydrochar and additional co-products from a green seaweed Ulva sp. with subcritical water hydrolysis. Energy Convers Manage 2020; 225: 113380.
[http://dx.doi.org/10.1016/j.enconman.2020.113380]

[946] Prabha SP, Nagappan S, Rathna R, Viveka R, Nakkeeran E. Blue biotechnology: A vision for future marine biorefineries.Refining Biomass Residues for Sustainable Energy and Bioproducts. Academic Press 2020; pp. 463-80.
[http://dx.doi.org/10.1016/B978-0-12-818996-2.00021-1]

[947] Sharma S, Basu S, Shetti NP, Kamali M, Walvekar P, Aminabhavi TM. Waste-to-energy nexus: A sustainable development. Environ Pollut 2020; 267: 115501.
[http://dx.doi.org/10.1016/j.envpol.2020.115501] [PMID: 32892013]

[948] Shobana S, Kumar G, Bakonyi P, *et al.* A review on the biomass pretreatment and inhibitor removal methods as key-steps towards efficient macroalgae-based biohydrogen production. Bioresour Technol 2017; 244(Pt 2): 1341-8.
[http://dx.doi.org/10.1016/j.biortech.2017.05.172] [PMID: 28602665]

[949] Ubando AT, Felix CB, Chen WH. Biorefineries in circular bioeconomy: A comprehensive review. Bioresour Technol 2020; 299: 122585.
[http://dx.doi.org/10.1016/j.biortech.2019.122585] [PMID: 31901305]

[950] Vo HNP, Ngo HH, Guo W, *et al.* A critical review on designs and applications of microalgae-based photobioreactors for pollutants treatment. Sci Total Environ 2019; 651(Pt 1): 1549-68.
[http://dx.doi.org/10.1016/j.scitotenv.2018.09.282] [PMID: 30360283]

[951] Boraso A, Ciancia M, Cerezo A, Piriz M L, Casas G, Eyras MC. Utilización de las macroalgas marinas de la costa argentina y sus hidrocoloides. 2015; 3-60.

[952] Akila V, Manikandan A, Sahaya Sukeetha D, Balakrishnan S, Ayyasamy PM, Rajakumar S. Biogas and biofertilizer production of marine macroalgae: An effective anaerobic digestion of Ulva sp. Biocatal Agric Biotechnol 2019; 18: 101035.
[http://dx.doi.org/10.1016/j.bcab.2019.101035]

[953] Jeong GT, Kim SK. Thermochemical conversion of defatted microalgae Scenedesmus obliquus into levulinic and formic acids. Fuel 2021; 283: 118907.
[http://dx.doi.org/10.1016/j.fuel.2020.118907]

[954] Khoo CG, Dasan YK, Lam MK, Lee KT. Algae biorefinery: Review on a broad spectrum of downstream processes and products. Bioresour Technol 2019; 292: 121964.
[http://dx.doi.org/10.1016/j.biortech.2019.121964] [PMID: 31451339]

[955] Liu X, Chen Y, Zhong M, Chen W, Lin Q, Du H. Isolation and pathogenicity identification of bacterial pathogens in bleached disease and their physiological effects on the red macroalga Gracilaria lemaneiformis. Aquat Bot 2019; 153: 1-7.
[http://dx.doi.org/10.1016/j.aquabot.2018.11.002]

[956] Nasaruddin RR, Chen T, Yao Q, Zang S, Xie J. Toward greener synthesis of gold nanomaterials: From biological to biomimetic synthesis. Coord Chem Rev 2021; 426: 213540.
[http://dx.doi.org/10.1016/j.ccr.2020.213540]

[957] Vinosha M, Palanisamy S, Muthukrishnan R, *et al.* Biogenic synthesis of gold nanoparticles from Halymenia dilatata for pharmaceutical applications: Antioxidant, anti-cancer and antibacterial activities. Process Biochem 2019; 85: 219-29.
[http://dx.doi.org/10.1016/j.procbio.2019.07.013]

[958] Dwi Kurniasari K, Arsianti A, Astika Nugrahayning Aziza Y, *et al.* Phytochemical Analysis and Anticancer Activity of Seaweed Gracilaria verrucosa against Colorectal HCT-116 Cells. Orient J Chem 2018; 34(3): 1257-62.
[http://dx.doi.org/10.13005/ojc/340308]

[959] Cano-Díaz C, Mateo P, Muñoz-Martín MÁ, Maestre FT. Diversity of biocrust-forming cyanobacteria in a semiarid gypsiferous site from Central Spain. J Arid Environ 2018; 151: 83-9.
[http://dx.doi.org/10.1016/j.jaridenv.2017.11.008] [PMID: 30038450]

[960] Krings M, Sergeev VN. A coccoid, colony-forming cyanobacterium from the Lower Devonian Rhynie chert that resembles Eucapsis (Synechococcales) and Entophysalis (Chroococcales). Rev Palaeobot Palynol 2019; 268: 65-71.
[http://dx.doi.org/10.1016/j.revpalbo.2019.06.002]

[961] Manickam N, Saravana Bhavan P, Santhanam P, *et al.* Phytoplankton biodiversity in the two perennial lakes of Coimbatore, Tamil Nadu, India. Acta Ecol Sin 2020; 40(1): 81-9.
[http://dx.doi.org/10.1016/j.chnaes.2019.05.014]

[962] Ortiz JE, Marcarelli AM, Juneau KJ, Huckins CJ. Invasive Myriophyllum spicatum and nutrients interact to influence algal assemblages. Aquat Bot 2019; 156: 1-9.
[http://dx.doi.org/10.1016/j.aquabot.2019.03.003]

[963] Taylor TN, Krings M. A colony-forming microorganism with probable affinities to the Chroococcales (cyanobacteria) from the Lower Devonian Rhynie chert. Rev Palaeobot Palynol 2015; 219: 147-56.
[http://dx.doi.org/10.1016/j.revpalbo.2015.04.003]

[964] Dvorak P, Casamatta D, Hasler P, Jahodářová E, Norwich A, Poulíčková A. Diversity of the cyanobacteria. 2017; 3-46.

[965] Choudhary P, Bhattacharya A, Prajapati SK, Kaushik P, Malik A. Phycoremediation-Coupled Biomethanation of Microalgal Biomass.Handbook of Marine Microalgae. Boston: Academic Press 2015; pp. 483-99.
[http://dx.doi.org/10.1016/B978-0-12-800776-1.00032-7]

[966] Priyanka; Kumar C, Chatterjee A, Wenjing W, Yadav D, Singh PK. Cyanobacteria: potential and role for environmental remediation. In: Singh P, Kumar A, Borthakur A, Eds. Abatement of Environmental Pollutants. Elsevier 2020; pp. 193-202.

[967] Belfield W, Dearden M. Variety of organisms. A Practical Course in Biology. Pergamon 1971; pp. 11-4.

[968] Crow WB. Epiphytes and climbers.A Synopsis of Biology. Butterworth-Heinemann 1960; pp. 512-5.
[http://dx.doi.org/10.1016/B978-1-4832-0029-3.50089-8]

[969] Crow WB. Epiphytes and climbers.A Synopsis of Biology. 2nd ed. Butterworth-Heinemann 1964; pp. 512-5.
[http://dx.doi.org/10.1016/B978-1-4831-6826-5.50091-1]

[970] Cwalina B. Biodeterioration of concrete, brick and other mineral-based building materials.Understanding Biocorrosion. Oxford: Woodhead Publishing 2014; pp. 281-312.
[http://dx.doi.org/10.1533/9781782421252.3.281]

[971] Hinton HE. PsocopteraBiology of Insect Eggs. Pergamon 1981; pp. 546-8.
[http://dx.doi.org/10.1016/B978-1-4832-8401-9.50024-6]

[972] John DM. Filamentous and plantlike green algae.Freshwater Algae of North America. Burlington: Academic Press 2003; pp. 311-52.
[http://dx.doi.org/10.1016/B978-012741550-5/50009-X]

[973] Nissenbaum A. Amber and the Direct Observation of Paleomicrobiota.Developments in Geochemistry. Elsevier 1991; Vol. 6: pp. 15-21.

[974] Shubert E, Gärtner G. Nonmotile Coccoid and Colonial Green Algae. In: Wehr JD, Sheath RG, Kociolek JP, Eds. Freshwater Algae of North America. Boston: Academic Press 2015; pp. 315-73.

[975] Shubert LE. Nonmotile Coccoid And Colonial Green Algae.Freshwater Algae of North America. Burlington: Academic Press 2003; pp. 253-309.
[http://dx.doi.org/10.1016/B978-012741550-5/50008-8]

[976] Jamieson BGM, Reynolds JF. Algae.Tropical Plant Types. Pergamon 1967; pp. 32-62.
[http://dx.doi.org/10.1016/B978-0-08-012119-2.50005-0]

[977] Beraldi-Campesi H, Cevallos-Ferriz SRS, Chacón-Baca E. Microfossil algae associated with Cretaceous stromatolites in the Tarahumara Formation, Sonora, Mexico. Cretac Res 2004; 25(2): 249-65.
[http://dx.doi.org/10.1016/j.cretres.2003.12.002]

[978] Golubić S, Pietrini AM, Ricci S. Euendolithic activity of the cyanobacterium Chroococcus lithophilus Erc. In biodeterioration of the Pyramid of Caius Cestius, Rome, Italy. Int Biodeterior Biodegradation 2015; 100: 7-16.
[http://dx.doi.org/10.1016/j.ibiod.2015.01.019]

[979] Reed JD, Illich HA, Horsfield B. Biochemical evolutionary significance of Ordovician oils and their sources. Org Geochem 1986; 10(1-3): 347-58.
[http://dx.doi.org/10.1016/0146-6380(86)90035-5]

[980] Khan I, Zhong N, Luo Q, Ai J, Yao L, Luo P. Maceral composition and origin of organic matter input in Neoproterozoic–Lower Cambrian organic-rich shales of Salt Range Formation, upper Indus Basin, Pakistan. Int J Coal Geol 2020; 217: 103319.
[http://dx.doi.org/10.1016/j.coal.2019.103319]

[981] Dar R, Bandh SA, Shafi S, Shameem N. Bacterial diversity of the rock-water interface in freshwater ecosystem.Freshwater Microbiology. Academic Press 2019; pp. 73-104.
[http://dx.doi.org/10.1016/B978-0-12-817495-1.00002-5]

[982] Heimann K, Cirés S. N2-fixing cyanobacteria: Ecology and biotechnological applications.Handbook of Marine Microalgae. Boston: Academic Press 2015; pp. 501-15.
[http://dx.doi.org/10.1016/B978-0-12-800776-1.00033-9]

[983] Kathiresan K. Bioprospecting potential of mangrove resources.Biotechnological Utilization of Mangrove Resources. Academic Press 2020; pp. 225-41.
[http://dx.doi.org/10.1016/B978-0-12-819532-1.00008-1]

[984] Mishra S. Cyanobacterial imprints in diversity and phylogeny.Advances in Cyanobacterial Biology. Academic Press 2020; pp. 1-15.
[http://dx.doi.org/10.1016/B978-0-12-819311-2.00001-2]

[985] Priya M, Gurung N, Mukherjee K, Bose S. Microalgae in removal of heavy metal and organic pollutants from soil.Microbial Biodegradation and Bioremediation. Oxford: Elsevier 2014; pp. 519-37.
[http://dx.doi.org/10.1016/B978-0-12-800021-2.00023-6]

[986] Rastogi RP, Sonani RR, Madamwar D. UV photoprotectants from algae—synthesis and bio-functionalities.Algal Green Chemistry. Amsterdam: Elsevier 2017; pp. 17-38.
[http://dx.doi.org/10.1016/B978-0-444-63784-0.00002-3]

[987] Singh DK, Pathak J, Pandey A, Singh V, Ahmed H. Ultraviolet-screening compound mycosporine-like amino acids in cyanobacteria: biosynthesis, functions, and applications. In: Singh PK, Kumar A, Singh VK, Shrivastava AK, Eds. Advances in Cyanobacterial Biology. Academic Press 2020; pp. 219-33.

[988] Sharma P, Slathia PS, Raina N, Bhagat D. Microbial diversity in freshwater ecosystems and its industrial potential.Freshwater Microbiology. Academic Press 2019; pp. 341-92.
[http://dx.doi.org/10.1016/B978-0-12-817495-1.00009-8]

[989] Bravakos P, Kotoulas G, Skaraki K, Pantazidou A, Economou-Amilli A. A polyphasic taxonomic approach in isolated strains of Cyanobacteria from thermal springs of Greece. Mol Phylogenet Evol 2016; 98: 147-60.
[http://dx.doi.org/10.1016/j.ympev.2016.02.009] [PMID: 26899923]

[990] Debiagi PEA, Trinchera M, Frassoldati A, Faravelli T, Vinu R, Ranzi E. Algae characterization and multistep pyrolysis mechanism. J Anal Appl Pyrolysis 2017; 128: 423-36.
[http://dx.doi.org/10.1016/j.jaap.2017.08.007]

[991] Markou G, Nerantzis E. Microalgae for high-value compounds and biofuels production: A review with focus on cultivation under stress conditions. Biotechnol Adv 2013; 31(8): 1532-42.
[http://dx.doi.org/10.1016/j.biotechadv.2013.07.011] [PMID: 23928208]

[992] Zia KM, Noreen A, Zuber M, Tabasum S, Mujahid M. Recent developments and future prospects on bio-based polyesters derived from renewable resources: A review. Int J Biol Macromol 2016; 82: 1028-40.
[http://dx.doi.org/10.1016/j.ijbiomac.2015.10.040] [PMID: 26492854]

[993] Burkholder JM. Harmful algal blooms.Encyclopedia of Inland Waters. Oxford: Academic Press 2009; pp. 264-85.
[http://dx.doi.org/10.1016/B978-012370626-3.00239-8]

[994] Haberle I, Hrustić E, Petrić I, *et al.* Adriatic cyanobacteria potential for cogeneration biofuel production with oil refinery wastewater remediation. Algal Res 2020; 50: 101978.
[http://dx.doi.org/10.1016/j.algal.2020.101978]

[995] Obrdlik P. The Longola Hot Springs of Zambia: The need for conservation. Biol Conserv 1988; 43(2): 81-6.
[http://dx.doi.org/10.1016/0006-3207(88)90083-3]

[996] Parent S, Morin A, Gagno D. Are meiofauna transient or resident in sand filters of marine aquariums? Water Res 2001; 35(15): 3625-34.
[http://dx.doi.org/10.1016/S0043-1354(01)00081-1] [PMID: 11561623]

[997] Sprachta S, Camoin G, Golubic S, Le Campion T. Microbialites in a modern lagoonal environment: nature and distribution, Tikehau atoll (French Polynesia). Palaeogeogr Palaeoclimatol Palaeoecol 2001; 175(1-4): 103-24.
[http://dx.doi.org/10.1016/S0031-0182(01)00388-1]

[998] Szubert K, Wiglusz M, Mazur-Marzec H. Bioactive metabolites produced by Spirulina subsalsa from the Baltic Sea. Oceanologia 2018; 60(3): 245-55.
[http://dx.doi.org/10.1016/j.oceano.2017.11.003]

[999] Titlyanov EA, Titlyanova TV, Yakovleva IM, Nakano Y, Bhagooli R. Regeneration of artificial injuries on scleractinian corals and coral/algal competition for newly formed substrate. J Exp Mar Biol Ecol 2005; 323(1): 27-42.
[http://dx.doi.org/10.1016/j.jembe.2005.02.015]

[1000] Włodarska-Kowalczuk M, Balazy P, Kobos J, Wiktor J, Zajączkowski M, Moskal W. Large red cyanobacterial mats (Spirulina subsalsa Oersted ex Gomont) in the shallow sublittoral of the southern

Baltic. Oceanologia 2014; 56(3): 661-6.
[http://dx.doi.org/10.5697/oc.55-3.661]

[1001] Prasad RN, Sanghamitra K, Antonia G-M, *et al.* Isolation, Identification and Germplasm Preservation of Different Native <i>Spirulina</i> Species from Western Mexico. Am J Plant Sci 2013; 4(12): 65-71.
[http://dx.doi.org/10.4236/ajps.2013.412A2009]

[1002] Ballot A, Krienitz L, Kotut K, Wiegand C, Pflugmacher S. Cyanobacteria and cyanobacterial toxins in the alkaline crater lakes Sonachi and Simbi, Kenya. Harmful Algae 2005; 4(1): 139-50.
[http://dx.doi.org/10.1016/j.hal.2004.01.001]

[1003] Krienitz L, Ballot A, Kotut K, *et al.* Contribution of hot spring cyanobacteria to the mysterious deaths of Lesser Flamingos at Lake Bogoria, Kenya. FEMS Microbiol Ecol 2003; 43(2): 141-8.
[http://dx.doi.org/10.1111/j.1574-6941.2003.tb01053.x] [PMID: 19719674]

[1004] Gonzales KN, Troncoso OP, Torres FG, López D. Molecular α-relaxation process of exopolysaccharides extracted from Nostoc commune cyanobacteria. Int J Biol Macromol 2020; 161: 1516-25.
[http://dx.doi.org/10.1016/j.ijbiomac.2020.07.268] [PMID: 32755710]

[1005] Poveda J. Cyanobacteria in plant health: Biological strategy against abiotic and biotic stresses. Crop Prot 2020; •••: 105450.

[1006] Van Den Hoek C, Cortel-Breeman AM, Wanders JBW. Algal zonation in the fringing coral reef of curaçao, Netherlands antilles, in relation to zonation of corals and gorgonians. Aquat Bot 1975; 1: 269-308.
[http://dx.doi.org/10.1016/0304-3770(75)90028-5]

[1007] Rosen B, Mares J. Catalog of microscopic organisms of the everglades, part 1 - The cyanobacteria. 2015; 108.

[1008] Afonina EY, Tashlykova NA. Plankton community and the relationship with the environment in saline lakes of Onon-Torey plain, Northeastern Mongolia. Saudi J Biol Sci 2018; 25(2): 399-408.
[http://dx.doi.org/10.1016/j.sjbs.2017.01.003] [PMID: 29472798]

[1009] Heynig H. Interessante phytoplankter aus gewässern des Bezirks Halle (DDR). VI. Arch Protistenkd 1989; 137(1): 57-68.
[http://dx.doi.org/10.1016/S0003-9365(89)80041-7]

[1010] Woelfel J, Schumann R, Adler S, Hübener T, Karsten U. Diatoms inhabiting a wind flat of the Baltic Sea: Species diversity and seasonal succession. Estuar Coast Shelf Sci 2007; 75(3): 296-307.
[http://dx.doi.org/10.1016/j.ecss.2007.04.033]

[1011] Albertano P, Bruno L, Bellezza S, Paradossi G. Polysaccharides as a key step in stone bio-erosion. In: Fassina V, Ed. Proceedings of the 9th International Congress on Deterioration and Conservation of Stone. 425-32.
[http://dx.doi.org/10.1016/B978-044450517-0/50125-2]

[1012] Lee H, Depuydt S, Choi S, *et al.* Potential use of nuisance cyanobacteria as a source of anticancer agents.Natural Bioactive Compounds. Academic Press 2021; pp. 203-31.
[http://dx.doi.org/10.1016/B978-0-12-820655-3.00010-0]

[1013] Mehta A, Soni VK, Shukla D, Vishvakarma NK. Cyanobacteria: A potential source of anticancer drugs.Advances in Cyanobacterial Biology. Academic Press 2020; pp. 369-84.
[http://dx.doi.org/10.1016/B978-0-12-819311-2.00024-3]

[1014] Pathak J, Mishra S, Kumari N, *et al.* Bionanotechnology of cyanobacterial bioactive compounds.Natural Bioactive Compounds. Academic Press 2021; pp. 115-42.
[http://dx.doi.org/10.1016/B978-0-12-820655-3.00006-9]

[1015] Bai R, Silaban AG, Gutierrez-Wing MT, Benton MG, Negulescu II, Rusch KA. Silver nanofiber assisted lipid extraction from biomass of a Louisiana Chlorella vulgaris/Leptolyngbya sp. co-culture.

Chem Eng J 2013; 225: 100-8.
[http://dx.doi.org/10.1016/j.cej.2013.03.075]

[1016] Maity JP, Hou CP, Majumder D, *et al.* The production of biofuel and bioelectricity associated with wastewater treatment by green algae. Energy 2014; 78: 94-103.
[http://dx.doi.org/10.1016/j.energy.2014.06.023]

[1017] Moungmoon T, Chaichana C, Pumas C, Pathom-aree W, Ruangrit K, Pekkoh J. Quantitative analysis of methane and glycolate production from microalgae using undiluted wastewater obtained from chicken-manure biogas digester. Sci Total Environ 2020; 714: 136577.
[http://dx.doi.org/10.1016/j.scitotenv.2020.136577] [PMID: 31982736]

[1018] Nath Tiwari O, Bhunia B, Muthuraj M, Kanti Bandyopadhyay T, Ghosh D, Gopikrishna K. Optimization of process parameters on lipid biosynthesis for sustainable biodiesel production and evaluation of its fuel characteristics. Fuel 2020; 269: 117471.
[http://dx.doi.org/10.1016/j.fuel.2020.117471]

[1019] Prabaharan D, Arun Kumar D, Uma L, Subramanian G. Dark hydrogen production in nitrogen atmosphere – An approach for sustainability by marine cyanobacterium Leptolyngbya valderiana BDU 20041. Int J Hydrogen Energy 2010; 35(19): 10725-30.
[http://dx.doi.org/10.1016/j.ijhydene.2010.03.007]

[1020] Ramesh kumar V, Narendrakumar G, Thyagarajan R, Melchias G. A comparative analysis of biodiesel production and its properties from Leptolyngbya sp. BI-107 and Chlorella vulgaris under heat shock stress. Biocatal Agric Biotechnol 2018; 16: 502-6.
[http://dx.doi.org/10.1016/j.bcab.2018.09.007]

[1021] Sen S, Dutta S, Guhathakurata S, Chakrabarty J, Nandi S, Dutta A. Removal of Cr(VI) using a cyanobacterial consortium and assessment of biofuel production. Int Biodeterior Biodegradation 2017; 119: 211-24.
[http://dx.doi.org/10.1016/j.ibiod.2016.10.050]

[1022] Singh J, Tripathi R, Thakur IS. Characterization of endolithic cyanobacterial strain, Leptolyngbya sp. ISTCY101, for prospective recycling of CO2 and biodiesel production. Bioresour Technol 2014; 166: 345-52.
[http://dx.doi.org/10.1016/j.biortech.2014.05.055] [PMID: 24926608]

[1023] Tsolcha ON, Tekerlekopoulou AG, Akratos CS, *et al.* Biotreatment of raisin and winery wastewaters and simultaneous biodiesel production using a Leptolyngbya -based microbial consortium. J Clean Prod 2017; 148: 185-93.
[http://dx.doi.org/10.1016/j.jclepro.2017.02.026]

[1024] Ángeles Torres R, Marín D, Rodero MR, *et al.* Biogas treatment for H2S, CO2, and other contaminants removal. From Biofiltration to Promising Options in Gaseous Fluxes Biotreatment. Elsevier 2020; pp. 153-76.

[1025] Beriş FŞ, Alpay-Karaoğlu Ş. Microbial energy production.Comprehensive Energy Systems. Oxford: Elsevier 2018; pp. 521-37.
[http://dx.doi.org/10.1016/B978-0-12-809597-3.00328-X]

[1026] Chen CY. Future perspective of algae: energy production and environmental cleanup. N Biotechnol 2018; 44: S127.
[http://dx.doi.org/10.1016/j.nbt.2018.05.1065]

[1027] Chiranjeevi P, Patil SA. Microbial fuel cell coupled with microalgae cultivation for wastewater treatment and energy recovery.Integrated Microbial Fuel Cells for Wastewater Treatment. Butterworth-Heinemann 2020; pp. 213-27.
[http://dx.doi.org/10.1016/B978-0-12-817493-7.00010-2]

[1028] Ellison CR, Overa S, Boldor D. Central composite design parameterization of microalgae/cyanobacteria co-culture pretreatment for enhanced lipid extraction using an external clamp-on ultrasonic transducer. Ultrason Sonochem 2019; 51: 496-503.

[http://dx.doi.org/10.1016/j.ultsonch.2018.05.006] [PMID: 29793838]

[1029] Marín D, Posadas E, Cano P, *et al*. Seasonal variation of biogas upgrading coupled with digestate treatment in an outdoors pilot scale algal-bacterial photobioreactor. Bioresour Technol 2018; 263: 58-66.
[http://dx.doi.org/10.1016/j.biortech.2018.04.117] [PMID: 29730519]

[1030] Nagarajan D, Lee DJ, Chang JS. Integration of anaerobic digestion and microalgal cultivation for digestate bioremediation and biogas upgrading. Bioresour Technol 2019; 290: 121804.
[http://dx.doi.org/10.1016/j.biortech.2019.121804] [PMID: 31327690]

[1031] Ng FL, Phang SM, Thong CH, *et al*. Integration of bioelectricity generation from algal biophotovoltaic (BPV) devices with remediation of palm oil mill effluent (POME) as substrate for algal growth. Environmental Technology & Innovation 2021; 21: 101280.
[http://dx.doi.org/10.1016/j.eti.2020.101280]

[1032] Rajneesh , Singh SP, Pathak J, Sinha RP. Cyanobacterial factories for the production of green energy and value-added products: An integrated approach for economic viability. Renew Sustain Energy Rev 2017; 69: 578-95.
[http://dx.doi.org/10.1016/j.rser.2016.11.110]

[1033] Sahu A, Pancha I, Jain D, *et al*. Fatty acids as biomarkers of microalgae. Phytochemistry 2013; 89: 53-8.
[http://dx.doi.org/10.1016/j.phytochem.2013.02.001] [PMID: 23453131]

[1034] Cho HU, Kim YM, Park JM. Enhanced microalgal biomass and lipid production from a consortium of indigenous microalgae and bacteria present in municipal wastewater under gradually mixotrophic culture conditions. Bioresour Technol 2017; 228: 290-7.
[http://dx.doi.org/10.1016/j.biortech.2016.12.094] [PMID: 28081527]

[1035] Elshobary ME, Zabed HM, Yun J, Zhang G, Qi X. Recent insights into microalgae-assisted microbial fuel cells for generating sustainable bioelectricity. Int J Hydrogen Energy 2020.

[1036] Ganesan R, Manigandan S, Samuel MS, *et al*. A review on prospective production of biofuel from microalgae. Biotechnol Rep (Amst) 2020; 27: e00509.
[http://dx.doi.org/10.1016/j.btre.2020.e00509] [PMID: 32775233]

[1037] García D, Alcántara C, Blanco S, Pérez R, Bolado S, Muñoz R. Enhanced carbon, nitrogen and phosphorus removal from domestic wastewater in a novel anoxic-aerobic photobioreactor coupled with biogas upgrading. Chem Eng J 2017; 313: 424-34.
[http://dx.doi.org/10.1016/j.cej.2016.12.054]

[1038] Konur O. Cyanobacterial bioenergy and biofuels science and technology: A scientometric overview.Cyanobacteria. Academic Press 2019; pp. 419-42.
[http://dx.doi.org/10.1016/B978-0-12-814667-5.00021-0]

[1039] Sharma J, Kumar V, Kumar SS, *et al*. Microalgal consortia for municipal wastewater treatment – Lipid augmentation and fatty acid profiling for biodiesel production. J Photochem Photobiol B 2020; 202: 111638.
[http://dx.doi.org/10.1016/j.jphotobiol.2019.111638] [PMID: 31733613]

[1040] Srinuanpan S, Cheirsilp B, Kassim MA. Oleaginous microalgae cultivation for biogas upgrading and phytoremediation of wastewater.Microalgae Cultivation for Biofuels Production. Academic Press 2020; pp. 69-82.
[http://dx.doi.org/10.1016/B978-0-12-817536-1.00005-9]

[1041] Vo Hoang Nhat P, Ngo HH, Guo WS, *et al*. Can algae-based technologies be an affordable green process for biofuel production and wastewater remediation? Bioresour Technol 2018; 256: 491-501.
[http://dx.doi.org/10.1016/j.biortech.2018.02.031] [PMID: 29472123]

[1042] Yang C, Nie R, Fu J, Hou Z, Lu X. Production of aviation fuel *via* catalytic hydrothermal decarboxylation of fatty acids in microalgae oil. Bioresour Technol 2013; 146: 569-73.

[http://dx.doi.org/10.1016/j.biortech.2013.07.131] [PMID: 23973977]

[1043] Abrevaya XC, Sacco NJ, Bonetto MC, Hilding-Ohlsson A, Cortón E. Analytical applications of microbial fuel cells. Part II: Toxicity, microbial activity and quantification, single analyte detection and other uses. Biosens Bioelectron 2015; 63: 591-601.
[http://dx.doi.org/10.1016/j.bios.2014.04.053] [PMID: 24906984]

[1044] Bernstein HC, Kesaano M, Moll K, *et al.* Direct measurement and characterization of active photosynthesis zones inside wastewater remediating and biofuel producing microalgal biofilms. Bioresour Technol 2014; 156: 206-15.
[http://dx.doi.org/10.1016/j.biortech.2014.01.001] [PMID: 24508901]

[1045] Bundschuh J, Yusaf T, Maity JP, Nelson E, Mamat R, Indra Mahlia TM. Algae-biomass for fuel, electricity and agriculture. Energy 2014; 78: 1-3.
[http://dx.doi.org/10.1016/j.energy.2014.11.005]

[1046] Cheng J, Zhu Y, Zhang Z, Yang W. Modification and improvement of microalgae strains for strengthening CO_2 fixation from coal-fired flue gas in power plants. Bioresour Technol 2019; 291: 121850.
[http://dx.doi.org/10.1016/j.biortech.2019.121850] [PMID: 31358426]

[1047] Patel A, Matsakas L, Rova U, Christakopoulos P. A perspective on biotechnological applications of thermophilic microalgae and cyanobacteria. Bioresour Technol 2019; 278: 424-34.
[http://dx.doi.org/10.1016/j.biortech.2019.01.063] [PMID: 30685131]

[1048] Soru S, Malavasi V, Concas A, Caboni P, Cao G. A novel investigation of the growth and lipid production of the extremophile microalga Coccomyxa melkonianii SCCA 048 under the effect of different cultivation conditions: Experiments and modeling. Chem Eng J 2019; 377: 120589.
[http://dx.doi.org/10.1016/j.cej.2018.12.049]

[1049] Thakur IS, Kumar M, Varjani SJ, Wu Y, Gnansounou E, Ravindran S. Sequestration and utilization of carbon dioxide by chemical and biological methods for biofuels and biomaterials by chemoautotrophs: Opportunities and challenges. Bioresour Technol 2018; 256: 478-90.
[http://dx.doi.org/10.1016/j.biortech.2018.02.039] [PMID: 29459105]

[1050] Young P, Taylor MJ, Buchanan N, Lewis J, Fallowfield HJ. Case study on the effect continuous CO_2 enrichment, *via* biogas scrubbing, has on biomass production and wastewater treatment in a high rate algal pond. J Environ Manage 2019; 251: 109614.
[http://dx.doi.org/10.1016/j.jenvman.2019.109614] [PMID: 31563600]

[1051] Upendar G, Mistry A, Das R, *et al.* Carbon dioxide biofixation using microorganisms and assessment of biofuel production. Environ Prog Sustain Energy 2017; •••: 37.

[1052] Pathak J. Recent developments in green synthesis of metal nanoparticles utilizing cyanobacterial cell factories. Nanomaterials in Plants, Algae and Microorganisms. Academic Press 2019; pp. 237-65.

[1053] Falkiewicz-Dulik M, Janda K, Wypych G. Microorganism involved in biodegradation of materials. Handbook of Material Biodegradation, Biodeterioration, and Biostablization. 2015; pp. 7-32.

[1054] Regaldo L, Reno U, Romero N, *et al.* Multifunctional approach to evaluate the efficiency of landfill leachate treatments. Removal of Toxic Pollutants Through Microbiological and Tertiary Treatment. Elsevier 2020; pp. 179-209.

[1055] Ling Y, Sun L, Wang S, Lin CSK, Sun Z, Zhou Z. Cultivation of oleaginous microalga Scenedesmus obliquus coupled with wastewater treatment for enhanced biomass and lipid production. Biochem Eng J 2019; 148: 162-9.
[http://dx.doi.org/10.1016/j.bej.2019.05.012]

[1056] Schmack M, Chambers J, Dallas S. Evaluation of a bacterial algal control agent in tank-based experiments. Water Res 2012; 46(7): 2435-44.
[http://dx.doi.org/10.1016/j.watres.2012.02.026] [PMID: 22386889]

[1057] Sellami I, Hamza A, Alaoui Mhamdi M, Aleya L, Bouain A, Ayadi H. Abundance and biomass of

rotifers in relation to the environmental factors in geothermal waters in Southern Tunisia. J Therm Biol 2009; 34(6): 267-75.
[http://dx.doi.org/10.1016/j.jtherbio.2009.03.003]

[1058] Angelidaki I, Treu L, Tsapekos P, *et al.* Biogas upgrading and utilization: Current status and perspectives. Biotechnol Adv 2018; 36(2): 452-66.
[http://dx.doi.org/10.1016/j.biotechadv.2018.01.011] [PMID: 29360505]

[1059] Aoki J, Kawamata T, Kodaka A, *et al.* Biofuel production utilizing a dual-phase cultivation system with filamentous cyanobacteria. J Biotechnol 2018; 280: 55-61.
[http://dx.doi.org/10.1016/j.jbiotec.2018.04.011] [PMID: 29678391]

[1060] Basri RS, Rahman RNZRA, Kamarudin NHA, Ali MSM. Cyanobacterial aldehyde deformylating oxygenase: Structure, function, and potential in biofuels production. Int J Biol Macromol 2020; 164: 3155-62.
[http://dx.doi.org/10.1016/j.ijbiomac.2020.08.162] [PMID: 32841666]

[1061] Chakraborty R, Mukhopadhyay P. Green fuel blending: A pollution reduction approach. In: Hashmi S, Choudhury IA, Eds. Encyclopedia of Renewable and Sustainable Materials. Oxford: Elsevier 2020; pp. 487-500.
[http://dx.doi.org/10.1016/B978-0-12-803581-8.11019-7]

[1062] Fuad Hossain M, Ratnayake RR, Mahbub S, Kumara KLW, Magana-Arachchi DN. Identification and culturing of cyanobacteria isolated from freshwater bodies of Sri Lanka for biodiesel production. Saudi J Biol Sci 2020; 27(6): 1514-20.
[http://dx.doi.org/10.1016/j.sjbs.2020.03.024] [PMID: 32489288]

[1063] Hena S, Fatimah S, Tabassum S. Cultivation of algae consortium in a dairy farm wastewater for biodiesel production. Water Resour Ind 2015; 10: 1-14.
[http://dx.doi.org/10.1016/j.wri.2015.02.002]

[1064] Lu Y, Zhuo C, Li Y, *et al.* Evaluation of filamentous heterocystous cyanobacteria for integrated pig-farm biogas slurry treatment and bioenergy production. Bioresour Technol 2020; 297: 122418.
[http://dx.doi.org/10.1016/j.biortech.2019.122418] [PMID: 31761632]

[1065] Oliveira DT, Turbay Vasconcelos C, Feitosa AMT, *et al.* Lipid profile analysis of three new Amazonian cyanobacteria as potential sources of biodiesel. Fuel 2018; 234: 785-8.
[http://dx.doi.org/10.1016/j.fuel.2018.07.080]

[1066] Sitther V, Tabatabai B, Fathabad SG, Gichuki S, Chen H, Arumanayagam ACS. Cyanobacteria as a biofuel source: Advances and applications. In: Singh PK, Kumar A, Singh VK, Shrivastava AK, Eds. Advances in Cyanobacterial Biology. Academic Press 2020; pp. 269-89.
[http://dx.doi.org/10.1016/B978-0-12-819311-2.00018-8]

[1067] Xu P, Xiao E, Wu J, He F, Zhang Y, Wu Z. Enhanced nitrate reduction in water by a combined bio-electrochemical system of microbial fuel cells and submerged aquatic plant Ceratophyllum demersum. J Environ Sci (China) 2019; 78: 338-51.
[http://dx.doi.org/10.1016/j.jes.2018.11.013] [PMID: 30665653]

[1068] Couto E, Calijuri ML, Assemany P. Biomass production in high rate ponds and hydrothermal liquefaction: Wastewater treatment and bioenergy integration. Sci Total Environ 2020; 724: 138104.
[http://dx.doi.org/10.1016/j.scitotenv.2020.138104] [PMID: 32408433]

[1069] Posadas E, Serejo ML, Blanco S, Pérez R, García-Encina PA, Muñoz R. Minimization of biomethane oxygen concentration during biogas upgrading in algal–bacterial photobioreactors. Algal Res 2015; 12: 221-9.
[http://dx.doi.org/10.1016/j.algal.2015.09.002]

[1070] Toledo-Cervantes A, Serejo ML, Blanco S, Pérez R, Lebrero R, Muñoz R. Photosynthetic biogas upgrading to bio-methane: Boosting nutrient recovery *via* biomass productivity control. Algal Res 2016; 17: 46-52.
[http://dx.doi.org/10.1016/j.algal.2016.04.017]

[1071] Yuan S, Chen X, Li W, Liu H, Wang F. Nitrogen conversion under rapid pyrolysis of two types of aquatic biomass and corresponding blends with coal. Bioresour Technol 2011; 102(21): 10124-30.
[http://dx.doi.org/10.1016/j.biortech.2011.08.047] [PMID: 21903383]

[1072] Nogueira IS, Leandro-Rodrigues NC. Algas planctônicas de um lago artificial do Jardim Botânico Chico Mendes, Goiânia, Goiás: florística e algumas considerações ecológicas. Rev Bras Biol 1999; 59(3): 377-95.
[http://dx.doi.org/10.1590/S0034-71081999000300003]

[1073] Gkelis S, Rajaniemi P, Vardaka E, Moustaka-Gouni M, Lanaras T, Sivonen K. Limnothrix redekei (Van Goor) Meffert (Cyanobacteria) strains from Lake Kastoria, Greece form a separate phylogenetic group. Microb Ecol 2005; 49(1): 176-82.
[http://dx.doi.org/10.1007/s00248-003-2030-7] [PMID: 15688255]

[1074] Borges GCP, Aquino EP, Eskinazi-Leça E, *et al.* Cell biovolume and carbon biomass of phytoplankton in degraded tropical estuaries in Northeastern Brazil. Reg Stud Mar Sci 2020; 40: 101522.
[http://dx.doi.org/10.1016/j.rsma.2020.101522]

[1075] Tonetto AF, Guillermo-Ferreira R, Cardoso-Leite R, Novaes MC, Peres CK. The relationship between water velocity and morphological complexity of stream dwellers. Limnologica 2018; 72: 22-7.
[http://dx.doi.org/10.1016/j.limno.2018.08.001]

[1076] Caires TA, da Silva AMS, Vasconcelos VM, *et al.* Biotechnological potential of Neolyngbya (Cyanobacteria), a new marine benthic filamentous genus from Brazil. Algal Res 2018; 36: 1-9.
[http://dx.doi.org/10.1016/j.algal.2018.10.001]

[1077] Caires TA, de Mattos Lyra G, Hentschke GS, de Gusmão Pedrini A, Sant'Anna CL, de Castro Nunes JM. Neolyngbya gen. nov. (Cyanobacteria, Oscillatoriaceae): A new filamentous benthic marine taxon widely distributed along the Brazilian coast. Mol Phylogenet Evol 2018; 120: 196-211.
[http://dx.doi.org/10.1016/j.ympev.2017.12.009] [PMID: 29246815]

[1078] Parker RDR, Drown DB. Effects of Phosphorus Enrichment and Wave Simulation on Populations of Ulothrix Zonata from Northern Lake Superior. J Great Lakes Res 1982; 8(1): 16-26.
[http://dx.doi.org/10.1016/S0380-1330(82)71937-9]

[1079] Conklin KY, Stancheva R, Otten TG, *et al.* Molecular and morphological characterization of a novel dihydroanatoxin-a producing Microcoleus species (cyanobacteria) from the Russian River, California, USA. Harmful Algae 2020; 93: 101767.
[http://dx.doi.org/10.1016/j.hal.2020.101767] [PMID: 32307065]

[1080] Monteagudo L, Moreno JL. Benthic freshwater cyanobacteria as indicators of anthropogenic pressures. Ecol Indic 2016; 67: 693-702.
[http://dx.doi.org/10.1016/j.ecolind.2016.03.035]

[1081] Shevchenko TF, Klochenko PD, Timchenko VM, Dubnyak SS. Epiphyton of a cascade plain reservoir under different hydrodynamic conditions. Ecohydrol Hydrobiol 2019; 19(3): 407-16.
[http://dx.doi.org/10.1016/j.ecohyd.2019.01.006]

[1082] Wood SA, Heath MW, Holland PT, Munday R, McGregor GB, Ryan KG. Identification of a benthic microcystin-producing filamentous cyanobacterium (Oscillatoriales) associated with a dog poisoning in New Zealand. Toxicon 2010; 55(4): 897-903.
[http://dx.doi.org/10.1016/j.toxicon.2009.12.019] [PMID: 20043936]

[1083] MubarakAli D, Arunkumar J, Nag KH, *et al.* Gold nanoparticles from Pro and eukaryotic photosynthetic microorganisms—Comparative studies on synthesis and its application on biolabelling. Colloids Surf B Biointerfaces 2013; 103: 166-73.
[http://dx.doi.org/10.1016/j.colsurfb.2012.10.014] [PMID: 23201734]

[1084] Murphin Kumar PS, MubarakAli D, Saratale RG, *et al.* Synthesis of nano-cuboidal gold particles for effective antimicrobial property against clinical human pathogens. Microb Pathog 2017; 113: 68-73.
[http://dx.doi.org/10.1016/j.micpath.2017.10.032] [PMID: 29056495]

[1085]Nies F, Wörner S, Wunsch N, *et al.* Characterization of Phormidium lacuna strains from the North Sea and the Mediterranean Sea for biotechnological applications. Process Biochem 2017; 59: 194-206.
[http://dx.doi.org/10.1016/j.procbio.2017.05.015]

[1086]Bauersachs T, Compaoré J, Hopmans EC, Stal LJ, Schouten S, Sinninghe Damsté JS. Distribution of heterocyst glycolipids in cyanobacteria. Phytochemistry 2009; 70(17-18): 2034-9.
[http://dx.doi.org/10.1016/j.phytochem.2009.08.014] [PMID: 19772975]

[1087]García ME, Aboal M. Environmental gradients and macroalgae in Mediterranean marshes: the case of Pego-Oliva marsh (East Iberian Peninsula). Sci Total Environ 2014; 475: 216-24.
[http://dx.doi.org/10.1016/j.scitotenv.2013.10.014] [PMID: 24238950]

[1088]Gugger M, Lenoir S, Berger C, *et al.* First report in a river in France of the benthic cyanobacterium Phormidium favosum producing anatoxin-a associated with dog neurotoxicosis. Toxicon 2005; 45(7): 919-28.
[http://dx.doi.org/10.1016/j.toxicon.2005.02.031] [PMID: 15904687]

[1089]Gurbuz F, Metcalf JS, Karahan AG, Codd GA. Analysis of dissolved microcystins in surface water samples from Kovada Lake, Turkey. Sci Total Environ 2009; 407(13): 4038-46.
[http://dx.doi.org/10.1016/j.scitotenv.2009.02.039] [PMID: 19395066]

[1090]Lilleheil G, Andersen RA, Skulberg OM, Alexander J. Effects of a homoanatoxin-a-containing extract from oscillatoria formosa (cyanophyceae/cyanobacteria) on neuromuscular transmission. Toxicon 1997; 35(8): 1275-89.
[http://dx.doi.org/10.1016/S0041-0101(97)00013-5] [PMID: 9278976]

[1091]Puddick J, van Ginkel R, Page CD, *et al.* Acute toxicity of dihydroanatoxin-a from Microcoleus autumnalis in comparison to anatoxin-a. Chemosphere 2021; 263: 127937.
[http://dx.doi.org/10.1016/j.chemosphere.2020.127937] [PMID: 32828056]

[1092]Catherine Q, Susanna W, Isidora ES, Mark H, Aurélie V, Jean-François H. A review of current knowledge on toxic benthic freshwater cyanobacteria – Ecology, toxin production and risk management. Water Res 2013; 47(15): 5464-79.
[http://dx.doi.org/10.1016/j.watres.2013.06.042] [PMID: 23891539]

[1093]Teneva I, Dzhambazov B, Koleva L, Mladenov R, Schirmer K. Toxic potential of five freshwater Phormidium species (Cyanoprokaryota). Toxicon 2005; 45(6): 711-25.
[http://dx.doi.org/10.1016/j.toxicon.2005.01.018] [PMID: 15804520]

[1094]Žegura B, Štraser A, Filipič M. Genotoxicity and potential carcinogenicity of cyanobacterial toxins – a review. Mutat Res Rev Mutat Res 2011; 727(1-2): 16-41.
[http://dx.doi.org/10.1016/j.mrrev.2011.01.002] [PMID: 21277993]

[1095]Martins MD, Branco LHZ, Werner VR. Cyanobacteria from coastal lagoons in southern Brazil: non-heterocytous filamentous organisms. Braz J Bot 2012; 35(4): 325-38.
[http://dx.doi.org/10.1590/S0100-84042012000400006]

[1096]Meunier CF, Dandoy P, Su BL. Encapsulation of cells within silica matrixes: Towards a new advance in the conception of living hybrid materials. J Colloid Interface Sci 2010; 342(2): 211-24.
[http://dx.doi.org/10.1016/j.jcis.2009.10.050] [PMID: 19944428]

[1097]Phoenix VR, Adams DG, Konhauser KO. Cyanobacterial viability during hydrothermal biomineralisation. Chem Geol 2000; 169(3-4): 329-38.
[http://dx.doi.org/10.1016/S0009-2541(00)00212-6]

[1098]Keshari N, Adhikary SP. Diversity of cyanobacteria on stone monuments and building facades of India and their phylogenetic analysis. Int Biodeterior Biodegradation 2014; 90: 45-51.
[http://dx.doi.org/10.1016/j.ibiod.2014.01.014]

CHAPTER 12

Lipid Production in Microalgae

Leyla Uslu[1,*] and **Oya Işık**[1]

[1] *Marine Biology Department, Fisheries Faculty, Cukurova University, 01330 Adana, Turkey*

Abstract: Microalgae, which are considered to be the living group that uses water and solar energy most effectively, have attracted the attention of researchers. Researches on efficient production of microalgae from starter cultures in the laboratory environment to outdoor ponds and photobioreactors continue in many countries. In addition to the known microalgae species, studies to search for new microalgae species that are rich in nutrient content and ease of production are ongoing. Of course, the researches also include more economical algae production studies. Algae and algae produced as larval food in aquaculture can be used as food support, as well as in the production of nutraceuticals and pharmaceuticals, like food coloring, soil fertilizer, and plant diseases. Algal oils and algal biomass have been gaining interest in renewable energy sources, especially in recent years, and studies in this area continue. The implementation of all these issues is based on well-known algal physiology and the realization of successful algae cultures.

Keywords: Algae Biomass, Biodiesel, Lipid Production, Microalgal Biotechnology.

INTRODUCTION

Energy is becoming one of the most expensive production inputs nowadays. Energy reserves are starting to run out and their polluting effects have been seen around the world. Therefore, there is an urgent need for renewable energies instead of fossil fuels. One of these energy sources is algae biomass, which is seen as promising for biofuel production.

Studies in microalgal biotechnology continue with modern biotechnology studies as well as traditional biotechnology studies. While genetic engineering researches are being carried out in microalgal lipid production, studies on the physiology of algae are also continuing. It is aimed to increase lipid production by creating

* **Correspondence author Leyla Uslu:** Marine Biology Department, Fisheries Faculty, Cukurova University, 01330, Adana, Turkey; Tel: +905064567826; Fax: +90 3223386439; E-mail: oyaisik@cu.edu.tr

Hüseyin Karaca and Cemil Koyunoğlu (Eds.)

stress in algae cells with different stress sources. Generally, with nitrogen deficiency in the environment, lipid enhancement studies are carried out in algal biomass.

Biodiesel can be produced from oils or lipids obtained from a variety of sources. Triglycerides, the major form of dietary lipid in fats and oils, are considered to be the main component essential for the production of biodiesel. The commonly used sources of fats for the production of biodiesel include pure vegetable oil, animal fats, and waste cooking oils. However, recently microalgae have been considered to be a potential source for biodiesel production because they can be harvested daily. Microalgae can multiply rapidly by dividing and can often grow over a wide temperature range and the oil content they produce can be very high (up to 80% of the dry weight) when compared to other traditional feedstocks.

Algae can be cultured in non-agricultural land, with high photosynthetic activity, harvested throughout the year with high biomass production. High lipid from algae is possible by reducing some elements of growth conditions from the nutrient medium. It is known that different nutrient sources and concentrations affect the growth and physiology of microalgae cells [1]. When nitrogen in the environment is reduced, lipid content increases while the amount of biomass decreases [2]. Production of protein is favored during periods of nitrogen sufficiency with limited carbohydrate synthesis; carbohydrates accumulate and protein production decreases whereas lipids usually increase [3, 4].

Our aim in our studies on microalgal lipids for more than ten years has been to carry out research to provide biomass with high lipid content for biodiesel that countries will need in the near or distant future.

We first carried out our studies to increase the microalgal lipid content in the laboratory conditions, then we tried the best results we achieved in outdoor photobioreactors and ponds.

The effect of nitrogen deficiency on the growth and lipid content of microalga *Isochrysis affinis galbana* in two photobioreactors was studied at the algal biotechnology outdoor unit. In this study, *Isochrysis affinis galbana* were cultured in two reactors: flat panel photobioreactors with different light paths (1, 3, 5, 7, and 10 cm) and tubular photobioreactors, with 50% nitrogen reduction and 20% inoculation densities. Biomass, lipid, and protein ratios were determined. The highest lipid content of 33.13% was obtained from *I. aff. galbana* with 12.11% protein in flat panel photobioreactors with 50% nitrogen reduction and 10 cm light path, and a 0.991 gL^{-1} biomass rate was obtained (Figs. **1** and **2**). The highest optical density was found in the 10 cm light path flat panel photobioreactor with a 50% nitrogen reduction [5].

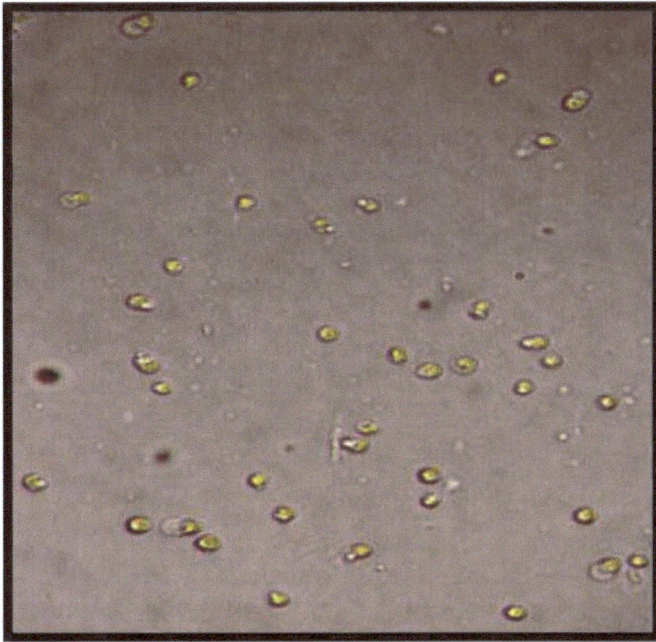

Fig. (1). Microalga *Isochrysis affinis galbana* (x40).

Fig. (2). *Isochrysis aff. galbana* cultures in tubular and flat-panel photobioreactors.

In the other study, we carried out, the effect of nitrogen limitation on the cell density, biomass, chlorophyll *a*, total carotene, protein, and lipid content of microalga *Phaeodactylum tricornutum* (Bohlin), Bacillariophyceae, cultured in photobioreactors outdoor was investigated. *Phaeodactylum tricornutum* (Fig. **3**) was cultivated in an appropriate medium as the control group, at the same time, it was cultured in a medium in which nitrogen was reduced to 50%. At the end of the study, it was determined that 35.04% lipid with 0.980±0.02 gL^{-1} biomass and

8.87% protein content, meanwhile, in the control group, 16.93% lipid, 1.036 ± 0.025 gL^{-1} biomass, and 31.05% protein were determined. As a result, it was observed that while nitrogen limitation increased the lipid amount, decreased the biomass and chlorophyll-a, total carotene and protein of the cell [6].

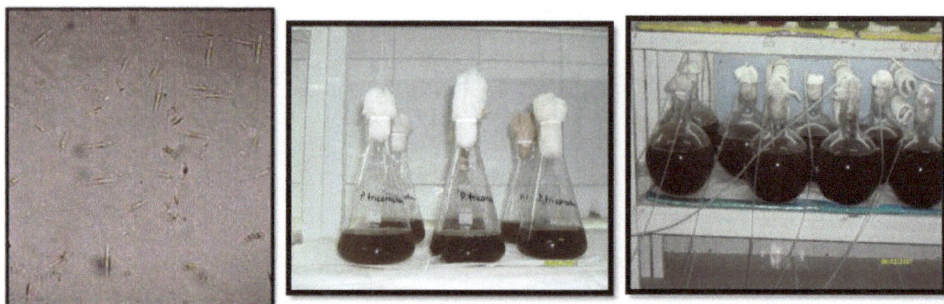

Fig. (3). Microalga *Phaeodactylum tricornutum* (x40) cultures.

In another work titled nitrogen limitation increases lipid content of microalga *Chlorella vulgaris* at the photobioreactor system (Fig. **4**). This study aimed to produce *C. vulgaris*, which is easy to culture with limited cost and whose outside culture proved practical by increasing its lipid content. *Chlorella vulgaris* was cultured at a photobioreactor in a non-nitrogen nutritive medium with different flow rates (0.3, 0.5, and 0.6 m sec^{-1}) and inoculation densities of 20 and 50%. The highest lipid content in the study was obtained from the culture with a biomass rate of $38.16\pm0.8\%$ and 0.409 ± 0.01 gL^{-1}, inoculation density of 50%, and a flow rate of 0.6 m sec^{-1}. The lowest lipid content was found as $21.34\pm0.5\%$ in the culture with an inoculation density of 50% and a flow rate of 0.3 m sec^{-1}. The highest protein content was found as $21.73\pm0.1\%$ in the culture with an inoculation density of 50% and a flow rate of 0.5 m sec^{-1}. The lowest biomass content was found as 0.035 ± 0.007 g L^{-1} in the culture with an inoculation density of 20% and a flow rate of 0.3 m sec^{-1}. The lipid content in that culture was found as $29.69\pm1\%$ and the protein content as $14\pm0.5\%$. These results suggest that N limitation increases *C. vulgaris* lipid content [7].

In the study carried out, *Phaeodactylum tricornutum* was cultured in a panel PBR system, with the different light paths (1, 3, 5, 7, and 10 cm), 50% nitrogen reduction, and with 20% inoculation densities. Lipid, protein, and biomass rates were determined. The highest lipid content (34.6%) for *P. tricornutum* in the study was obtained from the culture with a 1.064gL^{-1} biomass rate, 8.50% protein in the panel PBR system with 50% nitrogen reduction, and, 7 cm light path. The lowest lipid content and highest protein were found at 29.76%, 10.12% respectively, at a 1 cm light path PBR system with 50% nitrogen reduction. The highest OD and lowest *Chlorealla* were found in the 7 cm light path PBR system with a 50% nitrogen reduction [8].

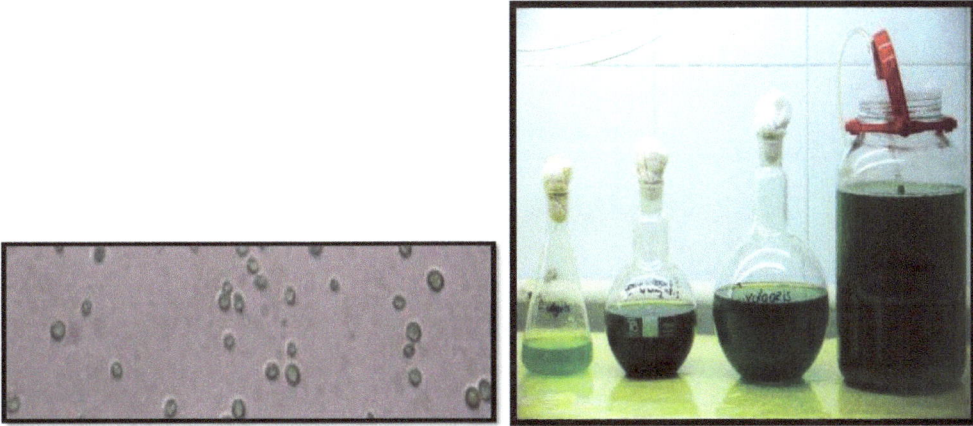

Fig. (4). Microalga *Chlorella vulgaris* (x40) cultures.

The effects of the changes in the composition of nutrients on the lipid content of microalgae were investigated. In this study, the effects of nitrogen deficit on the growth rates, lipid and *chlorophyll-a* content of the microalgae species of *Isochrysis affinis galbana* (Prymnesiophyceae), *Phaeodactylum tricornutum* (Bacillariophyceae), and *Porphyridium cruentum* (Rhodellophyceae) were examined (Fig. **5**). It was observed that nitrogen limitation increased the lipid content of *I. affinis galbana* and *P. tricornutum*. In the study subjected to 50% of nitrogen reduction, 30.91% lipid and 0.755±0.03 gL^{-1} dry matter were detected in *I. affinis galbana*, and 30.18% lipid and 0.978±0.02 gL^{-1} dry matter in the *P. tricornutum*. In the study in which nitrogen starvation (100% of nitrogen reduction) was applied to *P. cruentum*, the lipid content (11.23%) was lower than the other two species [9].

The effect of different nitrogen ratios on *Spirulina platensis* growth, protein, and fat content was studied in laboratory conditions (Fig. **6**). At the end of the study, 5.78% lipids were obtained with 1.12 gL^{-1} biomass in the control group, while lipid 17.05% and 1 gL^{-1} dry *Spirulina* were found in 100% N reduction in the nutrient medium [10].

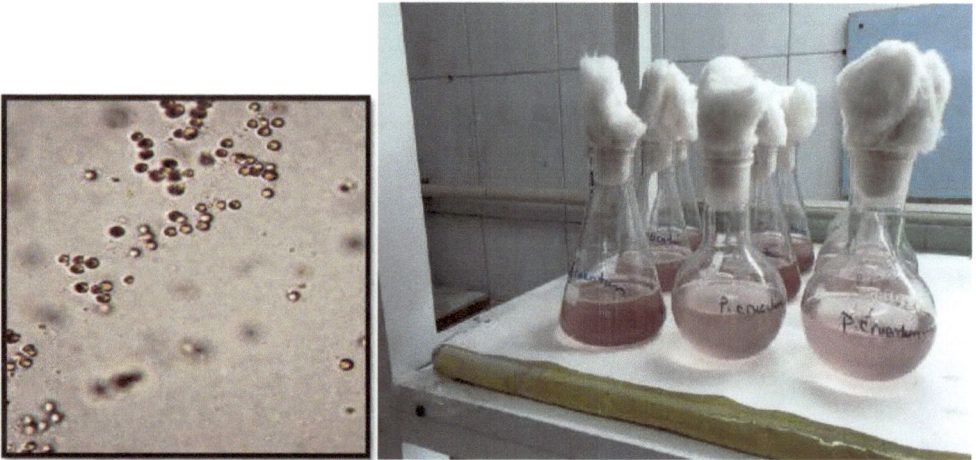

Fig. (5). Microalgae *Porphyridium cruentum* (x40) cultures.

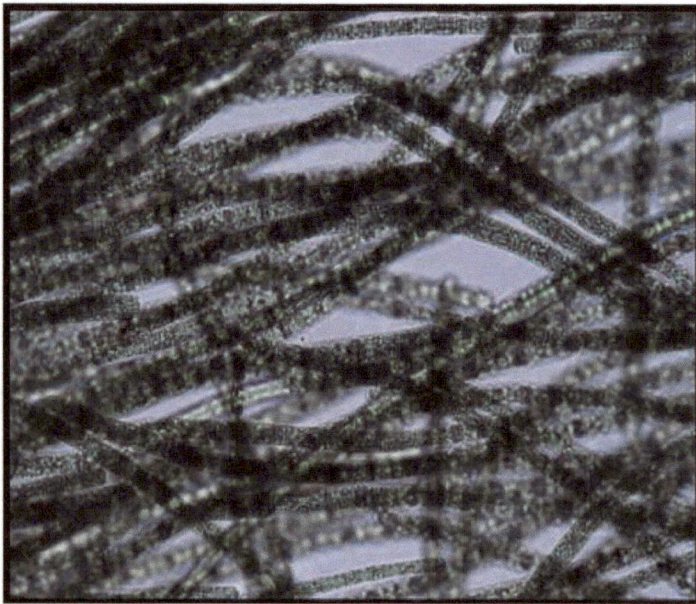

Fig. (6). Cyanobacteria *Spirulina platensis* (x40).

In another study with *Spirulina platensis*, conducted in ponds and outdoors, the effect of nitrogen deficiency (100% N (-)) and high salinity on the lipid amount of *Spirulina platensis* was investigated. The highest lipid levels were determined as 21.98% in winter in cultures without nitrogen added, while the lowest lipid amount in the control group was 4.5% in summer [11].

The effect of 50% N, 100% N, 50% N plus 50% P and 50% P deficiencies and nitrite addition were treated on *Chlorella vulgaris* (Chlorophyceae) and was studied in laboratory conditions to determine the effects of the deficient nutrient and different nitrogen sources on lipid and protein contents. Protein and lipid values of the biomass were found as 50.8 and 12.29% for the control group, 20.3 and 17.5% for 50% N(-), 13.01 and 35.6% for 100% N(-), 21.37 and 20.5% for 50% N(-) and 50% P(-), 38.16 and 16.7% for 50% P(-) and 41.03 and 13.04% for the nitrite group that was added. The highest lipid content was recorded with the culture in which 100% N(-) was treated with 0.18 g/L dry-weight (Fig. **7**) [12].

Fig. (7). Lipid analysis in microalgae.

To determine the effects of 50% and 100% deficient nitrogen on cell growth, lipid, and fatty acid composition under laboratory conditions, the effects of nitrogen deficiencies were investigated in cyanobacteria *Spirulina platensis* (Cyanophyceae). In this study carried out, it was determined that the stress of N deficiency increased the lipid content by 5.43%, 13.14% and 16.85% lipids were recorded for the groups of control, 50% N(-) and 100% N(-), respectively. The highest lipid content and biomass (1.05 gL^{-1}) were recorded from the culture treated with 100% N(-). The fatty acid composition of the *Spirulina platensis* cultured shows that (in order of abundance) palmitic, linolenic and linoleic acids were most prevalent. Nitrogen deficiencies increased the total lipid content while reducing the fatty acid content as a percentage of biomass. On the other hand, the composition of the fatty acids was unaffected, but *Spirulina platensis* showed an increase in C14:1 and C18:2 under nitrogen deficit conditions [13].

A more economical production can be made by supplying the carbon dioxide needed by algae from industrial flue gases and reducing the input cost of the nutrient element.

As a result of our research, lipids between 17 and 38% with dry matter amounts varying from 0.5 gram to 1 gram per liter were obtained from different microalgae species per day.

According to our results, 1 kg of oil is obtained from 3 kg of algae and 7.5 tons of liquid culture is required for 3 kg of algae.

Studies to discover new algae species with high lipid content, increased lipid production in algae, and, of course, reduced production costs with them will continue.

Another essential feature is the fatty acid distribution of algae oil. The fuel properties of biodiesel vary greatly depending on the fatty acid distribution [2].

The suitability of microalgal biomass as a biofuel feedstock is closely related to the length and degree of saturation of its fatty acids as specified by the four key figures iodine.

CONCLUSION

Microalgal lipid production is within the scope of algal biotechnology studies. Some algae species under stress can increase their cellular lipid content to the extent allowed by their genetic structures. For this purpose, stress factors such as nitrogen deficiency, high salinity, and high light intensity are generally applied. These studies are conducted by influencing the physiology of algal cells. However, in recent years, studies have been carried out to produce high levels of oil in fermenters with heterotrophic species. For example, the biomass and lipid concentrations obtained with the microalgae *Chlorella protothecoides* cultured with heterotrophic feeding using crude glycerol as the carbon source were 23.5 and 14.6 g/l for 6 days, respectively [14]. The important thing here is to determine and choose the most economical production model and technology. Could it be a more economical production model to focus on systems that use solar energy, which is a free energy source? Studies on these issues are ongoing.

CONSENT FOR PUBLICATION

Not applicable.

CONFLICT OF INTEREST

The authors declare no conflict of interest, financial or otherwise.

ACKNOWLEDGEMENT

Declared none.

REFERENCES

[1]   Fidalgo JP, Cid A, Abalde J, Herrero C. Culture of the marine diatom *Phaeodactylum tricornutum* with different nitrogen sources: Growth, nutrient conversion and biochemical composition. Cah Biol Mar 1995; 36: 165-73.

[2]   Xin L, Hong-ying H, Jia Y. Lipid accumulation and nutrient removal properties of a newly isolated freshwater microalga, *Scenedesmus* sp. LX1, growing in secondary effluent. N Biotechnol 2010; 27(1): 59-63.
[http://dx.doi.org/10.1016/j.nbt.2009.11.006] [PMID: 19969113]

[3]   Lourenço SO, Barbarino E, Marquez UML, Aidar E. Distribution of intracellular nitrogen in marine microalgae: Basis for the calculation of specific nitrogen☐to☐protein conversion factors. J Phycol 1998; 34(5): 798-811.
[http://dx.doi.org/10.1046/j.1529-8817.1998.340798.x]

[4]   Shifrin NS, Chisholm SW. Phytoplankton lipids: Interspecific differences and effects of nitrate, silicate and light–dark cycles. J Phycol 1981; 17(4): 374-84.
[http://dx.doi.org/10.1111/j.1529-8817.1981.tb00865.x]

[5]   Uslu L, Işık O, Ak Çimen B. The effect of nitrogen deficiency on the growth and lipid content of *Isochrysis affinis galbana* in two photobioreactor systems (PBR): Tubular and Flat Panel. J Agric Sci 2020; 26(3): 282-9.

[6]   Ak B, Işık O, Uslu L, Azgin C. The effect of stress due to nitrogen limitation on lipid content of *Phaeodactylum tricornutum* (bohlin) cultured outdoor in photobioreactor. Turk J Fish Aquat Sci 2015; 15(3): 647-52.
[http://dx.doi.org/10.4194/1303-2712-v15_3_09]

[7]   Uslu L, Durmaz Y, Işik O, Mutlu Y, Koç K, Ak B. Nitrogen limitation increases lipid content of *Chlorella vulgaris* at photobioreactor system. J Anim Vet Adv 2013; 12(1): 52-7.

[8]   Uslu L, Ak B, Işik O, Durmaz Y. Effect of light path length and nitrogen deficiency on the biochemical composition of *Phaeodactylum tricornutum*. Fresenius Environ Bull 2014; 23(6): 1309-13.

[9]   Uslu L, Işik O, Mutlu Y. Besleyici element kompozisyonundaki değişikliklerin mikroalglerde lipid içeriğine etkisi. J FisheriesSciences Com 2012; 6(3): 176-83.
[http://dx.doi.org/10.3153/jfscom.2012021]

[10]   Uslu L, Işik O, Koç K, Göksan T. The effects of nitrogen deficiencies on the lipid and protein contents of *Spirulina platensis*. Afr J Biotechnol 2011; 10(3): 386-9.

[11]   Işık O, Ak B, Uslu L, Uslu M. Azot Eksikliği ve Yüksek Tuzluluğun Spirulina platensis Lipid Miktarına Etkisi. 5. Doğu Anadolu Sempozyumu Su Ürünleri Sempozyumu 2014; 1: 364-5.

[12]   Mutlu YB, Isık O, Uslu L, Koç K, Durmaz Y. The effects of nitrogen and phosphorus deficiencies and nitrite addition on the lipid content of *Chlorella vulgaris* (Chlorophyceae). Afr J Biotechnol 2011; 10(3): 453-6.

[13]   AAk Çimen B, Işık O, Uslu L. The effects of nitrogen deficiencies on the growth, lipid and fatty acid composition of *Spirulina platensis*. In: Atik A, Ed. Innovative Approaches In Agriculture, Forestry And Aquaculture Sciences 2018; 137-50.

[14]   Chen YH, Walker TH. Biomass and lipid production of heterotrophic microalgae *Chlorella protothecoides* by using biodiesel-derived crude glycerol. Biotechnol Lett 2011; 33(10): 1973-83.
[http://dx.doi.org/10.1007/s10529-011-0672-y] [PMID: 21691839]

CHAPTER 13

The Science of Catalysts in Algae Oil Production

Cemil Koyunoğlu[1,2,*] and **Hüseyin Karaca**[3]

[1] *Energy Systems Engineering Department, Engineering Faculty, Yalova University, Yalova, Turkey*

[2] *Central Research Laboratory, Fuel-oil analysis laboratory, Yalova University, Yalova, Turkey*

[3] *Chemical Engineering Department, Engineering Faculty, Inonu University, Malatya, Turkey*

Abstract: CO_2, which is a gas produced as a result of combustion, has an important share in greenhouse gases such as SO_x, CO, and NO_x and causes global warming. As biodiesel produced from algae converts CO_2 within the biological carbon cycle, it accelerates the carbon cycle, as a result of which the greenhouse gas effect decreases. C16-C18 methyl esters in biodiesel whose degradability feature resembles dextrose (sugar) do not show any negative microbiological affect up to 10000 mg/L, they decompose rapidly in nature. 40% of diesel in water and 95% of biodiesel in 28 days can be degraded. In the production of biodiesel, catalysts that break the triglyceride bonds allow the esters to become free.

Keywords: Biocarbon cycle, Biodiesel production, Catalyst types, Greenhouse gases.

HOMOGEN CATALYSTS

Catalyst and reactant are present in the same phase in homogeneous reactions. They are in the form of base, acid, or liquid enzyme solutions. Generally, HCl and sulfuric acid (H_2SO_4) are used as homogeneous catalysts in the acidic transesterification reaction. Potassium hydroxide (KOH) and sodium hydroxide (NaOH) are used in the alkali transesterification reaction. Lipase enzyme in liquid form used in the transesterification reaction is a homogeneous enzyme catalyst. The temperature is high in the presence of acidic catalysts. The water in the environment negatively affects the reaction. The catalyst in the same homogeneous environment with glycerin is separated together, it cannot be recovered, and the water in the environment affects the reaction negatively. The temperature is high in the presence of acidic catalysts. The environment is strongly acidic because the equipment and materials used are resistant to pressure

---

* **Correspondence author Cemil Koyunoğlu:** Energy Systems Engineering Department, Faculty of Engineering, Yalova University, Yalova, Turkey; Tel: +902268155378; E-mail: cemil.koyunoglu@yalova.edu.tr

and corrosion and must be made of expensive materials. Cost increases and creates danger for the user. A base catalyst is 4000 times faster than an acidic catalyst. Alkaline catalysts are less corrosive than acids. Free fatty acid and moisture content are the main criteria affecting basic transesterification. With the hydrolysis of oils, fatty acids become free in the presence of water. However, under these conditions, when esterification is not performed, free fatty acids cause the formation of soap by depleting the basic catalyst by the reaction. Free fatty acid content should be less than 4% in a basic catalyzed reaction. Homogeneous enzyme catalysts in liquid lipase form are used in transesterification reactions. Aqueous enzyme solutions with stabilizers such as benzoate prevent microbial growth such as sorbitol and glycerol. Homogeneous catalysts take place in more suitable reaction environments. Their selectivity is high. Some of the disadvantages of homogeneous catalysts can be overcome by using a heterogeneous catalyst. There are many studies on the algae to oil process using various homogenous catalysts [1 - 10]. Various basic homogeneous catalyst reactions are summarized in Figs. (**1** - **7**).

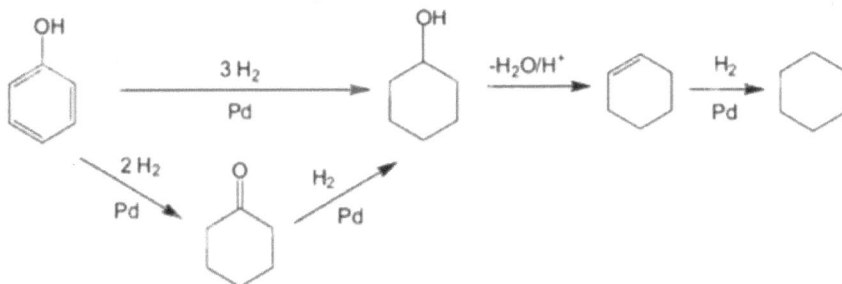

Fig. (1). Phenol hydrodeoxygenation over Pd/C catalyst [8].

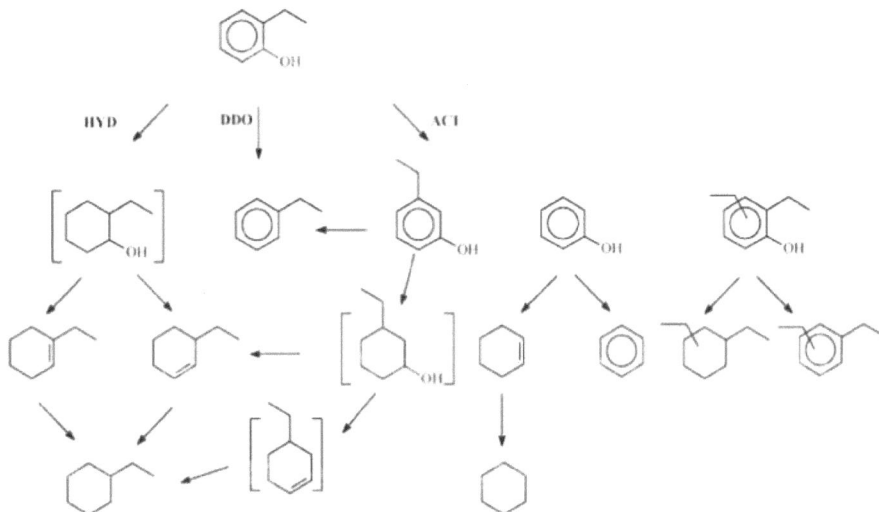

Fig. (2). 2-ethylphenol transformation over sulfided Mo-based catalysts [8].

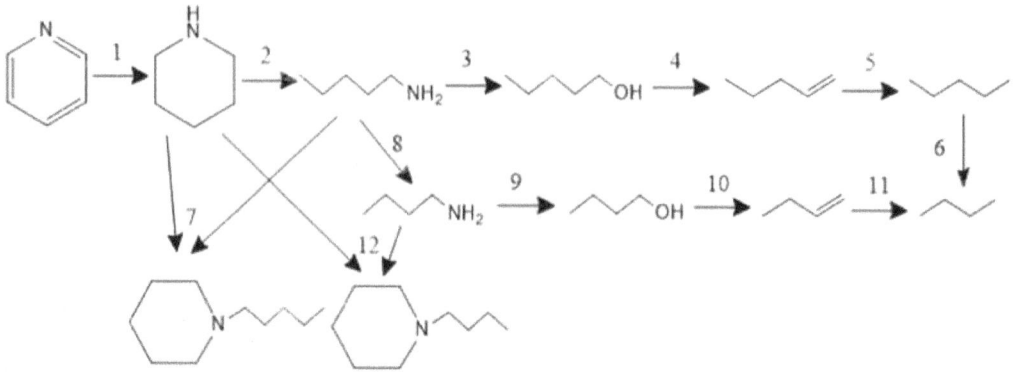

Fig. (3). Pyridine hydrodenitrogenation mechanism [8].

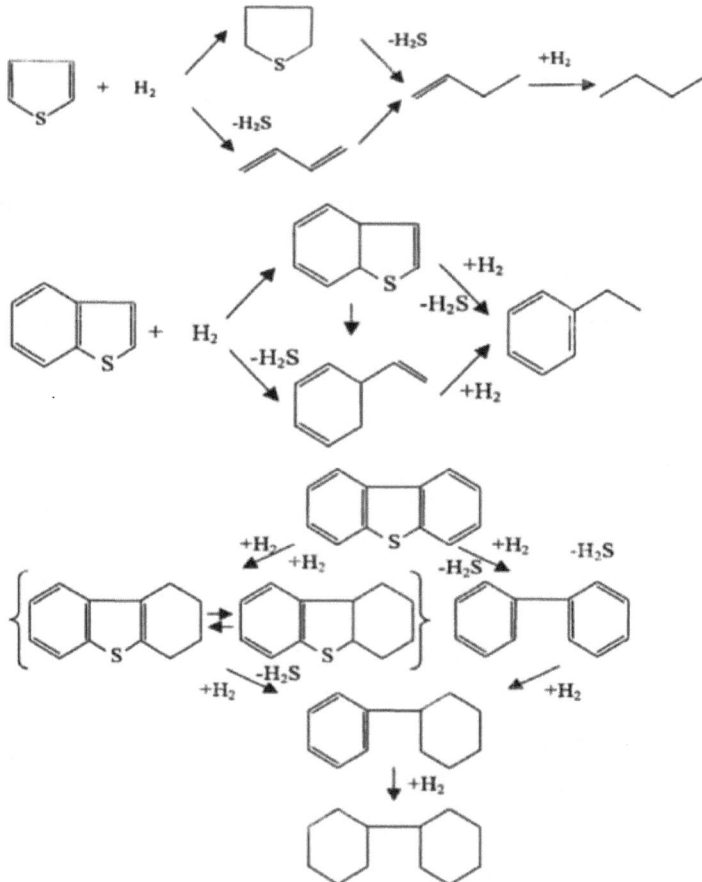

Fig. (4). Benzothiophene, thiophene, and dibenzothiophene desulfurization [8].

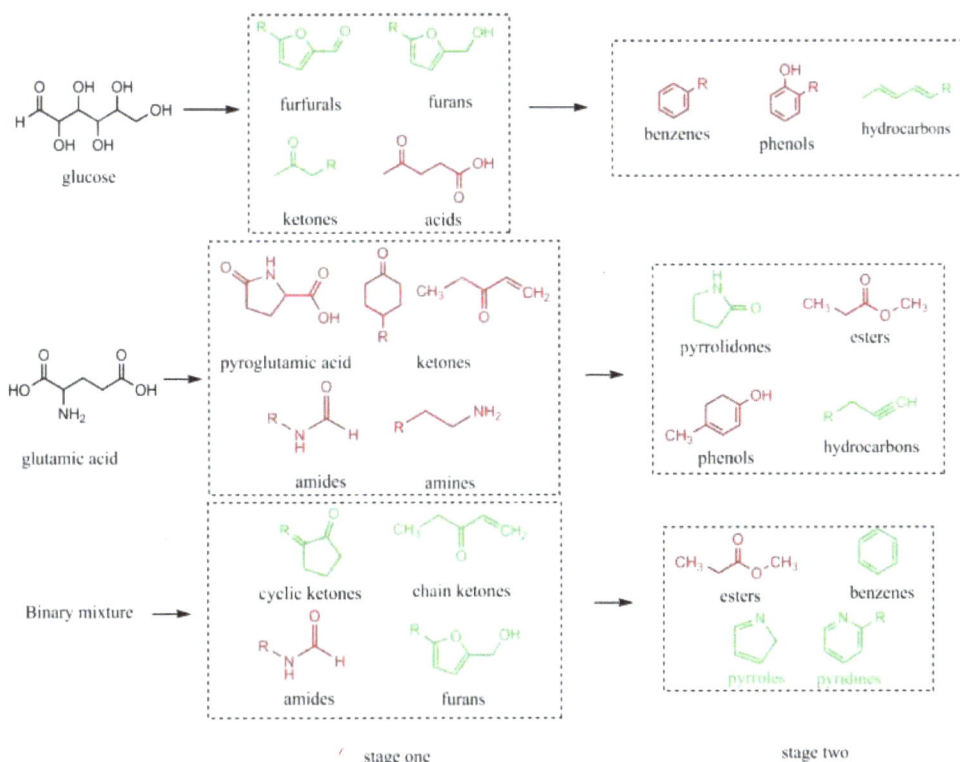

Fig. (5). Major reactions of glutamic acid, glucose, and their binary mixtures with HZSM-5 catalyst [10].

HETEROGEN CATALYSTS

Catalysts that are not in the same phase as reactants are called heterogeneous catalysts. The use of catalyst supports the reaction medium by providing a special porosity and surface area that allows the active groups to react with large triglyceride molecules, preventing mass transfer restriction. Supplements that increase catalyst activity by themselves do not have catalyzing properties.

Heterogeneous catalysts can be separated from the reaction mixture and reused in biodiesel production. Examples of heterogeneous catalysts are CaO, KOH/Al$_2$O$_3$ and ZnO-La$_2$O$_3$Mg/La. Heterogeneous catalysts can be in basic, acidic, or enzymatic forms [3, 11 - 19].

As observed in Fig. (9), when Devaraj *et al.* [20] increased the concentration of the heterogeneous catalyst they used in the biodiesel production experiment they carried out in an experimental setup seen in Fig. (8), they observed that the non-transformed triglyceride at the bottom of the tube decreased.

Step 1

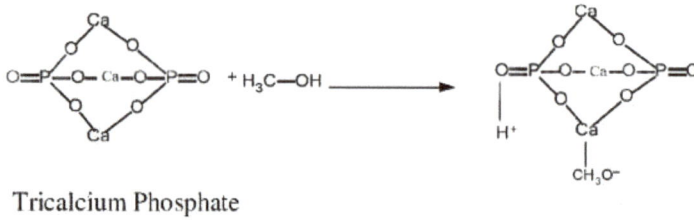

Tricalcium Phosphate

Step 2

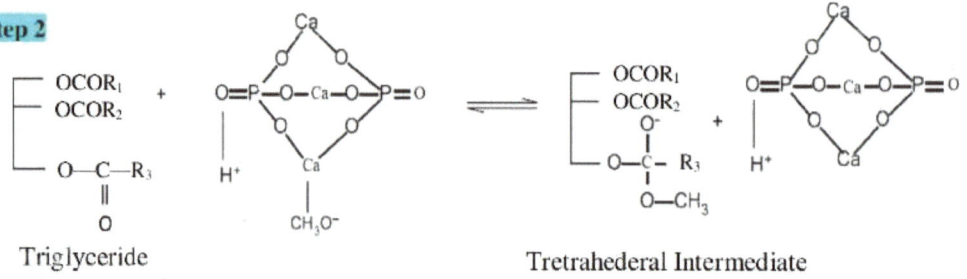

Triglyceride   Tretrahederal Intermediate

Step 3

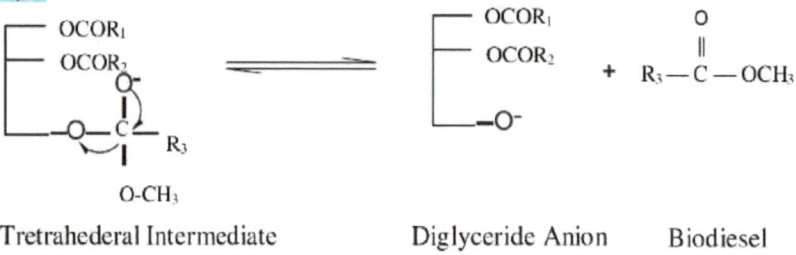

Tretrahederal Intermediate   Diglyceride Anion   Biodiesel

Step 4

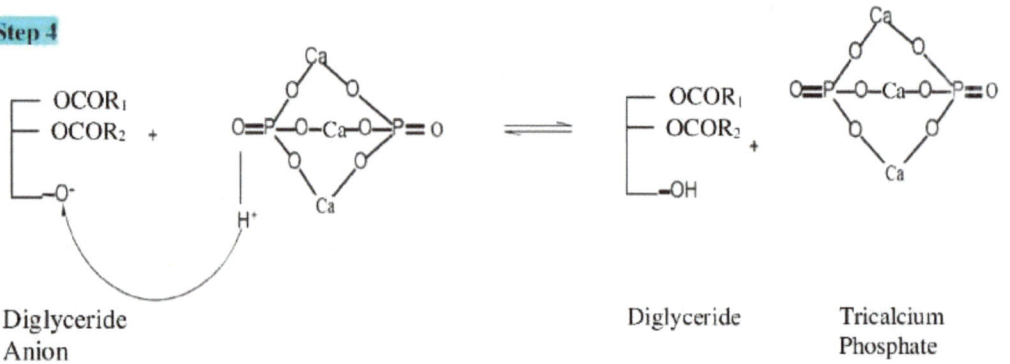

Diglyceride   Diglyceride   Tricalcium
Anion   Phosphate

Fig. (6). CaO catalyst reaction mechanism [7].

Step 1

Calcium oxide

Step 2

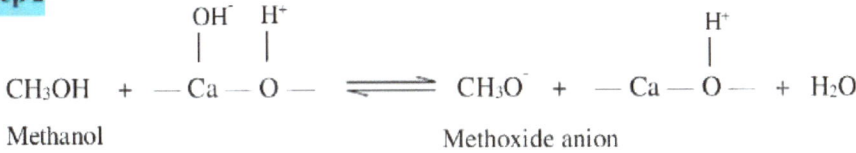

Methanol Methoxide anion

Step 3

Methoxide anion

Step 4

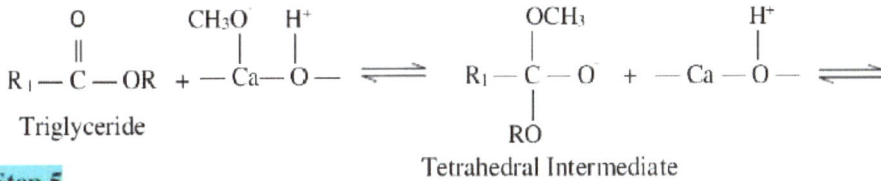

Triglyceride Tetrahedral Intermediate

Step 5

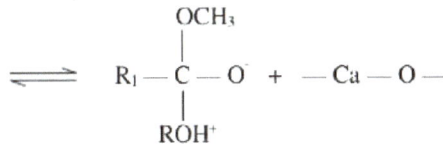

Step 6 **Step 7**

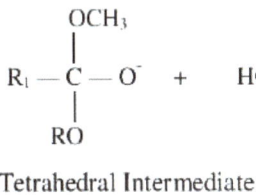 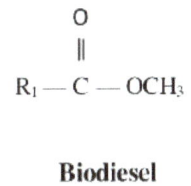

Tetrahedral Intermediate

Biodiesel

Fig. (7). β-tricalcium phosphate catalyst reaction mechanism [7].

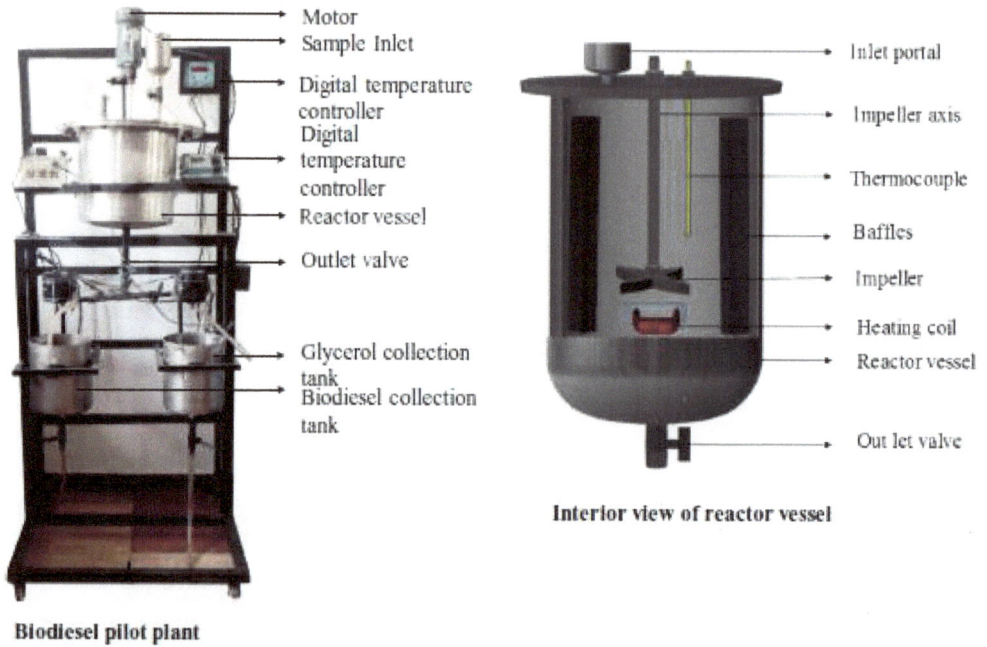

Fig. (8). Left; biodiesel production pilot plant facility, right; bioreactor inside view [20].

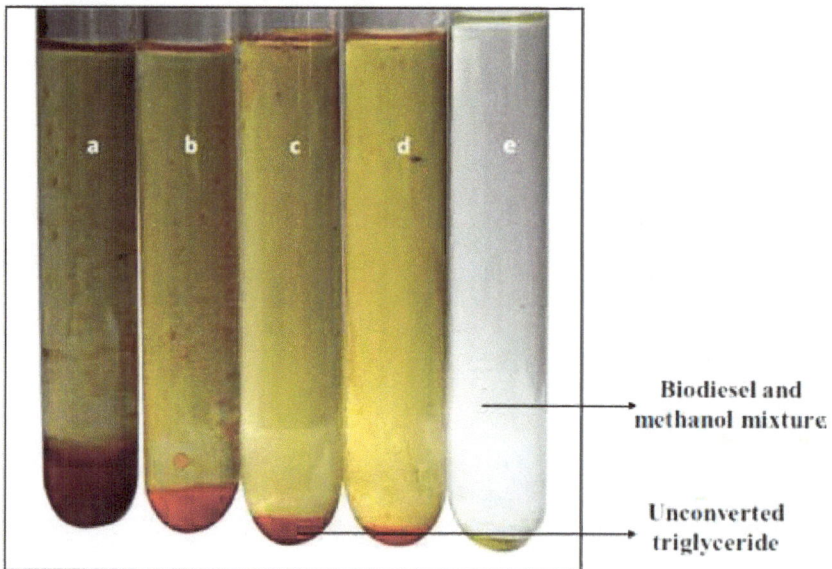

Fig. (9). a) Catalyst 1 wt%, **b)** Catalyst 2 wt%, **c)** Catalyst 3 wt%, **d)** Catalyst 1 wt%, **e)** Catalyst 5 wt%, used with 1:04 oil to methanol molar ratio [20].

Guo *et al.*, after preparing the slurry in Fig. (**10**) with the nickel catalyst they used in *Nannochloropsis gaditana* and *Chlorella vulgaris* liquefaction, performed a study with the catalysts in Fig. (**11**) [21].

Fig. (10). Microalgae strains (a, b) and algae slurry preparation [21].

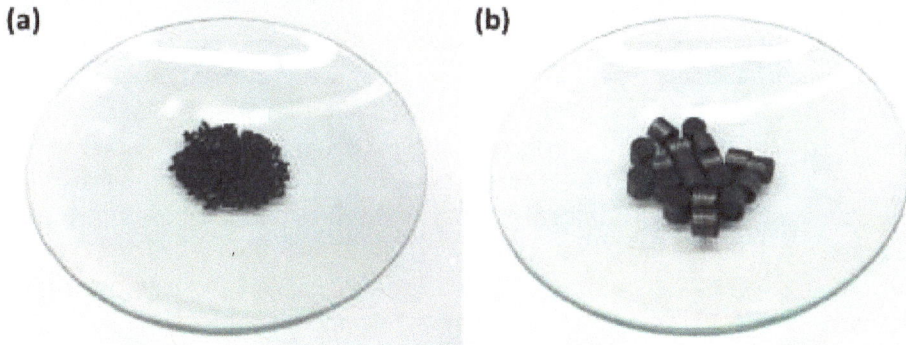

Fig. (11). a; NiMo/Al$_2$O$_3$ and b; NiW/Al$_2$O$_3$ catalysts [21].

Heterogeneous catalysts used in enzymatic biodiesel production are arrested enzymes. Without losing the catalytic activity of the enzyme or microorganism, porous or non-porous polymeric matrices are captured or bonded onto a carrier material such as inorganic matrices and gels by various methods. The use of captured lipases in the production of biodiesel provides reusability, stability, better control of the reaction, continuous operation, high product conversion, and high purity [22 - 25].

Lipase Catalysts

Enzymes that catalyze the reversible hydrolysis of lipids in living organisms are called natural lipase enzymes. Lipase enzymes, whose natural substrate is oil, have been preferred as catalysts in biodiesel production in recent years. Lipases belong to the triacylglycerol hydrolases group of the class of hydrolases. Lipases hydrolyze the ester bonds at the interface between the aqueous phase containing the enzyme and the organic phase containing the lipid. Bile salts, whose lipids are surfactants in the emulsion, facilitate hydrolysis. Lipase enzymes are obtained from bacterial cells such as, *Pseudomonas glumae*, *Rhizopus arrhizus*, and *Pseudomonas aeruginosa*. Lipase enzymes can also be obtained from yeast cells such as *Candida cylindracea*, *Candida antarctica*, *Geotrichum candidum*, *Saccharomycopsis lipolytica*, and *Trishosporon fermentas*.

Lipases break down the bonds between glycerides and fatty acids randomly. *Corynebacterium acnes*, *Candida rugosa*, *Staphylococcus aureus*, *Chromobacterium* spp. Lipases obtained from microorganisms such as can be given as examples of some enzymes. Lipases with position selectivity cleave the ester bonds at the Sn-1,3 position, ie the outer. Organisms such as *Rhizopus arrhizus*, *Mucor miehei*, *Aspergillus niger*, and *Rhizopus delemar* are examples of this group. Some lipases are selective according to the chain length of the fatty acid. *Aspergillus delemar* and *Aspergillus niger* lipases show selectivity to short-chain fatty acids. Lipases can be obtained from microbial, vegetable, and animal sources. *Penicillium cyclopium* lipase breaks down long-chain fatty acids. Industrial use of lipases in biodiesel production is still very limited. Particularly, the catalytic performances of microbial lipases are investigated in purified form or directly in studies conducted as microorganisms. Microbial lipases are derived from fungi, yeast, and other microorganisms. The fact that they are more resistant to environmental conditions than plant and animal cells makes it possible for microorganisms to reproduce rapidly and to produce lipase in large quantities and economically. With genetic regulation, microorganisms can be made to produce more enzymes. Lipases carry out transesterification, esterification, and similar reactions that are the opposite of hydrolysis reactions in systems with very low water activity. First, in the transesterification reaction, a covalent bond is established between the serine residues in the active part of the enzyme and the enzyme-substrate complex and triglyceride. Diglyceride and acyl-enzyme are released. The acylated enzyme interacts with methanol to form glycerin and fatty acid alkyl ester and start enzymatic biodiesel production. Since the reaction medium requires room temperature, neutral pH, and pressure in lipase-catalyzed biodiesel production, pressure, excessive temperature, and related corrosion do not occur. Therefore, process equipment or designs can be made from cheaper and simpler materials. This makes it possible to produce economical and safer biodiesel in terms of equipment and people. Besides, other advantages of this method are that it can catalyze free fatty acid esterification as well as triglycerides, short reaction time, low product inhibition, reusability, and easy production of the enzyme [26 - 30].

ENZYME DETENTION METHODS

Attaching it to a carrier material by various methods or capturing the catalytic activities of the microorganism enzyme in a matrix is called the arresting process [31 - 40].

ADVANTAGES OF CAPTURING ENZYMES

-The enzyme's resistance to environmental conditions (organic reagents, temperature, pH) increases.

-Products can be obtained pure.

-Enzymes can be separated from the reaction medium by simple methods.

-Production costs are low.

-They can be applied to processes running continuously.

Product inhibition can be prevented by keeping product formation under control.

-They are more stable than the free enzyme.

-They can show a better activity than the free enzyme.

-Applicable for multi-step and consecutive reactions.

-Enzymes can be arrested on synthetic (such as nylon, styrene) support materials and natural materials (such as perlite, clay, alumina, cellulose, silica gel). The most important factors in the selection of support for industrial production are the chemical and mechanical resistance of the support, its chemical structure (hydrophilic/hydrophobic) microbial resistance, loading cost, and capacity.

-The factors affecting the detention differ according to the method of detention. Generally, the concentration, pH, buffer type, retention time and temperature, amount of support, enzyme amount and type, type of crosslinking agent, and precipitator concentrations can be listed. The methods used for enzyme capture are divided into chemical and physical methods [31 - 40].

ADSORPTION

It is a method of arresting enzymes by adsorption on support in water dissolution. The forces responsible for arrest in physical adsorption are hydrophobic interactions, van der Waals forces, and hydrogen bonds. Less damage to the enzyme protein is caused by the adsorption process compared to chemical bonding methods. However, since the forces between the carrier and the enzyme are weak due to the adsorbed enzyme support, it can be peeled off. Another disadvantage is that substances and other proteins can also be adsorbed on the support [41 - 45].

TRAPPING ENZYMES

This method is based on trapping enzymes between the pores of the solid support material. Accordingly, the enzyme is dissolved in the polymer or aqueous monomer solution. The aqueous monomer is trapped within the enzyme polymer. Enzymes cannot exit the cage, while products and substrate can enter and exit the cage. The enzyme arrested by this method is more stable than the enzyme arrested by physical adsorption [46 - 49].

ENCAPSULATION

In this method, enzymes are confined within semi-permeable membranes. The semi-permeable membrane prevents enzymes or large proteins from getting out of the capsule, allowing free flow of product and small substrates [49 - 58].

IONIC BONDING

It occurs by ionic binding of the enzyme to the support material (insoluble in water). Synthetic polymers and polysaccharides containing ion exchange residues are widely used. Ionic bonding can be achieved since the capture of the enzyme does not cause significant changes in its conformation and active center with high efficiency. As in covalent bonding, the enzyme's activity is high by binding on the carrier surface. The binding forces formed between the support and the enzyme are not strong in ionic bonding as well as in covalent bonding. The enzyme carrier can be separated from the surface at pH changes and substrate solutions with high ionic strength [59 - 62].

COVALENT BONDING

In covalent bonding, which is the most durable method of arrest, the surface of the support material is activated in the first stage and then the enzyme is bound to the activated support material. Functional groups used for this attachment, are carboxyl, amino, phenolic, hydroxyl, sulfhydryl, thiol, imidazole, indole, threonine, and groups. Covalent bonding prevents the enzyme from passing into the solution as the bond is strong and ensures that the enzymes are stable. However, by affecting the active site of the enzyme, it may cause a loss of activity and change the conformation of the enzyme. Besides, covalent bonding is more complex and expensive than other methods of arrest [63 - 67].

Covalent Binding of Lipase to Perlite Support

Perlite is an inorganic volcanic rock composed of light gray, pearl-luster, small, amorphous, round glassy grains and these particles contain 2-6% water. Perlites, which are in an aluminum silicate structure, containing 70% silica. In special furnaces, unexpanded perlite is milled and heated rapidly up to the softening point between 760°C-1100°C. As a result of the evaporation of the water in them, these particles can expand 10 to 20 times their original volume. After the expansion, a large number of white, small softened glassy particles are formed. Perlite is porous, chemically inert, lightweight, and has good mechanical and thermal stability. It is resistant to organic solvents, cheap, and non-toxic. Due to these technical features, it attracts the attention of researchers in terms of using perlite as an enzyme arresting support. Most simply, one of the ways to modify the surface properties of silica gels is to completely or partially dehydroxylate the surface. Based on this, with the organic silaning technique, the perlite containing 70% silica is captured by the enzyme capture process, in the first step the silanol groups (Si-OH) on the perlite surface are covalently bonded with an alkyl amine (3-aminopropyl triethoxy silane). In the second step, one of the -CHO groups in glutaraldehyde is attached to the alkylamine group. In this way, it can be said that the enzyme molecules will bind to glutaraldehyde with one free -CHO group with -NH2 groups. The covalent binding of lipase to perlite then occurs as an arrest reaction [68 - 75].

CROSS-LINKING

In the small molecule enzyme capture method with cross-linking, soluble enzyme molecules are transformed into insoluble aggregates (aggregates) due to the formation of covalent bonds between the enzyme molecules of multi- or bifunctional reagents such as hexamethylene diisocyanate or glutaraldehyde. Glutaraldehyde is often used as a crosslinking agent. Glutaraldehyde, a bifunctional reactive agent, reacts between the amino groups on the enzyme surface and the carrier. Due to the strong binding of enzymes, their desorption is very low in the covalent binding method. As a result of cross-linking, high enzyme aggregates are obtained with their activity and stability. In this method, proteins must first be precipitated. Organic solvent or salt is added to the environment in the precipitation process. In the precipitation method, neutral salts are added to the environment by increasing the ionic strength. At 4°C, the loss of activity decreases, and ammonium sulfate is generally used as salt. Precipitation with organic solvents is carried out by adding water-miscible solvents (such as ethanol, acetone, PEG). The addition of organic solvent lowers the dissolving force and the dielectric constant of the solution. Thus, aggregation occurs with the effect of electrostatic attraction and the solubility of the protein is reduced. Precipitation with organic solvents should be performed at 0°C or below temperatures to minimize denaturation. Protein conformation changes rapidly at higher temperatures. This allows organic solvent molecules to reach the interior of the protein. To increase the applicability of the precipitation process, the freezing point of the mixtures of aqueous solutions and organic solvents must be below 0°C [37, 76 - 87].

CONCLUSION

This section, which describes homogeneous catalysts, is one of the most important issues in the production of fuel from algae. In cases where catalyst synthesis is possible, it is necessary to provide the necessary conditions for the preparation of algae and to produce fuel. We can say that the addition of a catalyst is generally important in the production of selective products. General information about lipase catalysts is given. Enzyme retention methods are described with their advantages and disadvantages. The adsorption mechanism has been mentioned. Covalently bound lipase and perlite promoters are mentioned. Finally, the importance of cross-links of catalysts used in algae technology is mentioned.

CONSENT FOR PUBLICATION

Not applicable.

CONFLICT OF INTEREST

The authors declare no conflict of interest, financial or otherwise.

ACKNOWLEDGEMENT

Declared none.

REFERENCES

[1]   Biswas B, Bhaskar T. Hydrothermal upgradation of algae into value-added hydrocarbons. In: Pandey A, Chang JS, Soccol CR, Lee D-J, Chisti Y, Eds. Biofuels from Algae. Elsevier 2019; pp. 435-59.

[2]   Chi NTL, Anto S, Ahamed TS, *et al.* A review on biochar production techniques and biochar based catalyst for biofuel production from algae. Fuel 2020; 119411.

[3]   Galadima A, Muraza O. Hydrothermal liquefaction of algae and bio-oil upgrading into liquid fuels: Role of heterogeneous catalysts. Renew Sustain Energy Rev 2018; 81: 1037-48.
        [http://dx.doi.org/10.1016/j.rser.2017.07.034]

[4]   Guan Q, Wei C, Ning P, tian S, Gu J. Catalytic gasification of algae *Nannochloropsis* sp. in sub/supercritical water. Procedia Environ Sci 2013; 18: 844-8.
        [http://dx.doi.org/10.1016/j.proenv.2013.04.113]

[5]   Guo Y, Yeh T, Song W, Xu D, Wang S. A review of bio-oil production from hydrothermal liquefaction of algae. Renew Sustain Energy Rev 2015; 48: 776-90.
        [http://dx.doi.org/10.1016/j.rser.2015.04.049]

[6]   Martín M, Grossmann I. Integrating glycerol to methanol *vs.* glycerol to ethanol within the production of biodiesel from algae. In: Klemeš JJ, Varbanov PS, Liew PY, Eds. Computer Aided Chemical Engineering. Elsevier 2014; 33: pp. 85-90.

[7]   Rahman MA. Valorization of harmful algae E. compressa for biodiesel production in presence of chicken waste derived catalyst. Renew Energy 2018; 129: 132-40.
        [http://dx.doi.org/10.1016/j.renene.2018.06.005]

[8]   Xu D, Lin G, Guo S, Wang S, Guo Y, Jing Z. Catalytic hydrothermal liquefaction of algae and upgrading of biocrude: A critical review. Renew Sustain Energy Rev 2018; 97: 103-18.
        [http://dx.doi.org/10.1016/j.rser.2018.08.042]

[9]   Yang W, Li X, Zhang D, Feng L. Catalytic upgrading of bio-oil in hydrothermal liquefaction of algae major model components over liquid acids. Energy Convers Manage 2017; 154: 336-43.
        [http://dx.doi.org/10.1016/j.enconman.2017.11.018]

[10]   Yang W, Wang Z, Song S, *et al.* Understanding catalytic mechanisms of HZSM-5 in hydrothermal liquefaction of algae through model components: Glucose and glutamic acid. Biomass Bioenergy 2019; 130: 105356.
        [http://dx.doi.org/10.1016/j.biombioe.2019.105356]

[11]   Anto S, Karpagam R, Renukadevi P, Jawaharraj K, Varalakshmi P. Biomass enhancement and bioconversion of brown marine microalgal lipid using heterogeneous catalysts mediated transesterification from biowaste derived biochar and bionanoparticle. Fuel 2019; 255: 115789.
        [http://dx.doi.org/10.1016/j.fuel.2019.115789]

[12]   Galadima A, Muraza O. Biodiesel production from algae by using heterogeneous catalysts: A critical review. Energy 2014; 78: 72-83.
        [http://dx.doi.org/10.1016/j.energy.2014.06.018]

[13]   Gohain M, Hasin M, Eldiehy KSH, *et al.* Bio-ethanol production: A route to sustainability of fuels using bio-based heterogeneous catalyst derived from waste. Process Saf Environ Prot 2021; 146: 190-200.
        [http://dx.doi.org/10.1016/j.psep.2020.08.046]

[14]   Jume BH, Gabris MA, Rashidi Nodeh H, Rezania S, Cho J. Biodiesel production from waste cooking oil using a novel heterogeneous catalyst based on graphene oxide doped metal oxide nanoparticles. Renew Energy 2020; 162: 2182-9.
        [http://dx.doi.org/10.1016/j.renene.2020.10.046]

[15]   Lin YC, Amesho KTT, Chen CE, Cheng PC, Chou FC. A cleaner process for green biodiesel synthesis from waste cooking oil using recycled waste oyster shells as a sustainable base heterogeneous catalyst under the microwave heating system. Sustain Chem Pharm 2020; 17: 100310.

[http://dx.doi.org/10.1016/j.scp.2020.100310]

[16] Ravi A, Gurunathan B, Rajendiran N, *et al.* Contemporary approaches towards augmentation of distinctive heterogeneous catalyst for sustainable biodiesel production. Environmental Technology & Innovation 2020; 19: 100906.
[http://dx.doi.org/10.1016/j.eti.2020.100906]

[17] Sahu O. Characterisation and utilization of heterogeneous catalyst from waste rice-straw for biodiesel conversion. Fuel 2021; 287: 119543.
[http://dx.doi.org/10.1016/j.fuel.2020.119543]

[18] Santha A, Varghese R, Joy Prabu H, Johnson I, Magimai Antoni Raj D, John Sundaram S. Production of sustainable biofuel from biogenic waste using CuO nanoparticles as heterogeneous catalyst. Mater Today Proc 2020.

[19] Singh R, Bux F, Sharma YC. Optimization of biodiesel synthesis from microalgal (Spirulina platensis) oil by using a novel heterogeneous catalyst, β-strontium silicate (β-Sr2SiO4). Fuel 2020; 280: 118312.
[http://dx.doi.org/10.1016/j.fuel.2020.118312]

[20] Devaraj K, Veerasamy M, Aathika S, *et al.* Study on effectiveness of activated calcium oxide in pilot plant biodiesel production. J Clean Prod 2019; 225: 18-26.
[http://dx.doi.org/10.1016/j.jclepro.2019.03.244]

[21] Guo B, Walter V, Hornung U, Dahmen N. Hydrothermal liquefaction of Chlorella vulgaris and Nannochloropsis gaditana in a continuous stirred tank reactor and hydrotreating of biocrude by nickel catalysts. Fuel Process Technol 2019; 191: 168-80.
[http://dx.doi.org/10.1016/j.fuproc.2019.04.003]

[22] Al-Sakkari EG, Abdeldayem OM, El-Sheltawy ST, *et al.* Esterification of high FFA content waste cooking oil through different techniques including the utilization of cement kiln dust as a heterogeneous catalyst: A comparative study. Fuel 2020; 279: 118519.
[http://dx.doi.org/10.1016/j.fuel.2020.118519]

[23] Hapońska M, Nurra C, Abelló S, Makkee M, Salvadó J, Torras C. Membrane reactors for biodiesel production with strontium oxide as a heterogeneous catalyst. Fuel Process Technol 2019; 185: 1-7.
[http://dx.doi.org/10.1016/j.fuproc.2018.11.010]

[24] Rabie AM, Shaban M, Abukhadra MR, Hosny R, Ahmed SA, Negm NA. Diatomite supported by CaO/MgO nanocomposite as heterogeneous catalyst for biodiesel production from waste cooking oil. J Mol Liq 2019; 279: 224-31.
[http://dx.doi.org/10.1016/j.molliq.2019.01.096]

[25] Ziembowicz S, Kida M, Koszelnik P. Reservoir bottom sediments as heterogeneous catalysts for effective degradation of a selected endocrine-disrupting chemical *via* a Fenton-like process. J Water Process Eng 2019; 32: 100950.
[http://dx.doi.org/10.1016/j.jwpe.2019.100950]

[26] Jayaraman J, Alagu K, Appavu P, Joy N, Jayaram P, Mariadoss A. Enzymatic production of biodiesel using lipase catalyst and testing of an unmodified compression ignition engine using its blends with diesel. Renew Energy 2020; 145: 399-407.
[http://dx.doi.org/10.1016/j.renene.2019.06.061]

[27] Makareviciene V, Gumbyte M, Sendzikiene E. Simultaneous extraction of microalgae Ankistrodesmus sp. oil and enzymatic transesterification with ethanol in the mineral diesel medium. Food Bioprod Process 2019; 116: 89-97.
[http://dx.doi.org/10.1016/j.fbp.2019.05.002]

[28] Nautiyal P, Subramanian KA, Dastidar MG. Recent advancements in the production of biodiesel from algae: A review. Reference Module in Earth Systems and Environmental Sciences. Elsevier 2014.
[http://dx.doi.org/10.1016/B978-0-12-409548-9.09380-5]

[29] Yasvanthrajan N, Sivakumar P, Muthukumar K, Murugesan T, Arunagiri A. Production of biodiesel

from waste bio-oil through ultrasound assisted transesterification using immobilized lipase. Environmental Technology & Innovation 2020; p. 101199.

[30]   Zarei A, Amin NAS, Talebian-Kiakalaieh A, Zain NAM. Immobilized lipase-catalyzed transesterification of Jatropha curcas oil: Optimization and modeling. J Taiwan Inst Chem Eng 2014; 45(2): 444-51.
[http://dx.doi.org/10.1016/j.jtice.2013.05.015]

[31]   Bhatia SK, Bhatia RK, Jeon JM, Kumar G, Yang YH. Carbon dioxide capture and bioenergy production using biological system-a review. Renew Sustain Energy Rev 2019; 110: 143-58.
[http://dx.doi.org/10.1016/j.rser.2019.04.070]

[32]   Hosoya-Matsuda N, Inoue K, Hisabori T. Roles of thioredoxins in the obligate anaerobic green sulfur photosynthetic bacterium Chlorobaculum tepidum. Mol Plant 2009; 2(2): 336-43.
[http://dx.doi.org/10.1093/mp/ssn077] [PMID: 19825618]

[33]   Karmakar A, Karmakar S, Mukherjee S. Properties of various plants and animals feedstocks for biodiesel production. Bioresour Technol 2010; 101(19): 7201-10.
[http://dx.doi.org/10.1016/j.biortech.2010.04.079] [PMID: 20493683]

[34]   Lan K, Park S, Yao Y. Key issue, challenges, and status quo of models for biofuel supply chain design.Biofuels for a More Sustainable Future. Elsevier 2020; pp. 273-315.
[http://dx.doi.org/10.1016/B978-0-12-815581-3.00010-5]

[35]   Li Y, Li C, Feng F, Wei W, Zhang H. Synthesis of medium and long-chain triacylglycerols by enzymatic acidolysis of algal oil and lauric acid. Lebensm Wiss Technol 2021; 136: 110309.
[http://dx.doi.org/10.1016/j.lwt.2020.110309]

[36]   Mine Y, Roy MK. 4.46-Egg components for heart health: Promise and progress for cardiovascular protective functional food ingredient. Comprehensive Biotechnology. 2nd ed.. Moo-Young M, Moo-Young M. ?Burlington: Academic Press 2011; pp. 553-65.

[37]   Niaounakis M. 1 - Introduction to Biopolymers. In: Niaounakis M, Ed. Biopolymers Reuse, Recycling, and Disposal. Oxford: William Andrew Publishing 2013; pp. 1-75.
[http://dx.doi.org/10.1016/B978-1-4557-3145-9.00001-4]

[38]   Park C, Heo K, Oh S, *et al.* Eco-design and evaluation for production of 7-aminocephalosporanic acid from carbohydrate wastes discharged after microalgae-based biodiesel production. J Clean Prod 2016; 133: 511-7.
[http://dx.doi.org/10.1016/j.jclepro.2016.05.168]

[39]   Reichle DE. Biological energy transformations by plants. In: Reichle DE, Ed. The Global Carbon Cycle and Climate Change. Elsevier 2020; pp. 43-53.
[http://dx.doi.org/10.1016/B978-0-12-820244-9.00004-4]

[40]   Urbieta MS, Donati ER, Chan KG, Shahar S, Sin LL, Goh KM. Thermophiles in the genomic era: Biodiversity, science, and applications. Biotechnol Adv 2015; 33(6): 633-47.
[http://dx.doi.org/10.1016/j.biotechadv.2015.04.007] [PMID: 25911946]

[41]   Bracco LF, Levin GJ, Urtasun N, *et al.* Covalent immobilization of soybean seed hull urease on chitosan mini-spheres and the impact on their properties. Biocatal Agric Biotechnol 2019; 18: 101093.
[http://dx.doi.org/10.1016/j.bcab.2019.101093]

[42]   Lee DS, Wi SG, Lee SJ, Lee YG, Kim YS, Bae HJ. Rapid saccharification for production of cellulosic biofuels. Bioresour Technol 2014; 158: 239-47.
[http://dx.doi.org/10.1016/j.biortech.2014.02.039] [PMID: 24607460]

[43]   Ma Y, Wang P, Wang Y, Liu S, Wang Q, Wang Y. Fermentable sugar production from wet microalgae residual after biodiesel production assisted by radio frequency heating. Renew Energy 2020; 155: 827-36.
[http://dx.doi.org/10.1016/j.renene.2020.03.176]

[44]   Shokrkar H, Ebrahimi S. Evaluation of different enzymatic treatment procedures on sugar extraction

from microalgal biomass, experimental and kinetic study. Energy 2018; 148: 258-68.
[http://dx.doi.org/10.1016/j.energy.2018.01.124]

[45] Zuccaro G, Pirozzi D, Yousuf A. Lignocellulosic biomass to biodiesel.Lignocellulosic Biomass to Liquid Biofuels. Academic Press 2020; pp. 127-67.
[http://dx.doi.org/10.1016/B978-0-12-815936-1.00004-6]

[46] Boone CD, McKenna R. Engineered mammalian carbonic anhydrases for CO_2 capture.Carbonic Anhydrases as Biocatalysts. Amsterdam: Elsevier 2015; pp. 291-309.
[http://dx.doi.org/10.1016/B978-0-444-63258-6.00016-0]

[47] Karmakar B, Halder G. Progress and future of biodiesel synthesis: Advancements in oil extraction and conversion technologies. Energy Convers Manage 2019; 182: 307-39.
[http://dx.doi.org/10.1016/j.enconman.2018.12.066]

[48] Satyawali Y, Vanbroekhoven K, Dejonghe W. Process intensification: The future for enzymatic processes? Biochem Eng J 2017; 121: 196-223.
[http://dx.doi.org/10.1016/j.bej.2017.01.016]

[49] Zhong L, Feng Y, Wang G, *et al.* Production and use of immobilized lipases in/on nanomaterials: A review from the waste to biodiesel production. Int J Biol Macromol 2020; 152: 207-22.
[http://dx.doi.org/10.1016/j.ijbiomac.2020.02.258] [PMID: 32109471]

[50] Dutta S. Lignin deconstruction: Chemical and biological approaches.Sustainable Catalytic Processes. Amsterdam: Elsevier 2015; pp. 125-55.
[http://dx.doi.org/10.1016/B978-0-444-59567-6.00005-4]

[51] Fabra MJ, Pérez-Bassart Z, Talens-Perales D, *et al.* Matryoshka enzyme encapsulation: Development of zymoactive hydrogel particles with efficient lactose hydrolysis capability. Food Hydrocoll 2019; 96: 171-7.
[http://dx.doi.org/10.1016/j.foodhyd.2019.05.026]

[52] Hasnain MS, Jameel E, Mohanta B, Dhara AK, Alkahtani S, Nayak AK. Alginates: sources, structure, and properties. Alginates in Drug Delivery. Nayak AK, Hasnain MS. Academic Press 2020; pp. 1-17.

[53] Jegannathan KR, Jun-Yee L, Chan ES, Ravindra P. Production of biodiesel from palm oil using liquid core lipase encapsulated in κ-carrageenan. Fuel 2010; 89(9): 2272-7.
[http://dx.doi.org/10.1016/j.fuel.2010.03.016]

[54] Kaur J, Choudhary S, Chaudhari R, Jayant RD, Joshi A. 9 - Enzyme-based biosensors. In: Pal K, Kraatz H-B, Khasnobish A, Bag S, Banerjee I, Kuruganti U, Eds. Bioelectronics and Medical Devices. Woodhead Publishing 2019; pp. 211-40.
[http://dx.doi.org/10.1016/B978-0-08-102420-1.00013-3]

[55] Lopez-Rubio A, Gavara R, Lagaron JM. Bioactive packaging: Turning foods into healthier foods through biomaterials. Trends Food Sci Technol 2006; 17(10): 567-75.
[http://dx.doi.org/10.1016/j.tifs.2006.04.012]

[56] Marx S. Glycerol-free biodiesel production through transesterification: A review. Fuel Process Technol 2016; 151: 139-47.
[http://dx.doi.org/10.1016/j.fuproc.2016.05.033]

[57] Parkhey P, Sahu R. Microfluidic microbial fuel cells: Recent advancements and future prospects. Int J Hydrogen Energy 2020.

[58] Russmayer H, Egermeier M, Kalemasi D, Sauer M. Spotlight on biodiversity of microbial cell factories for glycerol conversion. Biotechnol Adv 2019; 37(6): 107395.
[http://dx.doi.org/10.1016/j.biotechadv.2019.05.001] [PMID: 31071430]

[59] Khoo KS, Tan X, Ooi CW, *et al.* How does ionic liquid play a role in sustainability of biomass processing? J Clean Prod 2020; 124772.

[60] Kumar A, Sharma G, Naushad M, *et al.* Bio-inspired and biomaterials-based hybrid photocatalysts for

environmental detoxification: A review. Chem Eng J 2020; 382: 122937.
[http://dx.doi.org/10.1016/j.cej.2019.122937]

[61]   Rocha de Freitas G, Adeodato Vieira MG, Carlos da Silva MG. Characterization and biosorption of silver by biomass waste from the alginate industry. J Clean Prod 2020; 271: 122588.
[http://dx.doi.org/10.1016/j.jclepro.2020.122588]

[62]   Yusuf A, Giwa A, Mohammed EO, Mohammed O, Al Hajaj A, Abu-Zahra MRM. CO2 utilization from power plant: A comparative techno-economic assessment of soda ash production and scrubbing by monoethanolamine. J Clean Prod 2019; 237: 117760.
[http://dx.doi.org/10.1016/j.jclepro.2019.117760]

[63]   Cavalcante FTT, Neto FS, Rafael de Aguiar Falcão I, *et al.* Opportunities for improving biodiesel production *via* lipase catalysis. Fuel 2020; 119577.

[64]   Kant Bhatia S, Kant Bhatia R, Jeon JM, *et al.* An overview on advancements in biobased transesterification methods for biodiesel production: Oil resources, extraction, biocatalysts, and process intensification technologies. Fuel 2021; 285: 119117.
[http://dx.doi.org/10.1016/j.fuel.2020.119117]

[65]   Li Y, Tang W, Chen Y, Liu J, Lee CF. Potential of acetone-butanol-ethanol (ABE) as a biofuel. Fuel 2019; 242: 673-86.
[http://dx.doi.org/10.1016/j.fuel.2019.01.063]

[66]   Pan J, Muppaneni T, Sun Y, *et al.* Microwave-assisted extraction of lipids from microalgae using an ionic liquid solvent [BMIM][HSO4]. Fuel 2016; 178: 49-55. [BMIM]. [HSO4].
[http://dx.doi.org/10.1016/j.fuel.2016.03.037]

[67]   Shanmugam S, Hari A, Pandey A, Mathimani T, Felix L, Pugazhendhi A. Comprehensive review on the application of inorganic and organic nanoparticles for enhancing biohydrogen production. Fuel 2020; 270: 117453.
[http://dx.doi.org/10.1016/j.fuel.2020.117453]

[68]   Velgosova O, Mražíková A, Čižmárová E, Málek J. Green synthesis of Ag nanoparticles: Effect of algae life cycle on Ag nanoparticle production and long-term stability. Trans Nonferrous Met Soc China 2018; 28(5): 974-9.
[http://dx.doi.org/10.1016/S1003-6326(18)64732-6]

[69]   Xu C, Nasrollahzadeh M, Sajjadi M, Maham M, Luque R, Puente-Santiago AR. Benign-by-design nature-inspired nanosystems in biofuels production and catalytic applications. Renew Sustain Energy Rev 2019; 112: 195-252.
[http://dx.doi.org/10.1016/j.rser.2019.03.062]

[70]   Blundon D, Dale M, Goudey S, Hoddinott J. Effects Of Oil And Oil Spill Chemicals On Shoreline Plants Of Northern Freshwater Ecosystems. In: Vandermeulen JH, Hrudey SE, Eds. Oil in Freshwater: Chemistry, Biology, Countermeasure Technology. Pergamon 1987; pp. 403-9.
[http://dx.doi.org/10.1016/B978-0-08-031862-2.50033-2]

[71]   Flórez-Fernández N, Illera M, Sánchez M, *et al.* Integrated valorization of Sargassum muticum in biorefineries. Chem Eng J 2021; 404: 125635.
[http://dx.doi.org/10.1016/j.cej.2020.125635]

[72]   Wang X, Wang X, Zhao J, *et al.* Solar light-driven photocatalytic destruction of cyanobacteria by F-Ce-TiO 2 /expanded perlite floating composites. Chem Eng J 2017; 320: 253-63.
[http://dx.doi.org/10.1016/j.cej.2017.03.062]

[73]   Awad YM, Lee SE, Ahmed MBM, *et al.* Biochar, a potential hydroponic growth substrate, enhances the nutritional status and growth of leafy vegetables. J Clean Prod 2017; 156: 581-8.
[http://dx.doi.org/10.1016/j.jclepro.2017.04.070]

[74]   Gruda N, Bisbis M, Tanny J. Impacts of protected vegetable cultivation on climate change and adaptation strategies for cleaner production – A review. J Clean Prod 2019; 225: 324-39.

[http://dx.doi.org/10.1016/j.jclepro.2019.03.295]

[75]   Vikrant K, Kailasa SK, Tsang DCW, *et al.* Biofiltration of hydrogen sulfide: Trends and challenges. J Clean Prod 2018; 187: 131-47.
[http://dx.doi.org/10.1016/j.jclepro.2018.03.188]

[76]   Baadhe RR, Potumarthi R, Gupta VK. Lipase-catalyzed biodiesel production: Technical challenges.Bioenergy Research: Advances and Applications. Amsterdam: Elsevier 2014; pp. 119-29.
[http://dx.doi.org/10.1016/B978-0-444-59561-4.00008-5]

[77]   Kumar D, Das T, Giri BS, Verma B. Preparation and characterization of novel hybrid bio-support material immobilized from Pseudomonas cepacia lipase and its application to enhance biodiesel production. Renew Energy 2020; 147: 11-24.
[http://dx.doi.org/10.1016/j.renene.2019.08.110]

[78]   Muñoz J, Alfaro MC, Trujillo-Cayado LA, Santos J, Martín-Piñero MJ. Production of food bioactive-loaded nanostructures by microfluidization. In: Jafari SM, Ed. Nanoencapsulation of Food Ingredients by Specialized Equipment. Academic Press 2019; Vol. 3: pp. 341-90.
[http://dx.doi.org/10.1016/B978-0-12-815671-1.00007-X]

[79]   Nayak SN, Bhasin CP, Nayak MG. A review on microwave-assisted transesterification processes using various catalytic and non-catalytic systems. Renew Energy 2019; 143: 1366-87.
[http://dx.doi.org/10.1016/j.renene.2019.05.056]

[80]   Nematian T, Salehi Z, Shakeri A. Conversion of bio-oil extracted from Chlorella vulgaris micro algae to biodiesel *via* modified superparamagnetic nano-biocatalyst. Renew Energy 2020; 146: 1796-804.
[http://dx.doi.org/10.1016/j.renene.2019.08.048]

[81]   Noreen A, Jabeen M, Tabasum S, *et al.* Algae-derived polyester blends and composites. In: Zia KM, Zuber M, Ali M, Eds. Algae Based Polymers, Blends, and Composites. Elsevier 2017; pp. 459-97.
[http://dx.doi.org/10.1016/B978-0-12-812360-7.00012-4]

[82]   Patil CK, Jirimali HD, Mahajan MS, Paradeshi JS, Chaudhari BL, Gite VV. Functional anti-corrosive and anti-bacterial surface coatings based on mercaptosuccinic and thiodipropionic acids and algae oil as renewable feedstock. React Funct Polym 2019; 139: 142-52.
[http://dx.doi.org/10.1016/j.reactfunctpolym.2019.03.020]

[83]   Singh DP, Dwevedi A. Production of clean energy by green ways. In: Dwevedi A, Ed. Solutions to Environmental Problems Involving Nanotechnology and Enzyme Technology. Academic Press 2019; pp. 49-90.
[http://dx.doi.org/10.1016/B978-0-12-813123-7.00002-5]

[84]   Zuber M, Zia KM, Noreen A, *et al.* Algae-Based Polyolefins. In: Zia KM, Zuber M, Ali M, Eds. Algae Based Polymers, Blends, and Composites. Elsevier 2017; pp. 499-529.
[http://dx.doi.org/10.1016/B978-0-12-812360-7.00013-6]

[85]   Mansir N, Taufiq-Yap YH, Rashid U, Lokman IM. Investigation of heterogeneous solid acid catalyst performance on low grade feedstocks for biodiesel production: A review. Energy Convers Manage 2017; 141: 171-82.
[http://dx.doi.org/10.1016/j.enconman.2016.07.037]

[86]   Ong HC, Tiong YW, Goh BHH, *et al.* Recent advances in biodiesel production from agricultural products and microalgae using ionic liquids: Opportunities and challenges. Energy Convers Manage 2020; 113647.

[87]   Soltani S, Rashid U, Al-Resayes SI, Nehdi IA. Recent progress in synthesis and surface functionalization of mesoporous acidic heterogeneous catalysts for esterification of free fatty acid feedstocks: A review. Energy Convers Manage 2017; 141: 183-205.
[http://dx.doi.org/10.1016/j.enconman.2016.07.042]

CHAPTER 14

Economical Fundamentals for Algae to Fuel Technology

Cemil Koyunoğlu[1,2,*]

[1] *Energy Systems Engineering Department, Engineering Faculty, Cinarcik Road 5th km, 77200, Yalova University, Yalova, Turkey*

[2] *Fuel Oil Analysis Laboratory, Central Laboratory, Cinarcik Road 5th km, 77200, Yalova University, Yalova, Turkey*

Abstract: This chapter provides basic economic information that readers will need during the operation phase before establishing any algae business. The subject has been reinforced with simple examples.

Keywords: Capital recovery factor, Engineering economy basics, Present value factor.

LIFETIME COST CALCULATION OF ASSETS

The lifetime cost of an asset is the analysis of the initial purchase, scrap, operating, labor, material, interest, insurance, depreciation, and tax costs of an investment (equipment, plant, and/or service) throughout its life. The reliability of the lifetime cost calculation of the asset also depends on the good evaluation of economic data and making healthy predictions for the future.

In order to calculate the cost of the product, economic factors such as what will be inflation in the future, how much interest and fuel costs will increase must be estimated close to the truth. The future cost depends on the estimated life of the investment as well as economic pressures and political decisions. As a result, the lifetime cost calculation is quantitative and just an estimate. The parameters required for its calculation usually depend on the structural theories and the skill and experience of decision-making mechanisms. For the lifetime cost calculation to be reliable, expert experience or literature should be consulted on this subject. Lifetime cost calculation is usually made on a uil basis. Definition of Present

---
[*] **Correspondence author Cemil Koyunoğlu:** Energy Systems Engineering Department, Faculty of Engineering, Yalova University, Yalova, Turkey; Tel: +902268155378; E-mail: cemil.koyunoglu@yalova.edu.tr

Hüseyin Karaca and Cemil Koyunoğlu (Eds.)

Value Factor and Capital Recovery Factor concepts is needed in calculating lifetime cost every year [1 - 17].

PRESENT VALUE FACTOR

Let's assume that you borrow 100 $ with 25% interest per year. At the end of a year, the interest rate of 100 $ becomes 25 $. The sum of the debt and interest received is as follows:

$$100 \times (1 + 0.25) = 125 \text{ \$}$$

At the end of the second year, total debt will be

$$100 \times (1 + 0.25) \times (1 + 0.25) = 100 \times (1 + 0.25)^2 = 156.25 \text{ \$}$$

Total debt amount after n years (the value of 100 $ after n years) would be

$$100 \times (1 + 0.25)^n$$

It can be calculated from the formula. Each year the total value of money increases by a factor of $(1 + 0.25)$. If the annual interest rate is i, the total value TV (value of money over time) of money in n years with the present value of money PV would be,

$$TV = PV \times (1 + i)^n \tag{1}$$

If n years after Equation 1, the present value of the money whose total value will be TV is PV then,

$$PV = TV / (1 + i)^n \tag{2}$$

In other words, the value of the money whose present value will be ND decreases by $(1 + i)$ -n factor for each past year. This factor is also called the Present Value Factor [1 - 17].

$$PVF = (1 + i)^{-n} \tag{3}$$

CAPITAL RECOVERY FACTOR

Let's calculate how much money we can borrow with 25% interest, provided that we pay 100 $ at the end of each year and pay the whole in 3 years. The present value of 100 $ we will pay at the end of the first year would be,

$$100 / (1 + 0.25) = 80 \text{ \$}$$

The present value of 100 \$ we will pay at the end of the second and third year, respectively would be,

$100 / (1 + 0.25)^2 = 64$ \$

$100 / (1 + 0.25)^3 = 51.2$ \$

In that case, the total money we can receive in 3 years, provided that we pay 100 \$ at the end of each year, is $80 + 64 + 51.2 = 195.2$ \$.

The present value of the money we can receive with a maturity of n years provided that we pay the Annual interest rate (AIR) amount at the end of each year at the annual interest rate,

$$PV = AIR \sum_{m=1}^{n} \frac{1}{(1+i)^m} \tag{4}$$

This is the present value of the money received in n years, provided that they pay an equal AIR amount at the end of each year.

When

$$PV = AIR \sum_{m=1}^{n} \frac{1}{(1+i)^m} = \frac{1-(1+i)^{-n}}{i}$$ in this expression, its current value is PV the amount that is paid in n equal intervals

$$PV = AIR \left[\frac{i}{1-(1+i)^{-n}} \right] \tag{5}$$

will be, the $\left[\frac{i}{1-(1+i)^{-n}} \right]$

The capital Recovery Factor in this expression is called the CRF.

$$GKF = i / 1 - (1 + i)^{-n} \tag{6}$$

For the detailed explanation example -1 is given [18 - 36].

Example-1:

If we borrowed 100 \$ at equal intervals over 10 years with an annual interest of

25%, how many pounds are our annual payments, and what would be the total amount we paid at the end of 10 years?

Answer-1:

Since the debt received is BD = 1000 TL, from Equality 6, the amount we must pay each year is AIR,

AIR= PV [i/i-(1+i)-n] = 280.07 $,

After 10 years, the total amount of payments is calculated as 280.07 $ x 10 = 2800.7 $.

ANNUAL COST

Some costs arise immediately when the project starts. We call these costs the initial investment costs. If the initial investment cost is IC, the annual interest rate is i, the scrap value (or second-hand value) of the investment is SV at the end of n years, which is the life of the investment, then the current cost of this investment is expressed as,

$$PC = IC - SV \times PVF \text{ (present value factor) (i, n)} \tag{7}$$

Investment cost remains constant throughout the life of the investment, it does not change. In addition, operating, labor, insurance maintenance, and repair costs vary from year to year. If we show the cost of these expenses in the year they occur, with AEC_m, the current cost of the first year's operating expenses will be,

$$AEC_1 \times \frac{1}{(1+i)^1} \tag{8}$$

The current cost of the expenses in the second and subsequent years, respectively will be

$$AEC_2 \times \frac{1}{(1+i)^2}$$

$$AEC_3 \times \frac{1}{(1+i)^3} \tag{9}$$

$$AEC_4 \times \frac{1}{(1+i)^4}$$

Total cost of operating expenses today would be,

$$PTC = \sum_{m=1}^{n} AEC_m \, xPVC(i,m) \tag{10}$$

Present Value of Total Expenses in PVTE will be,

$$PVTE = IC - SV \times PVF(i,n) + \sum_{m=1}^{n} AEC_m \times PVF(i,m) \tag{11}$$

With these expenses, the annual cost of investment and operating expenses (annual investment and operating expenses cost AIOEC) can be formulated,

$$AIOEC = [IY - HR \times PVF(i,n) + \sum_{m=1}^{n} AECm \, xPVC(i,m)] \times RF(i,n) \tag{12}$$

For the detailed explanation example -2 is given [37 - 56].

Example-2:

The purchase price of the pump, which is estimated to have a life of 10 years, is $ 10000. It is assumed that the scrap price of the pump will also be 100 TL. It is estimated that the distribution of expenses for the operation of the pump will be as given below.

1. year 1000 $ 6. year 5750 $

2. year 1560 $ 7. year 7000 $

3. year 2100 $ 8. year 9800 $

4. year 2950 $ 9. year 13800 $

5. year 4100 $ 10. year 19300 $

Calculate the annual cost of the pump's product, assuming the annual interest rate is 40%.

Answer-2:

The present value of the pump scrap value of the investment will be

SV x PVF (i, n) = 100 x $(1 + i)^{-n}$ = 100 x $(1 + 0.4)^{-10}$ = 3.5 $

The current value of operating expenses estimated to be made each year is given below.

1. the year $AEC_1 x \dfrac{1}{(1+i)^1} \longrightarrow$ $\qquad$ $1000 \text{ x } \dfrac{1}{(1+0.40)^1} = 714.3\ \$$

2. the year $AEC_2 x \dfrac{1}{(1+i)^2} \longrightarrow$ $\qquad$ $1560 \text{ x } \dfrac{1}{(1+0.40)^2} = 796.0\ \$$

3. the year $AEC_3 x \dfrac{1}{(1+i)^3} \longrightarrow$ $\qquad$ $2100 \text{ x } \dfrac{1}{(1+0.40)^3} = 765.3\ \$$

4. the year $AEC_4 x \dfrac{1}{(1+i)^4} \longrightarrow$ $\qquad$ $2950 \text{ x } \dfrac{1}{(1+0.40)^4} = 768.0\ \$$

5. year $AEC_5 x \dfrac{1}{(1+i)^5} \longrightarrow$ $\qquad$ $4100 \text{ x } \dfrac{1}{(1+0.40)^5} = 762.3\ \$$

6. year $AEC_6 x \dfrac{1}{(1+i)^6} \longrightarrow$ $\qquad$ $5750 \text{ x } \dfrac{1}{(1+0.40)^6} = 763.7\ \$$

7. year $AEC_7 x \dfrac{1}{(1+i)^7} \longrightarrow$ $\qquad$ $7000 \text{ x } \dfrac{1}{(1+0.40)^7} = 664.1\ \$$

8. year $AEC_8 x \dfrac{1}{(1+i)^8} \longrightarrow$ $\qquad$ $9800 \text{ x } \dfrac{1}{(1+0.40)^8} = 664.1\ \$$

9. year $AEC_9 x \dfrac{1}{(1+i)^9} \longrightarrow$ $\qquad$ $13800 \text{ x } \dfrac{1}{(1+0.40)^9} = 668.0\ \$$

10. year $AEC_{10} x \dfrac{1}{(1+i)^{10}} \longrightarrow$ $\qquad$ $19300 \text{ x } \dfrac{1}{(1+0.40)^{10}} = 667.2\ \$$

The present value of total operating expenses in 10 years will be,

$$\text{TOC} = \sum_{m=1}^{10} AEC_m x PVF(i,m) = 7236\ \$$$

With this data, the annual investment and operating expenses cost will be calculated as,

$$\text{ACIOE} = [\text{IC-SC x } PVF(i,n) + \sum_{m=1}^{n} AEC_m x\ PVF\ (i,m)] x\ RF\ (i,n)$$

$$\text{ACIOE} = [10000\text{-} 3.5 + 7236]\ x\ \dfrac{0.4}{1-(1+0.4)^{-10}} = 7140\ \text{TL}/\$$$

EXERGY COST

Exergy analysis of systems in engineering is generally done by assuming that systems operate in steady-state continuous flow. In the exergy analysis of systems that exchange matter, heat and work with their environment through one or more channels, the energy and material exchanges with the environment of the systems are expressed in terms of the exergy equivalents of the energy, work, and heat exchanges transferred to the system by the material exchange. Exergy exchange is the reduction of the environment of the system to the "business potential" exchange between the system and its environment. Economic analysis of thermal systems according to exergy cost; is meaningful in terms of evaluating the energy transferred to the system by material heat and business exchanges - to a common base created - according to the exergy base. Exergy cost means that the economic analysis of the interactions of the system with its environment is reduced to a common base, it is built on exergy analysis [29, 57 - 68].

The unit flow exergy cost of substances flowing into the system through different channels can be expressed as $\Sigma_{in}c_{in}(ma_a)_{in}$ per unit time (year, hour, or second).

Here, the cost of the substances flowing into the c_g system per unit time (yearly) will be $\Sigma_{in}c_{in}(ma_a)_{in}$. If the unit exergy cost of the cw business exchange of the system is, the unit time (yearly) cost of the business exchange is cwW⋅. Similarly, since the exergy equivalent of heat exchange with the environment of the system will be $c_q(1-\frac{T_o}{T})Q_{kh}\cdot$; the unit exergy cost of heat exchange will be in c_q, and the cost per unit time of heat exchange will be

(1- The unit exergy cost of heat exchange can be written in and the cost per unit time of heat exchange can be written as $c_q(1-\frac{T_o}{T})Q_{kh}\cdot$.

Exergy cost is the cost balance for this system. With the above data, the cost balance for the system is also $c_q(1-\frac{T_o}{T})Q_{kh}\cdot$;

$$\Sigma_{in} c_{in}(\dot{m}a_a)_{in} + c_a(1 - \tfrac{\cdot o}{\cdot}) \dot{Q}_{kh} + \text{ACİOE} = \Sigma_{out} c_{out}(\dot{m}a_a)_{out} + c_w \dot{W}_{kh} \quad \textbf{(13)}$$

With the help of Equation 13, the exergy cost can be calculated separately for each item in the system. EIOPPY in this expression is the estimated investment and operating cost per unit time (per year) for that item.

For the detailed explanation example, -3 is given [29, 57 - 68].

Example-3:

In order to pump the same amount of water, two separate pumps with a power of 4.5 kW each are considered. The cost of the A pump is 3000 $, the second law efficiency is 80%, the cost of the B pump is 5000 $, and the second law efficiency is 90%. The physical life of both pumps will be 20 years, and the distribution of operating costs for both pumps by years will be as given in the table below:

1. year 100 $ 6. year 200 $ 11. year 300 $ 16. year 400 $

2. year 120 $ 7. year 220 $ 12. year 320 $ 17. year 420 $

3. year 140 $ 8 year 240 $ 13. year 340 $ 18. year 440 $

4. year 160 $ 9. year 260 $ 14. year 360 $ 19. year 460 $

5. year 180 $ 10. year 280 $ 15. year 380 $ 20. year 480 $

it is assumed that pumps are not costly scrap, annual interest is 20% and the price of electricity used is approximately 5 x 10-5 $ / kJ. Calculate the average cost of the annual products of the pumps per unit flow exergy.

Solution:

If we define the thermodynamic states of the fluid at the inlet and outlet of the adiabatic pump shown in Fig. (1) as 1 (inlet) and 2 (outlet), the equation for the pump is 13,

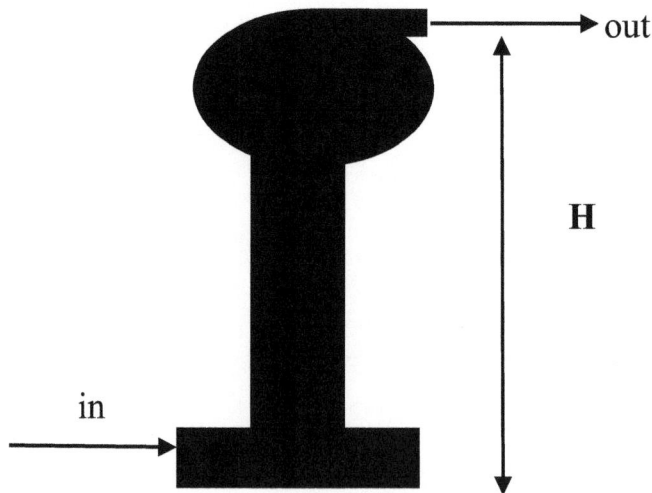

Fig. (1). Sample pump sketch.

$$c_{in}(\dot{m}a_a)_1 + EI\dot{O}PPY = c_{out}(\dot{m}a_a)_2 + c_w \dot{W}_{kh}$$

It can be written as. Since the cost of the unit flow exergy of the substances flowing into the system at the inlet and outlet of the pump will be equal;

$$c_{in} = c_{out}$$

and the expression of ˙ (PRODUCT) produced per year (per unit time) and ˙ consumed (FUEL) in terms of flow exergy, as defined in previous chapters;

$$PR\dot{O}DUCT = \dot{m}[(a_a)_2 - (a_a)_1] \quad F\dot{U}EL = -\dot{W}_{kh}$$

will be. Exergy cost equality for the pump,

$$EI\dot{O}PPY + c_{fuel}F\dot{U}EL = c_{product}PRODUCT \text{ is reduced.}$$

Here $c_{product} = c_{in} = c_{fuel}$ and $c_{fuel} = c_{cw}$ are defined as the cost of unit flow exergy of the product and fuel, respectively. Considering the definition of second law efficiency,

$$\varepsilon = \frac{PR\dot{O}DUCT}{F\dot{U}EL}$$

If the exergy cost equation is rearranged, the cost of the product for the pump is obtained as,

$$c_{product} = \frac{c_{fuel}}{\varepsilon} + \frac{[EI\dot{O}PPY]}{\varepsilon[F\dot{U}EL]} -$$

This result shows that since the second law efficiency is less than one, the cost of the product's unit flow exergy is greater than the cost of the unit flow exergy of the FUEL. If we increase the efficiency of the second law to reduce the cost of the product, the term c_{fuel}/ε in the equation that gives the product cost decreases, but the second law is to try to minimize the cost of the product by balancing the investment and operating costs made due to the increasing inefficiency.

The current value (CV) of operating expenses (OE) estimated to be made each year is given in Table **1** below.

Table 1. Estimation of CV and OE values.

1. year $AEC_1 x \frac{1}{(1+i)^1}$	83.3 $	11. year $AEC_1 x \frac{1}{(1+i)^{11}}$	40.4 $
2. year $AEC_2 x \frac{1}{(1+i)^2}$	83.3 $	12. year $AEC_{12} x \frac{1}{(1+i)^{12}}$	35.9 $
3. year $AEC_3 x \frac{1}{(1+i)^3}$	81.0 $	13. year $AEC_{13} x \frac{1}{(1+i)^{13}}$	31.8 $
4. year $AEC_4 x \frac{1}{(1+i)^4}$	77.2 $	14. year $AEC_{14} x \frac{1}{(1+i)^{14}}$	28.0 $
5. year $AEC_5 x \frac{1}{(1+i)^5}$	72.3 $	15. year $AEC_{15} x \frac{1}{(1+i)^{15}}$	24.7 $
6. year $AEC_6 x \frac{1}{(1+i)^6}$	67.0 $	16. year $AEC_{16} x \frac{1}{(1+i)^{16}}$	21.6 $
7. year $AEC_7 x \frac{1}{(1+i)^7}$	61.4 $	17. year $AEC_{17} x \frac{1}{(1+i)^{17}}$	18.9 $
8. year $AEC_8 x \frac{1}{(1+i)^8}$	55.8 $	18. year $AEC_{18} x \frac{1}{(1+i)^{18}}$	16.5 $
9. year $AEC_9 x \frac{1}{(1+i)^9}$	50.4 $	19. year $AEC_{19} x \frac{1}{(1+i)^{19}}$	14.4 $
10. year $AEC_{10} x \frac{1}{(1+i)^{10}}$	45.2 $	20. year $AEC_{20} x \frac{1}{(1+i)^{20}}$	12.5 $

Present value of the total operating expenses of the pumps will be

$$TOC = \sum_{m=1}^{10} AEC_m \, xPVF(i,m) = 921.7 \text{ \$}$$

With these values, the total annual investment and operating expenses of A and B pumps;

A pump's EIOPPY = $[3000 + 921.7]$ x $\dfrac{0.20}{1-(1+0.20)^{-20}}$ = 805.4 $/year

B pump's EIOPPY = $[5000 + 921. 7]$ x $\dfrac{0.20}{1-(1+0.20)^{-20}}$ = 1216.1 $/year

Since the powers of the A and B pumps are 4.5 kW, both pumps per year will use $[F\dot{U}EL]$,

$[F\dot{U}EL] = 4.5\ \dfrac{kj}{s}\ \dfrac{(60\ x\ 60\ x\ 24)s}{1\ day}\ \dfrac{360\ day}{1\ year}$ = 1.4 x 10^8 kJ/year.

The cost of products of A and B pumps is from Equation 20.

$$C_{product} = \dfrac{C_{fuel}}{\varepsilon} + \dfrac{[EI\dot{O}PPY]}{[F\dot{U}EL]}$$

A $\longrightarrow$ Pump

$$C_{product} = \dfrac{5\ x\ 10^{-5}\$/KJ}{0.8} + \dfrac{805.4\ \$/KJ}{0.8x1.4\ x\ 10^8\ kj/year} = 6.969\ x\ 10^{-5}\ \$/kJ$$

$C_{product}$ = 6.969 x 10^{-5} $/kj x 3600 kj/kWh = 0.25 $/kWh

B $\longrightarrow$ Pump

$$C_{product} = \dfrac{5\ x\ 10^{-5}\$/KJ}{0.8} + \dfrac{1216.1\ \$/KJ}{0.9x1.4\ x\ 10^8\ kj/year} = 7.215\ x\ 10^{-5}\ \$/kJ$$

$C_{product}$ = 7.215 x 10^{-5} $/kj x 3600 kj/kWh = 0.26 $/kWh.

CONCLUSION

The main features and requirements of the economic situation regarding algae technology are summarized below.

Microalgal biotechnology has attracted the attention of different sectors with the studies starting in the early 1950s. This is also related to the economic value of the found species and their wide usage area. Previously, simple systems that ensured certain needs were met, then more complex designs were needed with increasing and diversifying needs, and today, production systems called "photobioreactors" have been preferred. Photobioreactors are one of the important building blocks of microalgae production, and these systems increase the yield and economic quality by providing the best production environment according to the characteristics of microalgae.

CONSENT FOR PUBLICATION

Not applicable.

CONFLICT OF INTEREST

The authors declare no conflict of interest, financial or otherwise.

ACKNOWLEDGEMENT

Declared none.

REFERENCES

[1]   Chen Z, Ross TW. Refusals to deal and orders to supply in competitive markets. Int J Ind Organ 1999; 17(3): 399-417.
[http://dx.doi.org/10.1016/S0167-7187(97)00032-5]

[2]   Fufa SM, Labonnote N, Frank S, Rüther P, Jelle BP. Durability evaluation of adhesive tapes for building applications. Constr Build Mater 2018; 161: 528-38.
[http://dx.doi.org/10.1016/j.conbuildmat.2017.11.056]

[3]   Hast A, Ekholm T, Syri S. What is needed to phase out residential oil heating in Finnish single-family houses? Sustain Cities Soc 2016; 22: 49-62.
[http://dx.doi.org/10.1016/j.scs.2016.01.002]

[4]   Khazzoom JD. The demand for home insulation: A study in the household demand for conservation. In: Auer PL, Douglas D, Eds. Advances in Energy Systems and Technology. Academic Press 1986; pp. 167-312.
[http://dx.doi.org/10.1016/B978-0-12-014905-6.50006-X]

[5]   Loughlin DH, Dodder RS. Engineering economic assessment of whole-house residential wood heating in New York. Biomass Bioenergy 2014; 60: 79-87.
[http://dx.doi.org/10.1016/j.biombioe.2013.10.029]

[6]   Pieper H, Ommen T, Elmegaard B, Brix Markussen W. Assessment of a combination of three heat sources for heat pumps to supply district heating. Energy 2019; 176: 156-70.
[http://dx.doi.org/10.1016/j.energy.2019.03.165]

[7]   Rentizelas AA, Tolis AI, Tatsiopoulos IP. Investment planning in electricity production under CO2 price uncertainty. Int J Prod Econ 2012; 140(2): 622-9.
[http://dx.doi.org/10.1016/j.ijpe.2010.11.002]

[8]   Sahari A. Electricity prices and consumers' long-term technology choices: Evidence from heating investments. Eur Econ Rev 2019; 114: 19-53.
[http://dx.doi.org/10.1016/j.euroecorev.2019.02.002]

[9]   Weaver PM, Ashby MF, Burgess S, Shibaike N. Selection of materials to reduce environmental impact: A case study on refrigerator insulation. Mater Des 1996; 17(1): 11-7.
[http://dx.doi.org/10.1016/0261-3069(96)00024-6]

[10]   Xoubi N. Economic assessment of nuclear electricity from VVER-1000 reactor deployment in a developing country. Energy 2019; 175: 14-22.
[http://dx.doi.org/10.1016/j.energy.2019.03.071]

[11]   Adedipe T, Shafiee M, Zio E. Bayesian network modelling for the wind energy industry: An overview. Reliab Eng Syst Saf 2020; 202: 107053.
[http://dx.doi.org/10.1016/j.ress.2020.107053]

[12]   Casillas CE, Kammen DM. The delivery of low-cost, low-carbon rural energy services. Energy Policy 2011; 39(8): 4520-8.
[http://dx.doi.org/10.1016/j.enpol.2011.04.018]

[13]   Cui W, Caracoglia L. Simulation and analysis of intervention costs due to wind-induced damage on

tall buildings. Eng Struct 2015; 87: 183-97.
[http://dx.doi.org/10.1016/j.engstruct.2015.01.001]

[14]   Cui W, Caracoglia L. A unified framework for performance-based wind engineering of tall buildings in hurricane-prone regions based on lifetime intervention-cost estimation. Struct Saf 2018; 73: 75-86.
[http://dx.doi.org/10.1016/j.strusafe.2018.02.003]

[15]   Rentizelas A, Georgakellos D. Incorporating life cycle external cost in optimization of the electricity generation mix. Energy Policy 2014; 65: 134-49.
[http://dx.doi.org/10.1016/j.enpol.2013.10.023]

[16]   Sarmiento J, Iturrioz A, Ayllón V, Guanche R, Losada IJ. Experimental modelling of a multi-use floating platform for wave and wind energy harvesting. Ocean Eng 2019; 173: 761-73.
[http://dx.doi.org/10.1016/j.oceaneng.2018.12.046]

[17]   Tonmukayakul U, Shih STF, Bourke-Taylor H, *et al.* Systematic review of the economic impact of cerebral palsy. Res Dev Disabil 2018; 80: 93-101.
[http://dx.doi.org/10.1016/j.ridd.2018.06.012] [PMID: 29981952]

[18]   Farooq D, Thompson I, Ng KS. Exploring the feasibility of producing sustainable aviation fuel in the UK using hydrothermal liquefaction technology: A comprehensive techno-economic and environmental assessment. Cleaner Engineering and Technology 2020; 1: 100010.
[http://dx.doi.org/10.1016/j.clet.2020.100010]

[19]   Fózer D, Volanti M, Passarini F, Varbanov PS, Klemeš JJ, Mizsey P. Bioenergy with carbon emissions capture and utilisation towards GHG neutrality: Power-to-Gas storage *via* hydrothermal gasification. Appl Energy 2020; 280: 115923.
[http://dx.doi.org/10.1016/j.apenergy.2020.115923]

[20]   Gupta SS, Bhartiya S, Shastri Y. Model-based optimisation of integrated algae biorefinery. IFAC Proceedings Volumes 2014; 47(1)1011: 1018.

[21]   Hezi Z, Shpak S, Fliesher M, Gillerman L, Kasher R, Oron G. Optimal managing the coastal aquifer for seawater desalination and meeting nitrates level of drinking water. Desalination 2018; 436: 63-8.
[http://dx.doi.org/10.1016/j.desal.2018.02.014]

[22]   Hosseinizand H, Lim CJ, Webb E, Sokhansanj S. Economic analysis of drying microalgae Chlorella in a conveyor belt dryer with recycled heat from a power plant. Appl Therm Eng 2017; 124: 525-32.
[http://dx.doi.org/10.1016/j.applthermaleng.2017.06.047]

[23]   Kovacevic V, Wesseler J. Cost-effectiveness analysis of algae energy production in the EU. Energy Policy 2010; 38(10): 5749-57.
[http://dx.doi.org/10.1016/j.enpol.2010.05.025]

[24]   Lee JC, Lee B, Heo J, Kim HW, Lim H. Techno-economic assessment of conventional and direct-transesterification processes for microalgal biomass to biodiesel conversion. Bioresour Technol 2019; 294: 122173.
[http://dx.doi.org/10.1016/j.biortech.2019.122173] [PMID: 31586730]

[25]   Mahapatra DM, Chanakya HN, Ramachandra TV. Algae Derived Single-Cell Proteins: Economic Cost Analysis and Future Prospects. In: Singh Dhillon G, Ed. Protein Byproducts. Academic Press 2016; pp. 275-301.
[http://dx.doi.org/10.1016/B978-0-12-802391-4.00015-X]

[26]   Ng DKS, Ng KS, Ng RTL. Integrated Biorefineries. In: Abraham MA, Ed. Encyclopedia of Sustainable Technologies. Oxford: Elsevier 2017; pp. 299-314.
[http://dx.doi.org/10.1016/B978-0-12-409548-9.10138-1]

[27]   Vulsteke E, Van Den Hende S, Bourez L, Capoen H, Rousseau DPL, Albrecht J. Economic feasibility of microalgal bacterial floc production for wastewater treatment and biomass valorization: A detailed up-to-date analysis of up-scaled pilot results. Bioresour Technol 2017; 224: 118-29.
[http://dx.doi.org/10.1016/j.biortech.2016.11.090] [PMID: 27955865]

[28]   Belessiotis V, Kalogirou S, Delyannis E. Solar Distillation—Solar Stills. In: Belessiotis V, Kalogirou S, Delyannis E, Eds. Thermal Solar Desalination. Academic Press 2016; pp. 103-90.
[http://dx.doi.org/10.1016/B978-0-12-809656-7.00003-9]

[29]   Calise F, d'Accadia MD, Quiriti E, Vicidomini M, Piacentino A. Trigeneration and Polygeneration Configurations for Desalination and Other Beneficial Processes. In: Gude VG, Ed. Sustainable Desalination Handbook. Butterworth-Heinemann 2018; pp. 99-199.
[http://dx.doi.org/10.1016/B978-0-12-809240-8.00004-6]

[30]   Curcio E, Profio GD, Fontananova E, Drioli E. Membrane technologies for seawater desalination and brackish water treatment. In: Basile A, Cassano A, Rastogi NK, Eds. Advances in Membrane Technologies for Water Treatment. Oxford: Woodhead Publishing 2015; pp. 411-41.
[http://dx.doi.org/10.1016/B978-1-78242-121-4.00013-7]

[31]   El-Halwagi MM. Overview of Process Economics. In: El-Halwagi MM, Ed. Sustainable Design Through Process Integration. 2nd ed.. . Butterworth-Heinemann 2017; pp. 15-71.

[32]   Hetti RK, Karunathilake H, Chhipi-Shrestha G, Sadiq R, Hewage K. Prospects of integrating carbon capturing into community scale energy systems. Renew Sustain Energy Rev 2020; 133: 110193.
[http://dx.doi.org/10.1016/j.rser.2020.110193]

[33]   Koleva MN, Polykarpou EM, Liu S, Styan CA, Papageorgiou LG. Synthesis of water treatment processes using mixed integer programming. In: Gernaey KV, Huusom JK, Gani R, Eds. Computer Aided Chemical Engineering. Elsevier 2015; 37: pp. 1379-84.

[34]   Leperi KT, Snurr RQ, You F. Optimization of Pressure/Vacuum Swing Adsorption with Variable Dehydration Levels for Post Combustion Carbon Capture. In: Gernaey KV, Huusom JK, Gani R, Eds. Computer Aided Chemical Engineering. Elsevier 2015; 37: pp. 2447-52.

[35]   Pal P. Industry-specific water treatment: Case studies. In: Pal P, Ed. Industrial Water Treatment Process Technology. Butterworth-Heinemann 2017; pp. 243-511.
[http://dx.doi.org/10.1016/B978-0-12-810391-3.00006-0]

[36]   Voutchkov N. Comparison of granular media and membrane pretreatment. In: Voutchkov N, Ed. Pretreatment for Reverse Osmosis Desalination. Amsterdam: Elsevier 2017; pp. 221-38.
[http://dx.doi.org/10.1016/B978-0-12-809953-7.00010-3]

[37]   Acién Fernández FG, Fernández Sevilla JM, Molina Grima E. Costs analysis of microalgae production. In: Pandey A, Chang J-S, Soccol CR, Lee DJ, Chisti Y, Eds. Biofuels from Algae. Elsevier 2019; pp. 551-66.

[38]   Atta AP, Diango A, N'Guessan Y, Descombes G, Morin C, Jaecker-Voirol A. Life cycle assessment, technical and economical analyses of jatropha biodiesel for electricity generation in remote areas of côte d'Ivoire. In: Kumar RP, Gnansounou E, Raman JK, Baskar G, Eds. Refining Biomass Residues for Sustainable Energy and Bioproducts. Academic Press 2020; pp. 523-42.
[http://dx.doi.org/10.1016/B978-0-12-818996-2.00024-7]

[39]   Bost KL, Piller KJ. Plant-based mucosal immunotherapy: Challenges for commercialization. Mucosal Vaccines. 2nd ed.. Kiyono H, Pascual DW. Academic Press 2020; pp. 371-82.

[40]   Giraldo JAP, Ortiz Sanchez M, Solarte-Toro JC, Cardona Alzate CA. Economic and social aspects of biorefineries.Recent Advances in Bioconversion of Lignocellulose to Biofuels and Value-Added Chemicals within the Biorefinery Concept. Elsevier 2020; pp. 199-231.
[http://dx.doi.org/10.1016/B978-0-12-818223-9.00008-4]

[41]   Jeguirim M, Zorpas AA, Navarro Pedreno J, *et al.* Sustainability assessment for biomass-derived char production and applications. In: Jeguirim M, Limousy L, Eds. Char and Carbon Materials Derived from Biomass. Elsevier 2019; pp. 447-79.
[http://dx.doi.org/10.1016/B978-0-12-814893-8.00012-2]

[42]   Lan K, Park S, Yao Y. Key issue, challenges, and status quo of models for biofuel supply chain design. In: Ren J, Scipioni A, Manzardo A, Liang H, Eds. Biofuels for a More Sustainable Future.

Elsevier 2020; pp. 273-315.
[http://dx.doi.org/10.1016/B978-0-12-815581-3.00010-5]

[43] Mohamed A-MO, Maraqa MA, Howari FM, Paleologos EK. Outdoor air pollutants: sources, characteristics, and impact on human health and the environment. In: Mohamed A-MO, Paleologos EK, Howari FM, Eds. Pollution Assessment for Sustainable Practices in Applied Sciences and Engineering. Butterworth-Heinemann 2021; pp. 491-554.
[http://dx.doi.org/10.1016/B978-0-12-809582-9.00009-8]

[44] Rana MS, Bhushan S, Prajapati SK. Techno-economic analysis and environmental aspects of food waste management. In: Rana JR, Kumar R, Gunasekaran M, Kavitha S, Eds. Food Waste to Valuable Resources. Academic Press 2020; pp. 325-42.

[45] Rodenas P, Wardman C, Esteve-Nuñez A. Metals recovery from wastewater by microbial electrochemical technologies. In: Olivares JA, Puyol D, Melero JA, Dufour J, Eds. Wastewater Treatment Residues as Resources for Biorefinery Products and Biofuels. Elsevier 2020; pp. 281-307.
[http://dx.doi.org/10.1016/B978-0-12-816204-0.00013-8]

[46] Sanders JPM, Langeveld JWA. Development perspectives for the bio-based economy. In: Galanakis CM, Ed. Biobased Products and Industries. Elsevier 2020; pp. 41-78.
[http://dx.doi.org/10.1016/B978-0-12-818493-6.00002-6]

[47] Wypych G, Ed. solvent recycling, removal, and degradation. Handbook of Solvents. 3rd ed. ChemTec Publishing 2019; pp. 1635-727.
[http://dx.doi.org/10.1016/B978-1-927885-41-3.50011-5]

[48] Ball AS, Patil S, Soni S. Introduction into nanotechnology and microbiology. In: Gurtler V, Ball AS, Soni S, Eds. Methods in Microbiology. Academic Press 2019; 46: pp. 1-18.

[49] de Richter R, Caillol S, Ming T. Geoengineering: Sunlight reflection methods and negative emissions technologies for greenhouse gas removal. In: Letcher TM, Ed. Managing Global Warming. Academic Press 2019; pp. 581-636.
[http://dx.doi.org/10.1016/B978-0-12-814104-5.00020-X]

[50] Dimian AC, Bildea CS, Kiss AA. Bioethanol and biobutanol. In: Dimian AC, Bildea CS, Kiss AA, Eds. Applications in Design and Simulation of Sustainable Chemical Processes. Elsevier 2019; pp. 285-327.
[http://dx.doi.org/10.1016/B978-0-444-63876-2.00008-5]

[51] Dimian AC, Bildea CS, Kiss AA. Methanol. In: Dimian AC, Bildea CS, Kiss AA, Eds. Applications in Design and Simulation of Sustainable Chemical Processes. Elsevier 2019; pp. 101-45.
[http://dx.doi.org/10.1016/B978-0-444-63876-2.00003-6]

[52] Gnansounou E. Economic Assessment of Biofuels. In: Pandey A, Larroche C, Dussap C-G, Eds. Biofuels: Alternative Feedstocks and Conversion Processes for the Production of Liquid and Gaseous Biofuels. 2nd ed.. Academic Press 2019; pp. 95-121.

[53] Kiss AA, Pătrașcu I, Bîldea CS. From substrate to biofuel in the acetone–butanol–ethanol process. In: Basile A, Dalena F, Eds. Second and Third Generation of Feedstocks. Elsevier 2019; pp. 59-82.
[http://dx.doi.org/10.1016/B978-0-12-815162-4.00003-3]

[54] Kohli R. Applications of solid carbon dioxide (dry ice) pellet blasting for removal of surface contaminants. In: Kohli R, Mittal KL, Eds. Developments in Surface Contamination and Cleaning: Applications of Cleaning Techniques. Elsevier 2019; pp. 117-69.
[http://dx.doi.org/10.1016/B978-0-12-815577-6.00004-9]

[55] Lewis AJ, Borole AP. Bio-Electro-Refinery: Conversion, sustainability, and policy. In: Mohan SV, Varjani S, Pandey A, Eds. Microbial Electrochemical Technology. Elsevier 2019; pp. 1059-85.
[http://dx.doi.org/10.1016/B978-0-444-64052-9.00044-3]

[56] Poinern GEJ, Fawcett D. Sustainable utilization of renewable plant-based food wastes for the green synthesis of metal nanoparticles. In: Henry DJ, Ed. Harnessing Nanoscale Surface Interactions.

Elsevier 2019; pp. 1-39.
[http://dx.doi.org/10.1016/B978-0-12-813892-2.00001-X]

[57]   Dincer I, Bicer Y. Enhanced dimensions of integrated energy systems for environment and sustainability. In: Dincer I, Bicer Y, Eds. Integrated Energy Systems for Multigeneration. Elsevier 2020; pp. 403-40.
[http://dx.doi.org/10.1016/B978-0-12-809943-8.00007-8]

[58]   Kalıncı Y, Dincer I. 5.3 Waste Energy Management. In: Dincer I, Ed. Comprehensive Energy Systems. Oxford: Elsevier 2018; pp. 91-133.
[http://dx.doi.org/10.1016/B978-0-12-809597-3.00510-1]

[59]   Lu H, Chen J, Guo L. 5.7 Energy Quality Management. In: Dincer I, Ed. Comprehensive Energy Systems. Oxford: Elsevier 2018; pp. 258-314.
[http://dx.doi.org/10.1016/B978-0-12-809597-3.00521-6]

[60]   Sieniutycz S. Thermodynamics and Optimization of Practical Processes. In: Sieniutycz S, Ed. Thermodynamic Approaches in Engineering Systems. Amsterdam: Elsevier 2016; pp. 347-420.
[http://dx.doi.org/10.1016/B978-0-12-805462-8.00008-X]

[61]   Sieniutycz S. System analysis in energy engineering and ecology. In: Sieniutycz S, Ed. Complexity and Complex Thermo-Economic Systems. Elsevier 2020; pp. 117-73.
[http://dx.doi.org/10.1016/B978-0-12-818594-0.00006-4]

[62]   Sieniutycz S, Jeżowski J. Systems theory in thermal and chemical engineering. In: Sieniutycz S, Jeżowski J, Eds. Energy Optimization in Process Systems. Amsterdam: Elsevier 2009; pp. 391-426.
[http://dx.doi.org/10.1016/B978-0-08-045141-1.00011-1]

[63]   Sieniutycz S, Jeżowski J. Systems theory in thermal and chemical engineering. In: Sieniutycz S, Jeżowski J, Eds. Energy Optimization in Process Systems and Fuel Cells. 2nd ed.. Amsterdam: Elsevier 2013; pp. 429-64.

[64]   Sieniutycz S, Jeżowski J. Hamilton–Jacobi–Bellman theory and practical energy systems. Energy Optimization in Process Systems and Fuel Cells. 3rd ed. Sieniutycz S, Jeżowski J. Elsevier 2018; pp. 199-241.

[65]   Stanek W, Gładysz P, Czarnowska L, Simla T. Examples of application of TEC. In: Stanek W, Gładysz P, Czarnowska L, Simla T, Eds. Thermo-ecology. Academic Press 2019; pp. 67-189.

[66]   Acar C, Dincer I. 3.1 Hydrogen Production. In: Dincer I, Ed. Comprehensive Energy Systems. Oxford: Elsevier 2018; pp. 1-40.
[http://dx.doi.org/10.1016/B978-0-12-809597-3.00304-7]

[67]   Dincer I, Zamfirescu C. 1.4 Sustainability Dimensions of Energy. In: Dincer I, Ed. Comprehensive Energy Systems. Oxford: Elsevier 2018; pp. 101-52.
[http://dx.doi.org/10.1016/B978-0-12-809597-3.00104-8]

[68]   Dincer I, Zamfirescu C. 1.5 Thermodynamic Aspects of Energy. In: Dincer I, Ed. Comprehensive Energy Systems. Oxford: Elsevier 2018; pp. 153-211.
[http://dx.doi.org/10.1016/B978-0-12-809597-3.00105-X]

CHAPTER 15

Closing Remarks

Cemil Koyunoğlu[1,2,*] and **Hüseyin Karaca**[3]

[1] *Energy Systems Engineering Department, Engineering Faculty, Cinarcik Road 5th km, 77200, Yalova University, Yalova, Turkey*

[2] *Fuel Oil Analysis Laboratory, Central Laboratory, Cinarcik Road 5th km, 77200, Yalova University, Yalova, Turkey*

[3] *Chemical Engineering Department, Engineering Faculty, Inonu University, Malatya, Turkey*

Abstract: In order to better evaluate the water potential of our world, we will now take a look at the various aquatic products that have had a significant impact on the biotech industry, from research materials to nutritional additives.

Keywords: Biofuel from marine creatures, Biofuel production, Mechanical pressing process, Microalgae, Oil extraction, Solvent extraction.

INTRODUCTION

The process of combining water and sunlight by the plant and turning it into matter and oxygen is called photosynthesis. Photosynthesis, also known as carbon dioxide digestion, usually occurs in the green leaves of plants in lamellar structures called chloroplasts. With the help of sunlight, about 3 liters of oxygen per hour is produced per kilogram of a healthy green leaf. This phenomenon, which provides an energy flow of about 16 W, is obtained from a 1 m^2 leaf area. One of the most fundamental phenomena that occurs in nature and that plants interact with light in photosynthesis. In terms of the formation of nutrients, fuel, and the production of atmospheric oxygen, all living things are closely dependent on this event. A leaf area of 15-30 m^2 is required to produce the oxygen that a person needs for 24 hours. Calculations show that in an average climate zone, a large tree can only provide the oxygen consumed by an adult in a year [1 - 7].

It is known that the energy given by the sun to the world is approximately 1.5×10^{18} kWh/year and this issue is 10000 times greater than the total energy con-

* **Correspondence author Cemil Koyunoğlu:** Energy Systems Engineering Department, Engineering Faculty, Cinarcik Road 5th km, 77200, Yalova University, Yalova, Turkey; Tel: +902268155378; E-mail: cemil.koyunoglu@yalova.edu.tr

sumed in the world. Approximately 0.1% of this energy coming to the surface of the earth is converted into biomass by photosynthesis and stored. This is approximately 10 times more than the total energy used in the world. The average photosynthesis efficiency given above increases 0.5-1.3% in temperate regions and 0.5-2.5% in semi-tropical regions. In order to give an idea about the energy of biomass formed as a result of photosynthesis, it would be sufficient to say that this is equal to the power (9×10^7 MW) of a 100000 times larger nuclear power station [6, 8, 11].

For example, the total energy amount per one hectare area parallel to 40o is 1.47×10^{13} calories. If all of this energy was to be converted into carbohydrates, a yield of 2000 tons per hectare could be achieved. However, 43% of the total sunlight falling on the earth's surface is effective rays with photosynthesis. For willow and hybrid poplar, which can yield 10 tons of dry matter per hectare, 159 W of sunlight falls per m^2 and the plant can absorb 0.30-0.41% of it. In Thailand, where 186 W of sunlight falls per m^2, the amount of solar energy absorbed by green algae increases to 4.90%, and accordingly, the dry matter yield reaches 164 tons/ha per year. While the gas production range for algae species is 380-550 kg volatile matter per liter, the average gas production is 460 l / kg-UM [6, 8 -11].

Solid fermented fertilizers are used in algae production. The use of fermented fertilizer is the method of placing dry or green leaves in a pit with a filter-bed, reusing the bottom-filtered water in the reactor feed material, and storing the above dry matter for use. This fermented fertilizer contains bacteria heavily and is suitable for composting plant materials. During anaerobic fermentation, the nutritional value is not lost, and especially vitamin B12 is synthesized during this period. It is rich in protein. This material is dried under the sun or in a dryer. In this type of use, drying the fermented manure by using the produced biogas increases the total system efficiency. However, high temperatures should be avoided in order not to spoil the vitamin structure. This dried material is laid and can be used as an additive in animal feed or as fish feed after the venom is removed. The demilitarized fermented fertilizer is rich in nutrients and trace elements and can be used in algae production [12 - 16].

Microalgae

Microalgae, which is one of the sources of biomass energy, is also an important ecological cycle element. Blue, red, green, *etc.* microalgae production in colors is the most effective and economical way to convert solar energy into biomass [17 - 21].

Microalgae are used in the feed, food, cosmetic and pharmaceutical industries as well as in biotechnological production. Recently, it has been used in biofuel production [17 - 21].

Algae can be grown abundantly in much smaller plots. In addition, algae do not require special conditions such as fresh water and fertile soil and can be harvested many times a year. Algae can grow practically, wherever there is sufficient light. Some of them are also grown in saltwater [17 - 21].

Algae can grow indoors or outdoors under suitable conditions. Open systems include pools, channels, waterfalls, natural water environments, *etc.* systems. Closed systems consist of photobioreactors. The basic principle in this type of reactor design is to reduce the light path and thus increase the amount of light reaching each cell. In addition, there must be functions that increase gas exchange, provide the optimum amount of light to the cells, and create a good mixture. The necessary conditions for the production of microalgae are given in Table **1** [17 - 21].

Microalgae are much more advantageous than soy and corn plants for biodiesel production. Table **2** shows the comparison of microalgae and some other biodiesel sources [17 - 21].

Table 1. Production conditions for microalgae.

Parameters	Boundary Values	Optimum Conditions
Temperature (°C)	16-27	18-24
Saltiness (g/I)	12-40	20-24
Light density (Ix)	1000-10000	2500-5000
Lightness time (Morning: m night h)	-	16:8 minimum 24:0 maximum
pH	7-9	8.2-8.7

Table 2. Comparison of some biodiesel sources.

Plant	Oil Yield (I/ha)	Needed Area (Mha)*	Percentage of the Harvested Area*
Corn	172	1540	864
Soybean	446	594	326
Rape	1190	223	122
Jatropha	1892	140	77
Coconut	2689	99	54
Palm	5950	45	24

(Table 2) cont.....

Plant	Oil Yield (l/ha)	Needed Area (Mha)*	Percentage of the Harvested Area*
Microalgae **	136900	2	1.1
Microalgae ***	58700	4.5	2.5

*: To meet 50% of all transportation fuel (in the USA) **: 70% by weight of biomass is oil ***:30% of the weight in biomass is fat

According to the table; microalgae appear to be the only source of biodiesel, which has the potential to completely replace diesel fuel. Unlike other oil crops, microalgae grow extremely fast and some are very rich in excess and fat. Microalgae are the fastest-growing photosynthetic organisms. They can complete all growth cycles every few days. Approximately, 46 t/ha/year can be produced from unicellular marine algae. Different types of algae produce different amounts of oil [17 - 21].

Microalgae often double their biomass within 24 hours. Biomass doubling times during exponential growth are usually as short as 3.5 hours. The oil content of microalgae can exceed 80% of the weight of dry biomass. According to some evaluations, the return of oil from algae (per 0.404 ha) is 200 times higher than the yield of the best performing plant/vegetable oils [17 - 21].

In a study, biodiesel production was carried out from heterotroph microalgal oil. Table **3** shows the comparison of the properties of biodiesel produced from microalgal oil, with diesel and biodiesel in ASTM standards [17 - 21].

Table 3. Comparison of the properties of biodiesel produced from microalgal oil, Diesel, and biodiesel in the ASTM biodiesel standard.

Properties	Biodiesel Production From Microalgae	Diesel Fuel	ASTM Biodiesel Standart
Density (kg/l)	0.864	0.838	0.86-0.9
Viscosity (40 °C'de) mm²/s	5.2	1.9-4.1	3.5-5.0
Flash point (°C)	115	75	Min. 100
Solidification point (°C)	-12	-50-10	-
Cold filter plug point (°C)	-11	-3.0 (Max. -6.7)	Summer (max. 0) Winter max. < -15
Acid value (mg KOH /g)	0.374	Max. 0.5	Max. 0.5
Heating value (MJ/kg)	41	40-45	-
H/C_{ratio}	1.81	1.81	-

Biodiesel produced from microalgae oil appears to comply with most ASTM biodiesel quality standards. Biodiesel produced from microalgal oil showed a lower cold filter plugging point at -11 °C compared to diesel. Potentially, heterotrophic microorganisms that grow oil on natural organic carbon, such as sugar, instead of microalgae can be used to make biodiesel. However, heterotrophic microorganisms can be used for making biodiesel. However, heterotrophic production is not as efficient as using photosynthetic microalgae. Because the renewable organic carbon resources required for the development of heterotrophic microorganisms are ultimately produced by photosynthesis, usually in crop areas [22 - 26].

Biodiesel is produced in geothermal areas that receive carbon dioxide sources, using microalgae that can perform photosynthesis with high oil levels, in inefficient and inert lands and pools with treated wastewater (microalgae greenhouses heated with geothermal energy for four seasons) are produced in easy and economical ways [22 - 26].

With the use of microalgae in biodiesel production, carbon dioxide will become a useful waste that contributes to the economy of humans. This situation increases the importance of cheap and easily accessible carbon dioxide sources [22 - 26]. Geothermal fields are important sources of carbon dioxide production. Biodiesel production from microalgae in geothermal fields could be an important alternative for possible livable energy bottlenecks in the future. With biodiesel production from microalgae in geothermal fields;

- More biodiesel is produced in a shorter time, at less cost, and from algal oil.
- Carbon dioxide, which brings the biggest financial burden in microalgae production, is obtained simply.
- By using waste carbon dioxide in algae production ponds, environmental recycling is eliminated by geothermal power plants.
- Pools of square meters are established by utilizing idle lands. Thanks to the wastewater used in the pools, the water is recycled [22 - 26].

Fig. (**1**) shows the flow chart of biodiesel production from microalgae in the geothermal environment.

Microalgal oils differ from many vegetable oils in that they are highly rich in polyunsaturated fatty acids with four or more double bonds. For example, eicosapentaenoic acid (EPA, C20: 5n-3; five double bonds) and docosahexaenoic acid (DHA, C22: 6n-3; six double bonds) are usually formed in algae oils. Fatty acids with four or more double bonds and FAME are very susceptible to oxidation during storage, which reduces the acceptability of these oils for use in biodiesel production. Some vegetable oils also have this problem. For example, oils such as high oleic rapeseed oil contain large amounts of linoleic acid (C18: 2n-6, 2 double bonds) and linolenic acid (C18: 3n-3, 3 double bonds). Although the oxidative stability of these fatty acids is higher than EPA and DHA, EN 14214 limits the linolenic acid methyl ester content in biodiesel to 12 mol% for vehicle use. There are no such restrictions for the amount of oil in biodiesel used for heating purposes, but acceptable biodiesel must meet other criteria based on the degree of total unsaturation of the oil. The total unsaturation of oil is indicated by its iodine number. EN 14214 and EN 14213 standards should not exceed the iodine value of biodiesel 120 and 130 g iodine / 100 g biodiesel, respectively. In addition, European biodiesel standards limit the content of four or more double-bonded FAMEs to a maximum of 1 mol%. Given the composition of some of the microalgal oils, it is likely that many of them do not meet the European biodiesel standards; however, this situation need not be an important limitation. The degree of unsaturation of microalgae and fatty acids containing more than four double bonds can be easily reduced by partial catalytic hydrogenation, which is generally used to make margarine from vegetable oils [28 - 32].

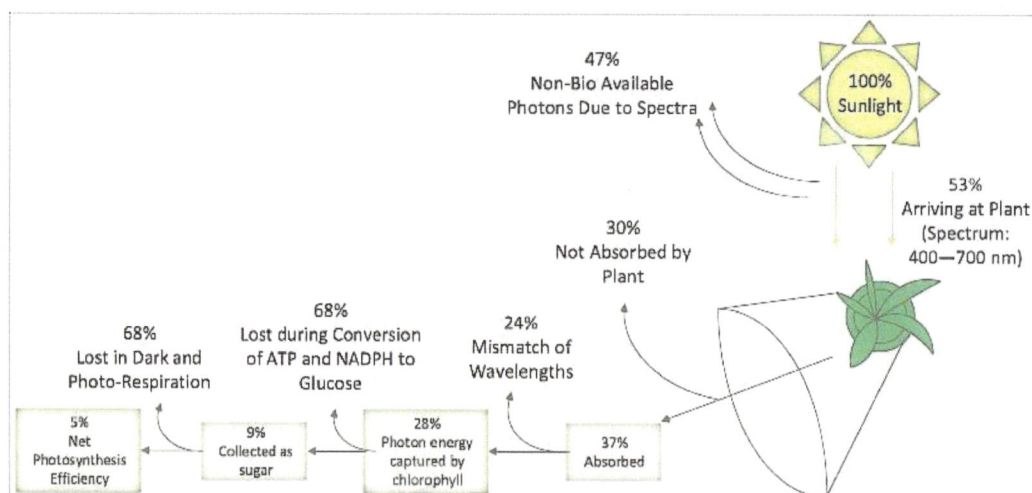

Fig. (1). Biodiesel production paths from microalgae [27].

OIL EXTRACTION FROM SEEDS

Oilseeds are subjected to preliminary processes such as cleaning (garbage separator), seed moistening and roasting (conditioning), shelling and separation, flaking, and so on before processing. The pre-treatments to be applied to vary according to the legal characteristics of the oilseed and the method to be used in crude oil production [33 - 42].

Oily seeds; Contains many impurities such as stone chips, metal particles, stalk, garbage. These substances will not only damage the equipment used during oil processing but also reduce the quality and efficiency of the oil obtained. For these reasons, oilseeds should be cleaned. Cleaning processes consist of systems that operate according to foreign matter, size, shape, density, and magnetism. These systems can be listed as sieves, triotors, pneumatic separators, magnet systems, brushing machines, and lintering machines (for cigite) [33 - 42].

Oilseeds purified from foreign substances: they are subjected to tempering and quenching under suitable conditions. This process allows the product to be taken easily without damaging the inside of the seed. After the annealing process, the oilseeds are peeled off. These processes are carried out in double-single-universal shell breaking and separation systems. The skins separated from the oilseeds leave the system as a by-product and are added to the meal when necessary to adjust the protein content of the meal [33 - 42].

Mechanical pressing, solvent extraction, pre-press-solvent extraction or two-stage solvent extraction, and supercritical fluid extraction methods are used in the production of crude oil from pre-processed oil seeds, depending on the oil content of the raw material [33 - 42].

Mechanical Pressing Process

Mechanical pressing can be defined as the solid-liquid phase separation method. Generally, mechanical pressing is applied to oil seeds with an oil content of more than 20%. As a result of mechanical pressing, crude oil is obtained as the main product and pulp as a by-product [43 - 51].

In order to produce fuel, the oil in the seeds is extracted by different methods. Among these methods, pressing has small or large capacities. Small scale (capacity) ones are preferred, especially since they do not impose a large financial burden on farmers. The most common of the presses in terms of working principle are screw presses. In this type of press driven by an electric motor, the material feeding amount is 4-8 kilograms per hour and the amount of oil obtained is 0.25-2 liters. Screw presses are also available for industrial purposes and their daily capacities are very high. These presses can be driven by electric motors or diesel engines [43 - 51].

Reciprocating presses, on the other hand, serve much larger capacities. The oil obtained as a result of pressing should be kept in the tank for a few days and the solid phase should settle to the bottom of the tank. Then the liquid part should be filtered [43 - 51].

In the commercial production of biodiesel, chemical extraction is used. Batch or continuous types are used in solvent extraction. These methods are suitable for large capacities. For example, the daily capacity of continuous types should be at least 200 tons for economic [43 - 51].

Solvent Extraction

Extraction methods; It is based on the principle of combining the oil solvent and the raw material, entering the oil into the solvent, and removing the solvent in the mixture from the oil. As the oil content of the seeds decreases in the production of crude oil, the efficiency of mechanical pressing decreases, and the ratio of the oil that cannot be taken to the total oil in the seed increases. Therefore, solvent extraction gives better results than mechanical pressing for seeds with low oil content [52 - 58].

In the pre-pressing-solvent extraction or direct solvent extraction method of oilseeds with a higher oil ratio than 20%, the oil rate is reduced to 14-20% in the first stage, and the remaining oil is extracted by solvent extraction in the second stage. In direct solvent extraction, two-stage solvent extraction is applied [52 - 58].

Supercritical Fluid Extraction

It is a technique in which important developments have been made in recent years. Hydrocarbons used as solvents in oil extraction have negative effects on human environmental health. Extraction of oil from oilseeds can be achieved with a liquid that acts as a solvent at temperatures and pressures above its critical point. Since the process is carried out under high pressure, the installation and operating costs of the system are quite high [53 - 62].

BIOFUEL FROM MARINE CREATURES

Taq polymerase, which is isolated from *Thermus aquaticus archaea* in hot spring waters and enables the development of PCR as a powerful tool in molecular biology, is important. Oceans have also been shown to be an excellent resource for enzymes and other products that play important roles in basic and applied research. For example, bacteria living near hydrothermal vents (ocean floor hot springs) have produced second-generation heat-resistant enzymes for PCR and enzymes that modify DNA, such as ligases and restriction enzymes [63 - 72].

Other enzymes produced by marine bacteria have interesting properties that will enable new applications that can be used in the future. Some enzymes are salt resistant, making them ideal for industrial-scale cultivation methods such as high salt solutions. *Bioluminescent Vibrio harveyi* plays a role in determining the environmental pollution of bacteria. Vibrio's marine species produce a variety of proteases; Some of these are unique proteases resistant to many detergents used in manufacturing processes. Consequently, these detergent-resistant proteases may have potential applications for cleaning processes as solvent proteins, such as use in laundry detergents that remove protein stains on clothes. Vibrio is also a good source for collagenase, a protease used in tissue culture. When scientists want to grow cells, such as cultured liver cells, the connective tissues that hold the cells together can use collagenases to cut the cells, allowing the cells to be dispersed separately in the culture plates [63 - 72].

Another product obtained from the sea; Carrageenan found in many preserved foods, toothpaste, and cosmetics. This sulfate-rich polysaccharide is derived from red waterweed and has been used in many products for over 50 years. There is a large family of carrageenan polysaccharides. Under different temperatures, they can form gels of varying densities. For this reason, carrageenans are used as thickening agents, to improve the "feel" of food in our mouth while chewing. Among the common applications of carrageenan; The use of this substance as a stabilizer and volume enhancer in chewing gum, chocolate milk, beer and wine, salad dressings, syrups, sauces, processed meats, adhesives, fabrics, polishes, and hundreds of other products can be shown. The Philippines is the world's largest producer of red algae (Rhodophyta) from which large amounts of carrageenan are obtained. Advances in a paper-thick algae cultivation used to enclose red algae strains have also increased the production of other algal polymers; these include important research materials such as agar used to make agar gels to grow bacteria and alginic acid used to make agarose gels for DNA electrophoresis [63 - 72].

Biofuel from Algae

Biofuel producer groups have recently been examining microscopic algae, rather than corn and frying oil, as a future alternative to oil. Microalgae naturally produce and store lipids (vegetable-like oils). If they can be changed biochemically, they can be more efficient in producing biofuels. Despite the U.S. Department of Energy (DOE) discovering the potential of algae before 1996, the same year the drop in crude oil prices from $ 50 to $ 20 a barrel reduced interest from the DOE, with the establishment of the Biofuel initiative in 2007 and the increasing price of crude oil. also back. The challenge is to produce at least a million barrels of fuel a day. The Solazyme company in San Francisco has developed a closed bioreactor system with algae that produce triacylglycerides and methanol by eating everything from waste glycerol to sugar pulp. After chemical modification and condensation, the alkanes produced are similar to biodiesel. While 50-60% fuel efficiency per gram of algae cell is considered excellent, Solazyme's algae produce 75% fuel per gram of dry weight. This company is the first to produce a barrel of microbial fuel oil and is confident that the target of one million barrels of fuel per day can be achieved [54, 73 - 81].

CONCLUSION

Such biomass-based resources are used in heating, cooking, and biofuel production such as ethanol, biodiesel, and electricity generation by going through different conversion processes. Biomass energy has an extremely important place as it is renewable and continuous, as well as being easily available, has low storage and transportation costs, contributes to socio-economic development, and is defined as environmentally friendly. Liquefaction of biomass is important in this sense. In other words, it is important that biomass is considered renewable energy and that it is in a position to meet the increasing energy supply.

CONSENT FOR PUBLICATION

Not applicable.

CONFLICT OF INTEREST

The authors declare no conflict of interest, financial or otherwise.

ACKNOWLEDGEMENT

Declared none.

REFERENCES

[1]   Allakhverdiev SI. Photosynthesis research for sustainability: From natural to artificial. Biochim Biophys Acta Bioenerg 2012; 1817(8): 1107-9.
       [http://dx.doi.org/10.1016/j.bbabio.2012.03.002] [PMID: 22687280]

[2]   Benemann JR, Weissman JC. biophotolysis: Problems and prospects. In: Schlegel HG, Barnea J, Eds. Microbial Energy Conversion. Pergamon 1977; pp. 413-26.

[3]   da Rosa AV. Biomass. In: da Rosa AV, Ed. Fundamentals of Renewable Energy Processes. 2nd ed.. Academic Press 2009; pp. 569-623.

[4]   Faunce T. Towards a global solar fuels project-artificial photosynthesis and the transition from anthropocene to sustainocene. Procedia Eng 2012; 49: 348-56.
       [http://dx.doi.org/10.1016/j.proeng.2012.10.147]

[5]   Restrepo-Coupe N, da Rocha HR, Hutyra LR, *et al.* What drives the seasonality of photosynthesis across the Amazon basin? A cross-site analysis of eddy flux tower measurements from the Brasil flux network. Agric For Meteorol 2013; 182-183: 128-44.
       [http://dx.doi.org/10.1016/j.agrformet.2013.04.031]

[6]   Sydney EB, Sydney ACN, de Carvalho JC, Soccol CR. Microalgal strain selection for biofuel production. In: Pandey A, Chang J-S, Soccol CR, Lee D-J, Chisti Y, Eds. Biofuels from Algae. 2nd ed.. Elsevier 2019; pp. 51-66.

[7]   Work VH, D'Adamo S, Radakovits R, Jinkerson RE, Posewitz MC. Improving photosynthesis and metabolic networks for the competitive production of phototroph-derived biofuels. Curr Opin Biotechnol 2012; 23(3): 290-7.
        [http://dx.doi.org/10.1016/j.copbio.2011.11.022] [PMID: 22172528]

[8]   Clark DP, Pazdernik NJ, McGehee MR. Molecular Evolution. In: Clark DP, Pazdernik NJ, McGehee MR, Eds. Molecular Biology. 3rd ed.. Academic Cell 2019; pp. 925-69.

[9]   Martínez DM, Ebenhack BW, Wagner TP. Energy–flow analyses and efficiency indicators. In: Martínez DM, Ebenhack BW, Wagner TP, Eds. Energy Efficiency. Elsevier 2019; pp. 101-27.
        [http://dx.doi.org/10.1016/B978-0-12-812111-5.00009-3]

[10]   Philp J, Atlas R. Microbial Resources for Global Sustainability. In: Kurtböke I, Ed. Microbial Resources. Academic Press 2017; pp. 77-101.
        [http://dx.doi.org/10.1016/B978-0-12-804765-1.00004-7]

[11]   Schlesinger WH, Bernhardt ES. The Biosphere: The Carbon Cycle of Terrestrial Ecosystems. In: Schlesinger WH, Bernhardt ES, Eds. Biogeochemistry. 3rd ed.. Boston: Academic Press 2013; pp. 135-72.

[12]   Avcioğlu AO, Türker U. Status and potential of biogas energy from animal wastes in Turkey. Renew Sustain Energy Rev 2012; 16(3): 1557-61.
        [http://dx.doi.org/10.1016/j.rser.2011.11.006]

[13]   Li J, Wang X, Kim HHM, Gates RS, Wang K. Optimal design of manure management for intensive swine feeding operation: A modeling method based on analytical target cascading. J Clean Prod 2021; 282: 124550.
        [http://dx.doi.org/10.1016/j.jclepro.2020.124550]

[14]   Lu X, Zhou Y, Zhang J, Ren Y. Determination of fluoroquinolones in cattle manure-based biogas residue by ultrasonic-enhanced microwave-assisted extraction followed by online solid phase extraction-ultra-high performance liquid chromatography-tandem mass spectrometry. J Chromatogr B Analyt Technol Biomed Life Sci 2018; 1086: 166-75.
        [http://dx.doi.org/10.1016/j.jchromb.2018.01.029] [PMID: 29680661]

[15]   Ma L, Liu Y, Xu J, *et al.* Mass loading of typical artificial sweeteners in a pig farm and their dissipation and uptake by plants in neighboring farmland. Sci Total Environ 2017; 605-606: 735-44.
        [http://dx.doi.org/10.1016/j.scitotenv.2017.06.027] [PMID: 28675883]

[16]   Wang T, Wu J, Yi Y, Qi J. Optimization of process conditions for infected animal tissues by alkaline hydrolysis technology. Procedia Environ Sci 2016; 31: 366-74.
        [http://dx.doi.org/10.1016/j.proenv.2016.02.049]

[17]   Abu Hasan H, Abu Bakar SNH, Takriff MS. Microalgae biofilms for the treatment of wastewater. In: Galanakis CM, Ed. Microalgae. Academic Press 2021; pp. 381-407.
        [http://dx.doi.org/10.1016/B978-0-12-821218-9.00012-8]

[18]   Choo M-Y, Phung YP, Juan JC. Catalytic conversion of microalgae oil to green hydrocarbon. In: Galanakis CM, Ed. Microalgae. Academic Press 2021; pp. 117-43.
        [http://dx.doi.org/10.1016/B978-0-12-821218-9.00005-0]

[19]   Hu Y, Bassi A. Extraction of biomolecules from microalgae. In: Jacob-Lopes E, Maroneze MM, Queiroz MI, Zepka LQ, Eds. Handbook of Microalgae-Based Processes and Products. Academic Press 2020; pp. 283-308.
        [http://dx.doi.org/10.1016/B978-0-12-818536-0.00011-7]

[20]   Manirafasha E, Jiao K, Zeng X, *et al.* Processing of Microalgae to Biofuels. In: Yousuf A, Ed. Microalgae Cultivation for Biofuels Production. Academic Press 2020; pp. 111-28.

[http://dx.doi.org/10.1016/B978-0-12-817536-1.00008-4]

[21]   Rebolledo-Oyarce J, Sáez-Navarrete C, Rodriguez-Cordova L. Transport phenomena models affecting microalgae growth. In: Galanakis CM, Ed. Microalgae. Academic Press 2021; pp. 63-90.
[http://dx.doi.org/10.1016/B978-0-12-821218-9.00003-7]

[22]   Chen J, Li J, Dong W, *et al.* The potential of microalgae in biodiesel production. Renew Sustain Energy Rev 2018; 90: 336-46.
[http://dx.doi.org/10.1016/j.rser.2018.03.073]

[23]   Mailaram S, Kumar P, Kunamalla A, Saklecha P, Maity SK. Biomass, biorefinery, and biofuels. In: Dutta S, Mustansar Hussain C, Eds. Sustainable Fuel Technologies Handbook. Academic Press 2021; pp. 51-87.
[http://dx.doi.org/10.1016/B978-0-12-822989-7.00003-2]

[24]   Makareviciene V, Skorupskaite V. Transesterification of microalgae for biodiesel production. In: Basile A, Dalena F, Eds. Second and Third Generation of Feedstocks. Elsevier 2019; pp. 469-510.
[http://dx.doi.org/10.1016/B978-0-12-815162-4.00017-3]

[25]   Teo SH, Islam A, Yusaf T, Taufiq-Yap YH. Transesterification of Nannochloropsis oculata microalga's oil to biodiesel using calcium methoxide catalyst. Energy 2014; 78: 63-71.
[http://dx.doi.org/10.1016/j.energy.2014.07.045]

[26]   Tüccar G, Özgür T, Aydın K. Effect of diesel–microalgae biodiesel–butanol blends on performance and emissions of diesel engine. Fuel 2014; 132: 47-52.
[http://dx.doi.org/10.1016/j.fuel.2014.04.074]

[27]   Dincer I, Bicer Y. Integration of renewable energy systems for multigeneration. In: Dincer I, Bicer Y, Eds. Integrated Energy Systems for Multigeneration. Elsevier 2020; pp. 287-402.
[http://dx.doi.org/10.1016/B978-0-12-809943-8.00006-6]

[28]   Atabani AE, Silitonga AS, Badruddin IA, Mahlia TMI, Masjuki HH, Mekhilef S. A comprehensive review on biodiesel as an alternative energy resource and its characteristics. Renew Sustain Energy Rev 2012; 16(4): 2070-93.
[http://dx.doi.org/10.1016/j.rser.2012.01.003]

[29]   Das M. Performance and emission characteristics of biodiesel–diesel blend. In: Hashmi S, Choudhury IA, Eds. Encyclopedia of Renewable and Sustainable Materials. Oxford: Elsevier 2020; pp. 202-11.
[http://dx.doi.org/10.1016/B978-0-12-803581-8.10579-X]

[30]   Straight vegetable oil as a diesel fuel? Bioenergy. 2nd ed.. Dahiya A, Ed. Academic Press 2020; pp. 439-42.

[31]   Elkelawy M, Alm-Eldin Bastawissi H, El Shenawy EA, Taha M, Panchal H, Sadasivuni KK. Study of performance, combustion, and emissions parameters of DI-diesel engine fueled with algae biodiesel/diesel/n-pentane blends. Energy Convers Manage 2020; X: 100058.

[32]   Ong HC, Mahlia TMI, Masjuki HH, Norhasyima RS. Comparison of palm oil, Jatropha curcas and Calophyllum inophyllum for biodiesel: A review. Renew Sustain Energy Rev 2011; 15(8): 3501-15.
[http://dx.doi.org/10.1016/j.rser.2011.05.005]

[33]   Ahmed SR, Shafique A, Azeem F, *et al.* Edible oil. Green Sustainable Process for Chemical and Environmental Engineering and Science. Inamuddin, Boddula R, Asiri AM. Elsevier 2021; pp. 99-126.

[34]   Cheikhyoussef N, Kandawa-Schulz M, Böck R, Cheikhyoussef A. Cold pressed Moringa oleifera seed oil. In: Ramadan MF, Ed. Cold Pressed Oils. Academic Press 2020; pp. 467-75.
[http://dx.doi.org/10.1016/B978-0-12-818188-1.00042-6]

[35]   Chew SC. Cold pressed rapeseed (Brassica napus) oil. In: Ramadan MF, Ed. Cold Pressed Oils. Academic Press 2020; pp. 65-80.
[http://dx.doi.org/10.1016/B978-0-12-818188-1.00007-4]

[36]   Chew SC. Cold-pressed rapeseed (Brassica napus) oil: Chemistry and functionality. Food Res Int

2020; 131: 108997.
[http://dx.doi.org/10.1016/j.foodres.2020.108997] [PMID: 32247493]

[37]   de Jesus SS, Filho RM. Recent advances in lipid extraction using green solvents. Renew Sustain Energy Rev 2020; 133: 110289.
[http://dx.doi.org/10.1016/j.rser.2020.110289]

[38]   López-Bascón MA, Luque de Castro MD. Soxhlet Extraction. In: Poole CF, Ed. Liquid-Phase Extraction. Elsevier 2020; pp. 327-54.

[39]   Musara C, Maroyi A, Cheikhyoussef N, Cheikhyoussef A. Cold pressed garden cress (Lepidium sativum L.) seed oil. In: Ramadan MF, Ed. Cold Pressed Oils. Academic Press 2020; pp. 477-89.
[http://dx.doi.org/10.1016/B978-0-12-818188-1.00043-8]

[40]   Mushtaq A, Hanif MA, Ayub MA, Bhatti IA, Jilani MI. Sesame. In: Hanif MA, Nawaz H, Khan MM, Byrne HJ, Eds. Medicinal Plants of South Asia. Elsevier 2020; pp. 601-15.
[http://dx.doi.org/10.1016/B978-0-08-102659-5.00044-6]

[41]   Peixouto YS, Pereira ACF, Ribeiro VS, Peixouto LS. Potential of Ricinus communis for the removal of toxic metals from mining dumping sites. In: Bauddh K, Korstad J, Sharma P, Eds. Phytorestoration of Abandoned Mining and Oil Drilling Sites. Elsevier 2021; pp. 263-86.
[http://dx.doi.org/10.1016/B978-0-12-821200-4.00004-2]

[42]   Ustun Argon Z, Celenk VU, Gumus ZP. Cold pressed grape (Vitis vinifera) seed oil. In: Ramadan MF, Ed. Cold Pressed Oils. Academic Press 2020; pp. 39-52.
[http://dx.doi.org/10.1016/B978-0-12-818188-1.00005-0]

[43]   Beg MS, Ahmad S, Jan K, Bashir K. Status, supply chain and processing of cocoa - A review. Trends Food Sci Technol 2017; 66: 108-16.
[http://dx.doi.org/10.1016/j.tifs.2017.06.007]

[44]   Cheng MH, Dien BS, Singh V. Economics of plant oil recovery: A review. Biocatal Agric Biotechnol 2019; 18: 101056.
[http://dx.doi.org/10.1016/j.bcab.2019.101056]

[45]   Ciuta S, Tsiamis D, Castaldi MJ. Chapter Two - Fundamentals of Gasification and Pyrolysis. In: Ciuta S, Tsiamis D, Castaldi MJ, Eds. Gasification of Waste Materials. Academic Press 2018; pp. 13-36.

[46]   de la Fuente B, Tornos A, Príncep A, *et al.* Chapter Six - Scaling-up processes: Patents and commercial applications. In: Lorenzo JM, Barba FJ, Eds. Advances in Food and Nutrition Research. Academic Press 2020; Vol. 92: pp. 187-223.

[47]   Hossain N, Chowdhury T, Chowdhury H, *et al.* Edible bio-oil production from microalgae and application of nano-technology. In: Galanakis CM, Ed. Microalgae. Academic Press 2021; pp. 91-116.
[http://dx.doi.org/10.1016/B978-0-12-821218-9.00004-9]

[48]   Lim JS, Abdul Manan Z, Hashim H, Wan Alwi SR. Towards an integrated, resource-efficient rice mill complex. Resour Conserv Recycling 2013; 75: 41-51.
[http://dx.doi.org/10.1016/j.resconrec.2013.04.001]

[49]   Uitterhaegen E, Evon P. Twin-screw extrusion technology for vegetable oil extraction: A review. J Food Eng 2017; 212: 190-200.
[http://dx.doi.org/10.1016/j.jfoodeng.2017.06.006]

[50]   Vandenbossche V, Candy L, Evon P, Rouilly A, Pontalier P-Y. Extrusion. In: Chemat F, Vorobiev E, Eds. Green Food Processing Techniques. Academic Press 2019; pp. 289-314.
[http://dx.doi.org/10.1016/B978-0-12-815353-6.00010-0]

[51]   Wang S, Zhu F, Kakuda Y. Sacha inchi (*Plukenetia volubilis* L.): Nutritional composition, biological activity, and uses. Food Chem 2018; 265: 316-28.
[http://dx.doi.org/10.1016/j.foodchem.2018.05.055] [PMID: 29884388]

[52]   Chamkalani A, Zendehboudi S, Rezaei N, Hawboldt K. A critical review on life cycle analysis of

algae biodiesel: current challenges and future prospects. Renew Sustain Energy Rev 2020; 134: 110143.
[http://dx.doi.org/10.1016/j.rser.2020.110143]

[53] Costa JAV, Freitas BCB, Moraes L, Zaparoli M, Morais MG. Progress in the physicochemical treatment of microalgae biomass for value-added product recovery. Bioresour Technol 2020; 301: 122727.
[http://dx.doi.org/10.1016/j.biortech.2019.122727] [PMID: 31983577]

[54] Cvejic JH, Langellotti AL, Bonnefond H, Verardo V, Bernard O. Microalgae as a source of edible oils. In: Galanakis CM, Ed. Lipids and Edible Oils. Academic Press 2020; pp. 175-211.
[http://dx.doi.org/10.1016/B978-0-12-817105-9.00005-7]

[55] Gullón B, Gagaoua M, Barba FJ, Gullón P, Zhang W, Lorenzo JM. Seaweeds as promising resource of bioactive compounds: Overview of novel extraction strategies and design of tailored meat products. Trends Food Sci Technol 2020; 100: 1-18.
[http://dx.doi.org/10.1016/j.tifs.2020.03.039]

[56] Jacob A, Ashok B, Alagumalai A, Chyuan OH, Le PTK. Critical review on third generation micro algae biodiesel production and its feasibility as future bioenergy for IC engine applications. Energy Convers Manage 2020; 113655.

[57] Patil RA, Kausley SB, Joshi SM, Pandit AB. Process intensification applied to microalgae-based processes and products. In: Jacob-Lopes E, Maroneze MM, Queiroz MI, Zepka LQ, Eds. Handbook of Microalgae-Based Processes and Products. Academic Press 2020; pp. 737-69.
[http://dx.doi.org/10.1016/B978-0-12-818536-0.00027-0]

[58] Rehmanji M, Suresh S, Nesamma AA, Jutur PP. Microalgal cell factories, a platform for high-valu--added biorenewables to improve the economics of the biorefinery. In: Das S, Dash HR, Eds. Microbial and Natural Macromolecules. Academic Press 2021; pp. 689-731.
[http://dx.doi.org/10.1016/B978-0-12-820084-1.00027-2]

[59] Deshmukh S, Kumar R, Bala K. Microalgae biodiesel: A review on oil extraction, fatty acid composition, properties and effect on engine performance and emissions. Fuel Process Technol 2019; 191: 232-47.
[http://dx.doi.org/10.1016/j.fuproc.2019.03.013]

[60] Khoo KS, Lee SY, Ooi CW, *et al.* Recent advances in biorefinery of astaxanthin from Haematococcus pluvialis. Bioresour Technol 2019; 288: 121606.
[http://dx.doi.org/10.1016/j.biortech.2019.121606] [PMID: 31178260]

[61] Li P, Sakuragi K, Makino H. Extraction techniques in sustainable biofuel production: A concise review. Fuel Process Technol 2019; 193: 295-303.
[http://dx.doi.org/10.1016/j.fuproc.2019.05.009]

[62] Li X, Liu J, Chen G, Zhang J, Wang C, Liu B. Extraction and purification of eicosapentaenoic acid and docosahexaenoic acid from microalgae: A critical review. Algal Res 2019; 43: 101619.
[http://dx.doi.org/10.1016/j.algal.2019.101619]

[63] Alalwan HA, Alminshid AH, Aljaafari HAS. Promising evolution of biofuel generations. Subject review. Renewable Energy Focus 2019; 28: 127-39.
[http://dx.doi.org/10.1016/j.ref.2018.12.006]

[64] Chu Van T, Ramirez J, Rainey T, Ristovski Z, Brown RJ. Global impacts of recent IMO regulations on marine fuel oil refining processes and ship emissions. Transp Res Part D Transp Environ 2019; 70: 123-34.
[http://dx.doi.org/10.1016/j.trd.2019.04.001]

[65] Devaney L, Iles A. Scales of progress, power and potential in the US bioeconomy. J Clean Prod 2019; 233: 379-89.
[http://dx.doi.org/10.1016/j.jclepro.2019.05.393]

[66]   Elie L, Granier C, Rigot S. The different types of renewable energy finance: A Bibliometric analysis. Energy Econ 2021; 93: 104997.
[http://dx.doi.org/10.1016/j.eneco.2020.104997]

[67]   Langle-Flores A, Quijas S. A systematic review of ecosystem services of Islas Marietas National Park, Mexico, an insular marine protected area. Ecosyst Serv 2020; 46: 101214.
[http://dx.doi.org/10.1016/j.ecoser.2020.101214]

[68]   Rehman Zia UU, Rashid T, Awan WN, Hussain A, Ali M. Quantification and technological assessment of bioenergy generation through agricultural residues in Punjab (Pakistan). Biomass Bioenergy 2020; 139: 105612.
[http://dx.doi.org/10.1016/j.biombioe.2020.105612]

[69]   Salame CT. Technologies and materials for renewable energy, environment and sustainability. Energy Rep 2020; 6: 1-3.
[http://dx.doi.org/10.1016/j.egyr.2020.11.068]

[70]   Taghipour A, Ramirez JA, Brown RJ, Rainey TJ. A review of fractional distillation to improve hydrothermal liquefaction biocrude characteristics; future outlook and prospects. Renew Sustain Energy Rev 2019; 115: 109355.
[http://dx.doi.org/10.1016/j.rser.2019.109355]

[71]   Usmani Z, Sharma M, Awasthi AK, *et al.* Bioprocessing of waste biomass for sustainable product development and minimizing environmental impact. Bioresour Technol 2021; 322: 124548.
[http://dx.doi.org/10.1016/j.biortech.2020.124548] [PMID: 33380376]

[72]   Zheng P, Chen T, Dong C, *et al.* Characteristics and sources of halogenated hydrocarbons in the Yellow River Delta region, Northern China. Atmos Res 2019; 225: 70-80.
[http://dx.doi.org/10.1016/j.atmosres.2019.03.039]

[73]   Dahiya A. Part 3 Biomass to liquid biofuels. In: Dahiya A, Ed. Bioenergy. 2nd ed. Academic Press 2020; pp. 129-31.
[http://dx.doi.org/10.1016/B978-0-12-815497-7.02003-0]

[74]   Dahiya A. Algae biomass cultivation for advanced biofuel production. In: Dahiya A, Ed. Bioenergy. 2nd ed. Academic Press 2020; pp. 245-66.
[http://dx.doi.org/10.1016/B978-0-12-815497-7.00013-0]

[75]   Ganguly P, Sarkhel R, Das P. The second- and third-generation biofuel technologies: Comparative perspectives. In: Dutta S, Mustansar Hussain C, Eds. Sustainable Fuel Technologies Handbook. Academic Press 2021; pp. 29-50.
[http://dx.doi.org/10.1016/B978-0-12-822989-7.00002-0]

[76]   Gounder H, Yaseen ZM, Deo R. Hybrid multilayer perceptron-firefly optimizer algorithm for modelling photosynthetic active solar radiation for biofuel energy exploration. In: Deo R, Samui P, Roy SS, Eds. Predictive Modelling for Energy Management and Power Systems Engineering. Elsevier 2021; pp. 191-232.
[http://dx.doi.org/10.1016/B978-0-12-817772-3.00007-0]

[77]   Konur O. The scientometric analysis of the research on the algal science, technology, and medicine. In: Konur O, Ed. Handbook of Algal Science, Technology and Medicine. Academic Press 2020; pp. 3-18.
[http://dx.doi.org/10.1016/B978-0-12-818305-2.00001-2]

[78]   Magda R, Szlovák S, Tóth J. The role of using bioalcohol fuels in sustainable development. In: Bochtis D, Achillas C, Banias G, Lampridi M, Eds. Bio-Economy and Agri-production. Academic Press 2021; pp. 133-46.
[http://dx.doi.org/10.1016/B978-0-12-819774-5.00007-2]

[79]   Rytter E, Hillestad M, Austbø B, Lamb JJ, Sarker S. Thermochemical production of fuels. In: Lamb JJ, Pollet BG, Eds. Hydrogen, Biomass and Bioenergy. Academic Press 2020; pp. 89-117.

[http://dx.doi.org/10.1016/B978-0-08-102629-8.00006-2]

[80]   Syrpas M, Venskutonis PR. Algae for the production of bio-based products. In: Galanakis CM, Ed. Biobased Products and Industries. Elsevier 2020; pp. 203-43.
[http://dx.doi.org/10.1016/B978-0-12-818493-6.00006-3]

[81]   Williams CL, Dahiya A, Porter P. Introduction to bioenergy and waste to energy. In: Dahiya A, Ed. Bioenergy. 2nd ed. Academic Press 2020; pp. 5-44.
[http://dx.doi.org/10.1016/B978-0-12-815497-7.00001-4]

SUBJECT INDEX

P

Palladium metal oxide semiconductors 19
Pavlova 29, 80
   lutheri 29, 80
   salina 80
Petro-chemistry technology laboratory 93
Petroleum gas, liquefied 33
Phaeodactylum tricornutum 80, 85, 117, 281,
   282, 283
Phormidium 177, 178, 198, 199, 200
   ambiguum 177, 178
   amphibium 200
   chlorinum 199
   formosum 200
   kützing 198
   terebriforme 199
   willei 199
Photosynthesis 73, 95, 113, 122, 127, 150,
   323, 324, 327
Physical-chemical properties of algae oil 69
Plant(s) 4, 11, 21, 25, 73, 81, 95, 146, 148,
   150, 154, 157, 158, 173, 195, 198, 199,
   279, 296, 323, 324, 327
   agricultural 95
   aquatic 195, 198, 199
   blue-green 173
   coal 146
   diseases 279
   geothermal power 327
   heat-generating 25
   terrestrial 81
   wood 4
Plectonema 182, 189
   boryanum 189
   notatum 182
Polymer electrolyte membrane 99, 105
Power generation 108
Pressure 21, 307
   economic 307
   generation 21
Process 15, 16, 17, 18, 20, 21, 38, 39, 40, 43,
   44, 47, 48, 61, 70, 99, 100, 101, 102,
   115, 125, 150, 329, 330

   aerobic biological 125
   anaerobic 15, 16, 20
   anaerobic digestion 18
   biological 15
   catalytic 38
   electrochemical 100
   hydrothermal conversion 38
   hydrothermal wave 147
   oxidation 100
   photosynthetic 150
   purification 115
   pyrolysis 61, 70
   thermochemical 39
Production systems 114, 152, 317
   light-receiving 152
Productivity 5, 113, 114, 125, 126, 162
   low biomass 113
Products 11, 15, 17, 41, 42, 43, 44, 45, 47, 52,
   100, 114, 115, 116, 123, 126, 145, 298,
   315, 323, 329, 331
   agricultural 126
   aquatic 323
   bioenergy 15
   carbon-related 52
   electrochemical 100
   gas 47
   hydrolysis 17
   nutritional 123
Properties 19, 20, 47, 61, 62, 68, 70, 73, 74,
   75, 76, 77, 81, 82, 93, 94, 141, 291
   catalyzing 291
   chemical 76
   chlorella-based biodiesel 76
   electrochemical 141
   spraying 61, 70
Proteases 331
   detergent-resistant 331
Pseudokirchneriella subcapitata 88
Pseudomonas 296
   aeruginosa 296
   glumae 296
Pumps 32, 117, 118, 139, 146, 155, 311, 314,
   315, 316
   peristaltic 155

Y

Yeast 158, 296
Yeast wastewater 136, 137
    discharge limits 136, 137

www.ingramcontent.com/pod-product-compliance
Lightning Source LLC
Chambersburg PA
CBHW050804220326
41598CB00006B/115